T0188908

Communications
in Computer and Information Science 1912

Rationale

The CCIS series is devoted to the publication of proceedings of computer science conferences. Its aim is to efficiently disseminate original research results in informatics in printed and electronic form. While the focus is on publication of peer-reviewed full papers presenting mature work, inclusion of reviewed short papers reporting on work in progress is welcome, too. Besides globally relevant meetings with internationally representative program committees guaranteeing a strict peer-reviewing and paper selection process, conferences run by societies or of high regional or national relevance are also considered for publication.

Topics

The topical scope of CCIS spans the entire spectrum of informatics ranging from foundational topics in the theory of computing to information and communications science and technology and a broad variety of interdisciplinary application fields.

Information for Volume Editors and Authors

Publication in CCIS is free of charge. No royalties are paid, however, we offer registered conference participants temporary free access to the online version of the conference proceedings on SpringerLink (http://link.springer.com) by means of an http referrer from the conference website and/or a number of complimentary printed copies, as specified in the official acceptance email of the event.

CCIS proceedings can be published in time for distribution at conferences or as post-proceedings, and delivered in the form of printed books and/or electronically as USBs and/or e-content licenses for accessing proceedings at SpringerLink. Furthermore, CCIS proceedings are included in the CCIS electronic book series hosted in the SpringerLink digital library at http://link.springer.com/bookseries/7899. Conferences publishing in CCIS are allowed to use Online Conference Service (OCS) for managing the whole proceedings lifecycle (from submission and reviewing to preparing for publication) free of charge.

Publication process

The language of publication is exclusively English. Authors publishing in CCIS have to sign the Springer CCIS copyright transfer form, however, they are free to use their material published in CCIS for substantially changed, more elaborate subsequent publications elsewhere. For the preparation of the camera-ready papers/files, authors have to strictly adhere to the Springer CCIS Authors' Instructions and are strongly encouraged to use the CCIS LaTeX style files or templates.

Abstracting/Indexing

CCIS is abstracted/indexed in DBLP, Google Scholar, EI-Compendex, Mathematical Reviews, SCImago, Scopus. CCIS volumes are also submitted for the inclusion in ISI Proceedings.

How to start

To start the evaluation of your proposal for inclusion in the CCIS series, please send an e-mail to ccis@springer.com.

Fazilah Hassan · Noorhazirah Sunar ·
Mohd Ariffanan Mohd Basri ·
Mohd Saiful Azimi Mahmud ·
Mohamad Hafis Izran Ishak ·
Mohamed Sultan Mohamed Ali
Editors

Methods and Applications for Modeling and Simulation of Complex Systems

22nd Asia Simulation Conference, AsiaSim 2023
Langkawi, Malaysia, October 25–26, 2023
Proceedings, Part II

Springer

Editors
Fazilah Hassan ⓘ
Universiti Teknologi Malaysia
Johor, Malaysia

Noorhazirah Sunar ⓘ
Universiti Teknologi Malaysia
Johor, Malaysia

Mohd Ariffanan Mohd Basri ⓘ
Universiti Teknologi Malaysia
Johor, Malaysia

Mohd Saiful Azimi Mahmud ⓘ
Universiti Teknologi Malaysia
Johor, Malaysia

Mohamad Hafis Izran Ishak ⓘ
Universiti Teknologi Malaysia
Johor, Malaysia

Mohamed Sultan Mohamed Ali ⓘ
Universiti Teknologi Malaysia
Johor, Malaysia

ISSN 1865-0929 ISSN 1865-0937 (electronic)
Communications in Computer and Information Science
ISBN 978-981-99-7242-5 ISBN 978-981-99-7243-2 (eBook)
https://doi.org/10.1007/978-981-99-7243-2

This Springer imprint is published by the registered company Springer Nature Singapore Pte Ltd.
The registered company address is: 152 Beach Road, #21-01/04 Gateway East, Singapore 189721, Singapore

Paper in this product is recyclable.

Preface

The 22nd Asia Simulation Conference (AsiaSim 2023) was held on Langkawi Island, Malaysia for two days on 25–26th October 2023. This Asia Simulation Conference is annually organized by the following Asian Simulation societies: the Korea Society for Simulation (KSS), the China Simulation Federation (CSF), the Japan Society for Simulation Technology (JSST), the Society of Simulation and Gaming of Singapore, and the Malaysia Simulation Society (MSS). This conference provides a platform for scientists, academicians, and professionals from around the world to present and discuss their work and new emerging simulation technologies with each other. AsiaSim 2023 aimed to serve as the primary venue for academic and industrial researchers and practitioners to exchange ideas, research findings, and experiences. This would encourage both domestic and international research and technical innovation in these sectors.

Bringing the concept of 'Experience the Power of Simulation', the papers presented at this conference have been compiled in this volume of the series Communications in Computer and Information Science. The papers in these proceedings tackle complex modelling and simulation problems in a variety of domains. The total of submitted papers to the conference was 164 papers. All submitted papers were reviewed by at least three (3) reviewers based on single-blind peer review. The papers presented in AsiaSim 2023 are included in two (2) volumes (1911 and 1912) containing 77 full papers divided into seven (7) main tracks, namely: Artificial Intelligence and Simulation, Digital Twins (Modelling), Simulation and Gaming, Simulation for Engineering, Simulation for Industry 4.0, Simulation for Sustainable Development, and Simulation for Social Sciences.

As editorial members of the AsiaSim 2023 conference, we would like to express our gratitude to all the authors who chose this conference as a venue for their publications and to all reviewers for providing professional reviews and comments. We are also very grateful for the support and tremendous effort of the Program and Organizing Committee members for soliciting and selecting research papers with a balance of high quality and new ideas and applications. Thank you from us.

Fazilah Hassan
Noorhazirah Sunar
Mohd Ariffanan Mohd Basri
Mohd Saiful Azimi Mahmud
Mohamad Hafis Izran Ishak
Mohamed Sultan Mohamed Ali

Organization

Advisor

Yahaya Md. Sam   Universiti Teknologi Malaysia, Malaysia

Chair

Zaharuddin bin Mohamed   Universiti Teknologi Malaysia, Malaysia

Deputy Chair

Norhaliza Abdul Wahab   Universiti Teknologi Malaysia, Malaysia

Secretariats

Mohd Saiful Azimi Mahmud   Universiti Teknologi Malaysia, Malaysia
Nurulaqilla Khamis   Universiti Teknologi Malaysia, Malaysia

Treasurer

Noorhazirah Sunar   Universiti Teknologi Malaysia, Malaysia

Technical and Publication

Mohamed Sultan Mohamed Ali   Universiti Teknologi Malaysia, Malaysia
Mohd. Ridzuan Ahmad   Universiti Teknologi Malaysia, Malaysia
Fazilah Hassan   Universiti Teknologi Malaysia, Malaysia

Program

Salinda Buyamin   Universiti Teknologi Malaysia, Malaysia
Mohd Ariffanan Mohd Basri   Universiti Teknologi Malaysia, Malaysia

Local Arrangement

Anita Ahmad   Universiti Teknologi Malaysia, Malaysia
Nurul Adilla Mohd Subha   Universiti Teknologi Malaysia, Malaysia
Mohamad Hafis Izran Ishak   Universiti Teknologi Malaysia, Malaysia

Website and Publicity

Herman Wahid   Universiti Teknologi Malaysia, Malaysia
Hazriq Izzuan Jaafar   Universiti Teknikal Malaysia Melaka, Malaysia

Sponsorship

Abdul Rashid Husain   Universiti Teknologi Malaysia, Malaysia
Leow Pei Ling   Universiti Teknologi Malaysia, Malaysia

International Program Committee

Abul K. M. Azad   Northern Illinois University, USA
Andi Andriansyah   Universitas Mercu Buana, Indonesia
Axel Lehmann   Universitat de Bundeswehr München, Germany
Bidyadhar Subudhi   National Institute of Technology Rourkela, India
Bohu Li   Beijing University of Aeronautics and
                                 Astronautics, China
Byeong-Yun Chang   Ajou University, South Korea
Cai Wentong   National University of Singapore, Singapore
Gary Tan   National University of Singapore, Singapore
He Chen   Hebei University, China
Jie Huang   Beijing Institute of Technology, China
Le Anh Tuan   Vietnam Maritime University, Vietnam
Liang Li   Ritsumeikan University, Japan
Lin Zhang   Beihang University, China
Mehmet Önder Efe   Hacettepe University, Turkey

Mohammad Hasan Shaheed	Queen Mary University of London, UK
Muhammad Abid	COMSATS University Islamabad, Pakistan
Mukarramah Yusuf	Universitas Hasanuddin, Indonesia
Ning Sun	Nankai University, China
Ramon Vilanova Arbos	University of Barcelona, Spain
Sarvat M. Ahmad	King Fahd University of Petroleum and Minerals, Saudi Arabia
Satoshi Tanaka	Ritsumeikan University, Japan
Sumeet S. Aphale	University of Aberdeen, UK
Teo Yong Meng	National University of Singapore, Singapore
Walid Aniss	Aswan University, Egypt
Xiao Song	Beihang University, China
Yahaya Md Sam	Universiti Teknologi Malaysia, Malaysia
Yuanjun Laili	Beihang University, China
Yun Bae Kim	Sungkyunkwan University, South Korea

Contents – Part II

Contents – Part I

Identification of First Order Plus Dead Time for a pH Neutralization Process Using Open Loop Test

Azavitra Zainal[1,2] (iD), Norhaliza Abdul Wahab[1(✉)] (iD), Mohd Ismail Yusof[2] (iD), and Mashitah Che Razali[1,3] (iD)

[1] Control and Mechatronics Engineering Department, Faculty of Electrical Engineering, Universiti Teknologi Malaysia, Johor Bahru, Johor, Malaysia
norhaliza@utm.my
[2] Instrumentation and Control Engineering Section, Malaysian Institute of Industrial Technology, Universiti Kuala Lumpur, Johor Bahru, Johor, Malaysia
[3] Faculty of Electrical Engineering, Universiti Teknikal Malaysia Melaka, Hang Tuah Jaya, Durian Tunggal, Melaka, Malaysia

Abstract. In this paper, a systematic method for determining the mathematical model for the pH neutralization process is proposed by applying system identification techniques to the controller design requirements. The identification is conducted experimentally using an actual pilot plant and analysis is performed using MATLAB toolbox for system identification. The identification is based on the first order plus dead time (FOPDT) method and subsequent application of model estimation and validation tests. In this study, pH is a controlled variable and alkali dosing pump stroke rate is a manipulated variable. The acid dosing pump stroke rate is fixed at 10%. Four open-loop step response experiments were conducted in which the alkali dosing pump stroke rate was set at 30% (Set A), 40% (Set B), 50% (Set C), and 60% (Set D). Based on the best fit performance criteria, Set B, Set C and Set D show approximately the same results. Frequency response analysis was performed using Bode plot and Nyquist diagram to determine the stability. The results show that Set B is more stable than Set C and Set D. The stability criterion is important because the obtained FOPDT model is used for controller design.

Keywords: System Identification · First order plus dead time · Palm oil mill effluent · pH neutralization

1 Introduction

At the palm oil mill wastewater treatment plant, the pH is neutralized to treat the highly acidic wastewater from the palm oil mill and to ensure that the wastewater discharged into the river meets environmental standards. It is important to keep the pH within the desired range because it is an inherent process with strong nonlinearity, whereby a small change in inputs resulting a large change in output. The exact acid-alkali ratio is critical to achieve the desired value [1–3].

F. Hassan et al. (Eds.): AsiaSim 2023, CCIS 1912, pp. 1–12, 2024.
https://doi.org/10.1007/978-981-99-7243-2_1

System identification in control engineering is used to determine the process dynamics and represent it in mathematical form based on empirical data [4, 5]. After data collection, the model structure is selected based on the criteria for model estimation and validation. Model estimation is used to minimize the error function between measured and predicted output. Model validation determines whether the model can reproduce the measured data. The obtained model is then used to design the controller. Due to the nonlinear behavior of the pH neutralization process, the process dynamics are usually represented by the first order plus dead time (FOPDT) model. Due to the process parameters such as process gain, time constant and dead time are different based on the process, the proper tuning method is required for controller design to achieve good closed-loop performance [6–8]. The study was conducted by Reshma et.al [9] and Aksay et. al [10] was used FOPDT model obtained from open loop response to tune the PID controller for pH neutralization process. Meng et.al [11] was used FOPDT model to design BP-PID-Smith based controller for the liquid fertilizer pH regulation process.

In this study, pH neutralization process is a single-input single-output (SISO), in which the controlled variable is pH and the manipulated variable is the alkali dosing pump stroke rate. The focus of this study is to determine the FOPDT model for the pH neutralization process utilizing offline identification by using experimental data. Then, the obtained model is compared based on the frequency response to determine the most suitable experimental setup.

2 Methodology

2.1 The Experimental Setup

Figure 1 shows the schematic diagram of the pilot plant for the pH neutralization process. The diluted palm oil mill effluent (POME) is prepared in the T-83 analysis tank by adding 1 L POME with 29 L of 0.1M hydrochloric acid, HCl. The POME pH value after diluted is 3.8. The POME is continuously circulated in the analysis tank by the P-803A pump to ensure that the acid and alkali are well mixed with the POME. Tank T-81 consists of the alkali, 0.1 M natrium hydroxide, NaOH. The alkali and acid dosing pump stroke rate range are between 0% dan 100%, which corresponds to a current signal of 420 mA.

The open-loop response was performed experimentally to determine the appropriate setting of the manipulated variable for the system identification. The experimental data is used to define the mathematical model through system identification. The experiment was designated into four data sets, the alkali dosing pump stroke rate was set at 30% (set A), 40% (set B), 50% (set C), and 60% (set D), respectively. The alkali dosing pump stroke rate is adjusted using a square wave signal with a 100 s phase and 100 s period. The acid dosing pump stroke rate is constantly set to 10% for all data sets. The sampling time is 1 s. The acid and alkali are continuously added until the pH reaches steady state based on the pH titration curve. The System Identification Toolbox in MATLAB version R2021b was used to determine the process dynamics, which provides a simple simulation with multiple iteration functions to select the best model that exhibit the model estimation and validation compared to the real experiment data. Table 1 summarizes the experimental setup.

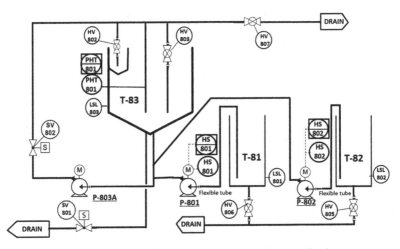

Fig. 1. Schematic diagram of the pilot plant for the pH neutralization process.

Table 1. The open loop experiment setting for the pH neutralization process.

Data set	Experiment setting	
	Alkali dosing pump stroke rate (%)	Total no. of data
Set A	30	2064
Set B	40	2068
Set C	50	2065
Set D	60	2061

2.2   System Identification

The pH neutralization process data set, which includes the alkali dosing pump stroke rate data and output pH value, is loaded into MATLAB. Figure 2 and 3 shows the time series plot of the experimental data. Both the pH output value and the alkali dosing pump stroke rate are plotted against the sampling time. In the titration curve in Fig. 3, the pH increases slowly as the alkali is added to an acidic solution (POME with HCl) at the beginning of the titration. Near the equivalence point, which is equal to 7, the pH of Set A, Set B, Set C, and Set D begins to increase rapidly. After exceeding the equivalence point, the rate of change of pH slows down. The conclusion from Fig. 3 is that the relationship between input and output is non-linear. However, for Set A, the neutralization process still does not reach the equivalence point because more alkali must be added to the process.

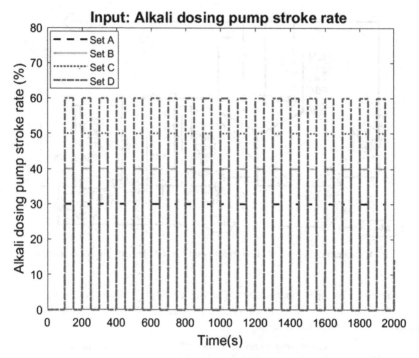

Fig. 2. The input: Alkali dosing pump stroke rate.

The measured data were divided into two parts, the first reserved for model estimation and the other for testing, and randomly divided. The percentage of data division for training and testing are 50%-50%, 60%-40%, and 70%-30% for each set. In this study, the first order plus dead time (FOPDT) is used to determine the best model for the pH neutralization process.

$$G(s) = \left(\frac{k}{(Ts+1)}\right)e^{-T_d s} \qquad (1)$$

Equation (1) shows the FOPDT equation, where K is the process gain, T is the time constant and T_d is the time delay. Each model's performance is evaluated based on the highest percentage of the model's best fits. Although the pH neutralization process is a higher-order system, the system needs to be represented in the FOPDT model to simplify the controller design.

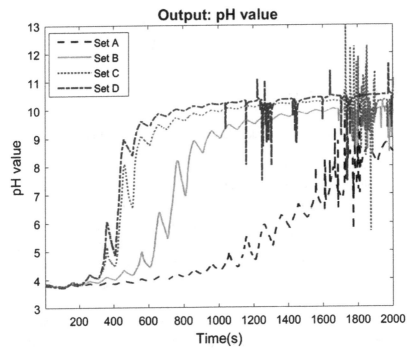

Fig. 3. The open-loop experimental data: pH value.

3   Results and Discussions

3.1   FOPDT Model

Table 2 shows the results of the system identification with FOPDT using the measurement data in Fig. 2 and Fig. 3 for each data set. For data sets B, C and D, the 50%-50% performs better than other split percentages. Set A cannot be used for system identification with the FOPDT model because the data obtained do not represent the S titration curve for the strong acid and strong alkali.

For the process gain, K value, Set B, Set C, and Set D show no significant difference, with a value of 0.55437, 0.14223, and 0.35831, respectively. As for the time constant, T, Set B takes more time for the process response that Set C and Set D, which show a value of 397.54 s. The values of time constants for Set C and Set D are almost the same, 203.65 s and 180.18 s, respectively. For the time delay, all data sets have almost the same value, about 24 s. To determine the fit of the model, the percentage of the output that the model reproduces is used as the best fit criterion. The best fit of the model is calculated as follows,

$$\text{Bestfit} = \left(1 - \frac{\sum_{i=1}^{n} |y_i - \hat{y}_i|}{\sum_{i=1}^{n} |y_i - \hat{y}|}\right) \times 100\% \tag{2}$$

where $\bar{y}$ is the mean of measured output.

Table 2. The system identification results using FOPDT models.

Data set	System Identification (FOPDT)			
	Training (%)	Testing (%)	FOPDT	Best fits (%)
Set A	50	50	$G(s) = (0.18322/(1 + (1e\text{-}06)s))e^{-15s}$	−143.5
	60	40	$G(s) = (0.18379/(1 + (1e\text{-}06)s))e^{-8s}$	−142.2
	70	30	$G(s) = (0.18272/(1 + (1e\text{-}06)s))e^{-1s}$	−139.9
Set B	50	50	$G(s) = (0.55437/(1 + 397.54s))e^{-24.95s}$	66.06
	60	40	$G(s) = (0.5797/(1 + 534.79s))e^{-26.8s}$	60.39
	70	30	$G(s) = (0.60951/(1 + 703.04s))e^{-17.23s}$	44.11
Set C	50	50	$G(s) = (0.14223/(1 + 203.65s))e^{-24.88s}$	70.13
	60	40	$G(s) = (0.17579/(1 + (1e\text{-}06)s)e^{-5s}$	−178.2
	70	30	$G(s) = (0.43727/(1 + 333.99s))e^{-19.83s}$	46.05
Set D	50	50	$G(s) = (0.35831/(1 + 180.18s))e^{-24.61s}$	69.16
	60	40	$G(s) = (0.3643/(1 + 233.81s))e^{-23.78s}$	67.08
	70	30	$G(s) = (0.36455/(1 + 285.36s))e^{-26.13s}$	45.09

Figure 4 shows the comparison between the experimental data and the FOPDT model approximation, and the results are quite good for the high order process model. The S-shape of the higher-order system is adequately represented by the first order dynamic characteristics. For the FOPDT model approximation [12], the processes can be divided into three groups based on the following equation.

$$\tau = T/(T + T_d) \tag{3}$$

where T and T_d value are obtained from Eq. (1). The system is defined based on the intervals, which are lag dominant system if $0 < \tau < 0.5$, the balance system if $\tau = 0.5$; and delay dominant system if $0.5 < \tau < 1$. Based on the FOPDT model in Table 2, the τ value for set B is 0.94, for set C is 0.89, and set D is 0.88. Thus, the system is a delay dominant system.

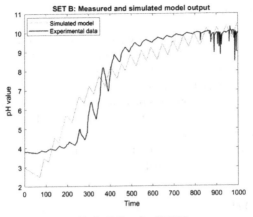

(a)   Set B (Best fits: 66.06 %).

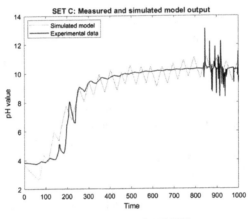

(b)   Set C (Best fits: 70.13%).

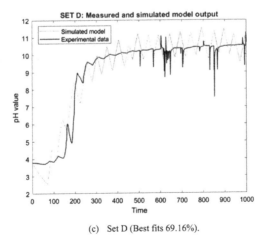

(c)   Set D (Best fits 69.16%).

Fig. 4. Comparison between FOPDT approximation with experimental data for (a) Set B, (b) Set C and (c) Set D.

3.2 Frequency Responses

The Bode plot is used to represent the frequency response of a system. It is used to analyze the stability margins (GM - gain margin and PM - phase margin) of a control system, which indicate the stability and robustness of the system. Table 3 shows the values for the gain margin and phase margin based on the FOPDT for Set B, Set C and Set D. The value of the gain margin was determined from the Bode plot (see Fig. 5), and it was found to be approximately same for Set B, Set C, and Set D. Since the value of the gain margin is more than 0 dB, the system is stable. The phase margin of all data sets is infinite, which means that the phase of the system never reaches $-180°$ or pi radians. This shows that the system is stable and has a very high degree of stability.

Table 3. Stability margins for Set B, Set C and Set D.

Data set	GM	PM
Set B	33.3 dB (at 0.0645 rad/s)	Infinity
Set C	39.5 dB (at 0.0661 rad/s)	Infinity
Set D	30.6 dB (at 0.0672 rad/s)	Infinity

For the FOPDT system, the Nyquist diagram can also be used to determine the stability and performance of the system. One of the methods for determining the stability is to analyze the number of encirclements around the point $(-1,0)$ in the Nyquist diagram. When the Nyquist diagram crosses the critical point $(-1,0)$ in the complex plane, the system is unstable. This critical point corresponds to the frequency at which the phase angle of the system is $-180°$. To draw the Nyquist diagram of the FOPDT model, the Nyquist (sys) command is used for the LTI model system, created with the tf command in MATLAB. Figure 6 shows the comparison of the Nyquist diagram for each data set. Due to the negative phase added by the time delay, the graph spirals toward the origin. From the magnified Nyquist diagram, set B intersects the imaginary axis twice, while sets C and D intersect the imaginary axis more than twice. Each intersection represents a pole at a particular frequency on the imaginary axis. From Fig. 3, sets C and D oscillate more than set B. Therefore, the model approximation of set B is more suitable for controller design.

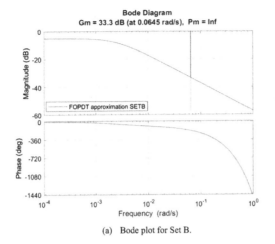

(a) Bode plot for Set B.

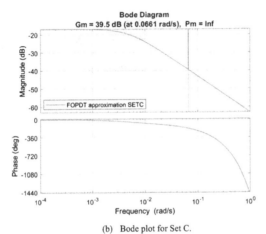

(b) Bode plot for Set C.

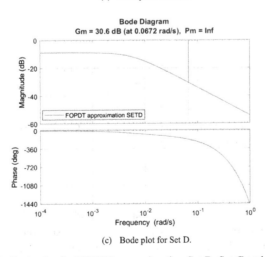

(c) Bode plot for Set D.

Fig. 5. Bode plot for FOPDT approximation Set B, Set C and Set D.

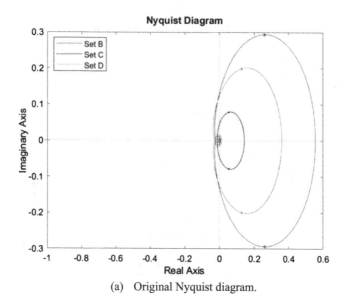

(a)   Original Nyquist diagram.

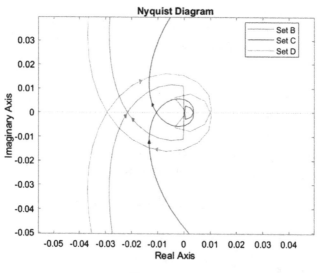

(b)   Enlarged image.

Fig. 6. Nyquist diagram for the open loop test for Set B, Set C and Set D (a) Nyquist diagram for each data set (b) Enlarged image at the origin.

4   Conclusions

In this work, the frequency response analysis was conducted to determine the FOPDT model of the pH neutralization process using open-loop experiment data. Four experimental settings were established, and the data are presented in Set A, Set B, Set C and

Set D. For Set A, more time was needed to show the dynamics of the process. The results show that the best fits value of the FOPDT approximation is approximately the same for Set B, Set C and Set D. On top of the time response analysis, the frequency response was performed using the Bode plot and Nyquist diagram. The frequency response results in terms of stability show that Set B was more stable compared to Set C and Set D. Therefore, the experimental setting for Set B represents the process dynamics in FOPDT. Further work will use the obtained FOPDT model to design the PID and model predictive controller for the pH neutralization plant.

Acknowledgement. The authors are grateful to the Universiti Teknologi Malaysia, UTM High Impact Research (UTMHR) & No Vot. Q.J130000.2451.08G74 and the Ministry of Higher Education (MOHE), for their partial financial support through their research funds. The first author wants to thank the Universiti Kuala Lumpur (UniKL) for the Academic Staff Higher Education Scheme (ASHES) and the Majlis Amanah Rakyat (MARA) for the "Skim Pinjaman Pelajaran MARA" scholarship.

References

1. Saad, M.S., Wirzal, M.D.H., Putra, Z.A.: Review on current approach for treatment of palm oil mill effluent: integrated system. J. Environ. Manage. **286**, 112209 (2021). https://doi.org/10.1016/j.jenvman.2021.112209
2. Kamyab, H., Chelliapan, S., Din, M.F.M., Rezania, S., Khademi, T., Kumar, A.: Palm oil mill effluent as an environmental pollutant. Palm Oil. InTech (2018). https://doi.org/10.5772/intechopen.75811
3. Kamal, S.A., et al.: Pre-treatment effect of palm oil mill effluent (POME) during hydrogen production by a local isolate clostridium butyricum. Int. J. Advanced Science, Eng. Information Technol. **2**(4), 325–331 (2012). https://doi.org/10.18517/ijaseit.2.4.214
4. Burn, K., Cox, C.: A hands-on approach to teaching system identification using first-order plus dead time modelling of step response data. Int. J. Electrical Eng. Educ. **57**(1), 24–40 (2020). https://doi.org/10.1177/0020720918813825
5. Burn, K., Maerte, L., Cox, C.: A matlab toolbox for teaching modern system identification methods for industrial process control. Int. J. Mech. Eng. Educ. **38**(4), 352–364 (2012). https://doi.org/10.7227/IJMEE.38.4.7
6. Ramachandran, R., Lakshminarayanan, S., Rangaiah, G.P.: Process identification using open-loop and closed-loop step responses. J. Institution of Engineers **45**(6), 1–13 (2005)
7. Haugen, F.: The good gain method for simple experimental tuning of PI controllers. MIC---Model. Identif. Control **33**(4), 141–152 (2012). https://doi.org/10.4173/mic.2012.4.3
8. Sakthiya Ram, S., Dinesh Kumar, D., Meenakshipriya, B.: Designing of PID controllers for pH neutralization process. Indian J. Science and Technol. **9**(12), 1–5(2016). https://doi.org/10.17485/ijst/2016/v9i12/89940
9. Babu, R., Swarnalatha, R.: Comparison of different tuning methods for ph neutralization in textile industry. J. Appl. Sci. **17**(3), 142–147 (2017). https://doi.org/10.3923/jas.2017.142.147
10. Akshay, N., Subbulekshmi, D.: Online auto selection of tuning methods and auto tuning PI Controller in FOPDT real time process-pH neutralization. Energy Procedia **117**, 1109–1116 (2017). https://doi.org/10.1016/j.egypro.2017.05.235

11. Meng, Z., et al.: Design and application of liquid fertilizer pH regulation controller based on BP-PID-smith predictive compensation algorithm. Applied Sci. **12**(12), (2022). https://doi.org/10.3390/app12126162
12. Muresan, C.I., Ionescu, C.M.: Generalization of the FOPDT model for identification and control purposes. Processes **8**(6), 682 (2020). https://doi.org/10.3390/pr8060682

Tapered Angle Microfluidic Device for Cell Separation Using Hydrodynamic Principle

Muhammad Asyraf Jamrus and Mohd Ridzuan Ahmad[✉]

Department of Control and Mechatronics Engineering, Faculty of Electrical Engineering, Universiti Teknologi Malaysia, 81310 Johor Bahru, Malaysia
mdridzuan@utm.my

Abstract. Metastasis is responsible for 90% of all cancer-related fatalities. CTCs are difficult to detect due to their rarity. Currently, the devices that aid in the detection of CTCs have limitations, such as a high price, complex apparatus, or low sample purity. Consequently, the purpose of this research is to construct and enhance a microfluidic device with a tapered angle. This device aims to outperform conventional microfluidic devices in terms of cost, sample purity, and apparatus. Using the finite element simulation software COMSOL Multiphysics, two studies are conducted, the first of which examines the effect of taper angle on particle separation and the second of which examines the effect of flow rate on particle separation. This design enables particle separation based on hydrodynamic theory and the sedimentation process. A mixture of 3 μm and 10 μm polystyrene (PS) microbeads were effectively separated when the taper angle approached 20 degrees, and separation continued until the taper angle reached 89 degrees. Using 12° and 6° taper angles, 3 μm, and 10 μm polystyrene microbeads were not effectively separated using finite element simulation. The proposed product is functionally enticing despite the fact that this design utilizes a passive separation technique. This technology provides uncomplicated, label-free, continuous separation of numerous particles in a self-contained device without the need for cumbersome equipment. Consequently, both point-of-care diagnostic instruments and micro total analytic systems could utilize this device.

Keywords: CTCs · microfluidic device · tapered angle

1 Introduction

Metastases are responsible for ninety percent of all cancer-related fatalities. Cancer's circulating tumor cells (CTC) are malignant cells that enter the bloodstream from the tumor; the overwhelming majority of CTCs die, but some survive and spread to other parts of the body, where they become metastases [2]. [3–5] High CTC levels are associated with decreased overall and progression-free survival. CTCs, on the other hand, are extremely rare; one CTC is ordinarily present for every million or more white blood cells, necessitating the use of highly sensitive technologies for detecting and analyzing CTCs.

© The Author(s), under exclusive license to Springer Nature Singapore Pte Ltd. 2024
F. Hassan et al. (Eds.): AsiaSim 2023, CCIS 1912, pp. 13–28, 2024.
https://doi.org/10.1007/978-981-99-7243-2_2

Despite the significance of CTCs, investigations into them continue to raise major concerns. Particularly in regards to how CTCs should be studied, as well as their actual metastatic potential and tumor heterogeneity as a consequence of diverse cancerous cell population groups. The clinical utility of CTCs has not been demonstrated; consequently, they cannot be used to guide therapy. There have been studies examining the predictive role of CTCs [6], but no definitive conclusions have been reached as of yet. Despite this, the results of these experiments could reveal the true potential of CTCs. CTC therapy will be determined by the research group's comprehension of CTCs and the metastatic stage. Only a comprehensive comprehension of CTCs and their role in tumor biology will enable us to overcome these obstacles.

There have been considerable efforts to resolve these issues. During the past decade, numerous CTC-study technology options have been investigated. Based on biological and/or physical factors, each technique possesses advantages and disadvantages. Currently, CellSearch is the only FDA-approved method for CTC detection. This procedure has been approved by the U.S. Food and Drug Administration. The initial release of the CellSearch system, designed to count CTC in 7.5 mL of blood, occurred in 2004. [7]. Despite the ground-breaking results obtained with this method [3, 4, 8, 9], it was unable to acquire all CTC subpopulations, including EpCAM-negative cells. This is due to the fact that this method only employs a detection strategy based on epithelial markers.

In addition, flow cytometry was used to detect CTCs. Flow cytometers pass a solution of single, monodisperse, unclamped cells through a laser beam, where each cell is counted, sorted, and classified [10]. This method has disadvantages, including limited sensitivity, no visual confirmation of cell specificity, and technical and analytic difficulty [11]. PCR-based approaches, EPISPOT, laser scanning cytometry, and CTC microchips are some of the few CTC detection technologies used in hospitals at present. Most of these techniques have either limited sensitivity or require the use of EpCam, which may miss some tumor cells.

CTCs were discovered and described in human blood nearly 150 years ago, which is quite intriguing. Recent technological advancements have significantly improved methods for detecting CTCs. There are a number of CTC detection methods, such as centrifugal sedimentation, magnetic cell separation, fluorescence-activated cell sorting, and microfluidics, that have not been clinically validated despite significant technological advances in this field, particularly over the past decade.

This investigation focuses on microfluidics [12]. Microfluidics is the study of the behavior, manipulation, and precise control of fluids on a microscopic scale (typically sub-millimeter), where surface forces dominate volumetric forces [13]. Microfluidics permits precise fluid control, reduced sample consumption, device minimization, minimal analytical cost, and nanoliter volume manipulation with ease. Biomedical applications of microfluidics include immobilization of C. aeruginosa, pH control, drug administration, gradient generation, and most importantly, cell analysis. A benefit of microfluidics is the ability to isolate uncommon cell populations without using cell adhesion or labeling techniques [14].

Microfluidics can be divided into two categories: active separation (using an external force field) and passive separation (using channel geometry and hydrodynamic force) [15]. Passive separation will be used in this endeavor, and passive cell separation microfluidic devices already exist, including pillar and weir, pinch flow fractionation (PFF), hydrodynamic filtration, inertial, and biomimetic.

Although passive separations of microfluidic devices are abundant, these devices have disadvantages. For instance, in pressure flow fractionation (PFF), the minimum separation efficiency is 90 percent. In this instance, I believe that a new design for cell separation with high sample purity and reduced complexity was required. Consequently, I wish to propose a passive microfluidic separation utilizing hydrodynamic principles and taper angles.

2 Literature Review

2.1 Pinch Flow Fractionation

In 2004, Yamada et al. presented the pressure flow fractionation (PFF) method. This device has two inlets: one will be filled with liquid, and the other will be filled with both liquid and particulates [16]. By modulating the flow rates from both inlets at this point, the liquid-containing particulates are concentrated on the wall of the constricted portion of the microchannel, and the fluid flow is laminar. This method aligns particles of varying sizes on a single sidewall. Then, at the intersection between constricted and widened microchannel segments, the spreading streamline exerts a force towards the center of the microchannel, primarily on larger particles, whereas the sidewall exerts a force primarily on smaller particles. Consequently, particulate sizes are segregated perpendicular to the flow direction in the widened channel segment.

2.2 Hydrodynamic Filtration

Yamada et al. introduced hydrodynamic filtration in 2005. This method isolates particles by consistently extracting a small volume of liquid from the main channel. A laminar fluid flow containing particles is continually injected from left to right into the microchannel [17]. As the primary microchannel is continuously purged of liquid, all particles align themselves along its sidewalls. Particles with a diameter greater than a specified value would never travel through the side channels, regardless of their proximity to the main channel's sidewalls. This is due to the fact that the centroid of the particle cannot be located within a predetermined distance from the sidewall. Past the branch point, particles become significantly more concentrated. The passage of particulates down the sidewalls is made possible by increased flow rates in the side channels. Because smaller particles are located closer to the sidewall than larger particles, they are filtered out first.

2.3 Flow Cytometry

Bonner et al. introduced flow cytometry in 1960, and the method has evolved over time. The flow cytometer instrument consists of three fundamental systems: fluidics, optics, and electronics. Flow cells are a component of the fluidic system. To transport and align cells or particles through a narrow channel and into the laser interceptor, sheath fluid (light beam) is required. This hydrodynamic focusing technique enables laser examination of a single cell [18]. The laser line, which generates a single wavelength of light at a predetermined frequency, is one of the most widely employed optical systems in use today. Subsequently, the particles are transmitted through one or more laser beams. Lasers are available in numerous wavelengths and intensities (photon output or time). A laser probe stimulates fluorescent probes connected to antibodies, causing them to fluoresce (or emit light) at specific wavelengths.

2.4 Inertial Microfluidic

In addition, for inertial microfluidics, numerous forces influence the mainline and lateral motion of suspended particles when a suspension of particles is infused into a microfluidic channel. Along the primary current, the viscous drag force accelerates particles to within a few percent of the speed of the fluid constituents in the surrounding area. When particles approach a channel wall, the wall slows them down, causing them to trail behind the fluid flow. In addition to the lateral lift forces induced by rotation and slip shear, a secondary flow drag force is also perpendicular to the fluid mainstream. Not all of these forces must be considered in inertial microfluidics research and applications. Depending on the medium, particulates, and channel structure, some are minimal [19].

2.5 Cell-affinity Chromatography

Cell-affinity chromatography is a technique for isolating cancer cells selectively from a heterogeneous cell population. It is typically affixed to a solid, insoluble matrix (such as agarose or polyacrylamide) that has been chemically modified to introduce reactive functional groups with which the ligand can react, resulting in the formation of stable covalent bonds. Those molecules in the stationary phase interact with the ligand. The target biomolecules can then be eluted using an elution buffer. Thus, the target molecule is recovered from the eluting solution, and its structure is preserved for further study [20].

2.6 Limitations

Even though these existing methods offer a number of benefits, we must examine their particular drawbacks and limitations. First, for pinched flow fractionation, the separation will be diminished when the Reynolds number (Re) is large, which is approximately 200 in a rectangular microchannel, due to the addition of equilibrium along the two sidewalls. Due to the fact that the constricting zone is dependent on the size of the targeted cell, it is necessary to create a unique microdevice for each desired cell type. This renders hydrodynamic filtration ineffective for separating cells not included in the design. In addition, flow cytometry is limited by the time required to separate cells. Flow cytometry works by injecting a single cell into liquid droplets, which makes sorting billions of cells a time-consuming, hundreds-of-hour-long procedure that is also expensive and requires highly skilled personnel. Lastly, the disadvantage of affinity-based separation is that it is limited by the specificity of biomarkers; therefore, it is essential to choose the appropriate chemical.

3   Methodology

3.1   Design Specifications

The microfluidic design with a tapered angle has one inlet and up to three outlets with a 2 mm diameter, a focusing zone with a length of 1 mm and a height of 100 μm, and a tapered area. This study employed a variety of tapered angles to determine the optimal angle for cell separation. The inlet will be supplied with laminar fluid flow at a flow rate stretching from 0.5 μL/min to 3 μL/min.

3.2   Proposed Solution

An optimal cell separation device should have high sample purity, be cost-effective, utilize straightforward equipment, and drastically reduce the time required to sort a large volume of samples. In light of this, the proposed design for a microfluidic device with a tapered angle is based on the hydrodynamic principle and will be described in greater detail below. This microfluidic device will have one inlet, one focusing region, one primary channel, and three outlets (Fig. 1). In addition, the angle of the taper will be determined once the desired results have been achieved.

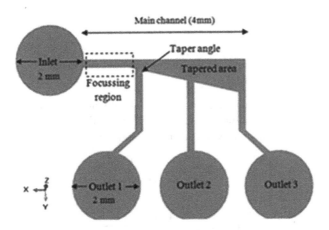

Fig. 1. Tapered angle microfluidic proposed design

3.3   Working Principle

Using hydrodynamic spreading to separate micro-objects required adjusting the hydro-dynamic spreading generated by the tapered device design. The width of a tapered microchannel increases progressively. This one-of-a-kind design maximizes velocity dispersion to generate a pressure differential between the primary channel and the three outputs. Increased velocities entail a decrease in hydrodynamic resistance, which is required for successful particle entrapment at the appropriate outlets. Because it converged via a brief focus zone, a buffer or sheath flow was unnecessary. After the sample flow is continuously conveyed to the inlet, the focusing zone concentrates the particles. As particles approach the end of the focusing region, they begin to follow their own trajectory, resulting in a difference in vertical position between particles of various sizes. When the velocity continues to decrease at the tapered zone, gravity takes over and sedimentation commences. Sedimentation is the separation by gravity of particulates suspended in a liquid [21]. In the carrier flow, larger particles with size-dependent trajectories were drawn in. Sedimentation is observable in microchannels with a tapered cross-section due to the microscopic dimensions employed, specifically the channel height [22]. Consequently, the largest particles are collected at Outlet 1, the smallest particles are collected at Outlet 3, and some particles are collected at Outlet 2.

3.4 Hydrodynamic Equations

The sample flow across a separation device can be described using an equivalent fluidic circuit with discrete resistive components R_{sense} and R_{h1-3} (for the tapered region), as illustrated in Fig. 2. The resistance to hydrodynamic flow for an incompressible Newtonian fluid is given by (1):

$$\Delta P = QR_h \tag{1}$$

where:

Q = Volumetric flow rate
ΔP = Pressure differential between the channel's outlet and inlet
R_h = Hydrodynamic resistance

The R_h outlet has the maximum velocity, which influences how suspended particles migrate within tapered microfluidics to exit through this outlet. Hydrodynamic resistance is entirely dependent on geometry; therefore, the equation is (2)(3):

$$R_{h,reCt} = \frac{32\mu L}{D^2 \omega d} \tag{2}$$

$$R_{h,tapered} = \frac{12\mu L}{h^3} \frac{\ln\left(\frac{\omega_1}{\omega_0}\right)}{\omega_1 - \omega_0} \tag{3}$$

where:

h = height
$\omega_{0,1}$ = width
L = linear dimension

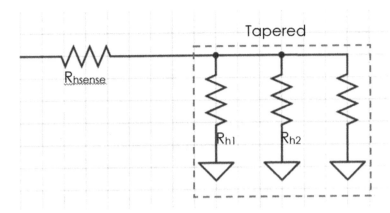

Fig. 2. Tapered microfluidic equivalent fluidic circuit

3.5 Tapered Angle Microfluidic Design

COMSOL Multiphysics was used to design the tapered-angle microfluidic device. Figure 3 depicts the first geometry design for the tapered angle microfluidic, in which the inlet and outflow diameters are fixed at 2 mm and the focusing region has dimensions of 1 mm in length and 100 μm in width. For the initial design, multiple tapered angles were utilized. Figures 4 and 5 depict, respectively, the designs with feedback connected from outlet 2 to the tapered area and the design with a length of 4 mm for the tapered area.

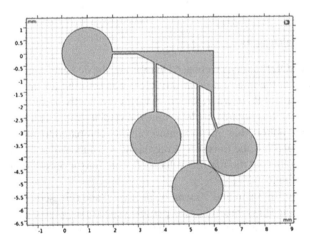

Fig. 3. First proposed geometry design for a microfluidic device with a tapered angle

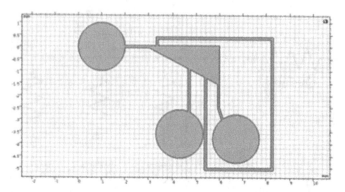

Fig. 4. Second proposed geometry design for a microfluidic device with a tapered angle

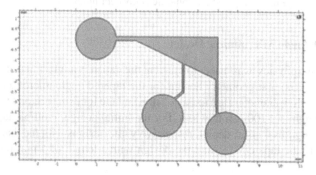

Fig. 5. Third proposed geometry design for tapered angle microfluidic device

3.6 Evaluate the Designs

Numerous COMSOL functions were used to examine the design. Laminar fluid flow was used in this study at the inlet. Laminar fluid flow was selected for this project because, in contrast to turbulent flow, where the fluid varies erratically, it moves smoothly or in predictable patterns. It is possible to determine the velocity value using probing features, and it appears that outlet 3 has the maximum velocity compared to outlets 1 and 2. This suggests that outlet 3 has the lowest hydrodynamic resistance. The hydrodynamic concept can be used to further modify this velocity feature for successful particle separation or particle entrapment.

With a diameter of 3 μm and 10 μm, respectively, and released continuously at the inlet from 0 to 5 s in 0.1-s intervals, two-particle trajectories were used to simulate the trajectory of PS microbeads. In order to count the particle traps at each outlet, the probe was also placed at each outlet.

4   Results and Discussions

4.1   The Effect of Tapering Angle on Cell Separation

In this work, the influence of tapering angle on cell separation was examined using a variety of tapered angles and a constant flow rate of 3 l/min. Figures 6 and 7, respectively, demonstrate how the streamline is impacted by the tapered angle for 6 and 27 tapering angles. Figure 8 shows the particles gathered at each outlet in the meantime. This is because sedimentation inside microfluidic devices is governed by the Stokes number, which is highly dependent on mass. The goal of this inquiry is to establish the ideal tapering angle at which the cells should divide.

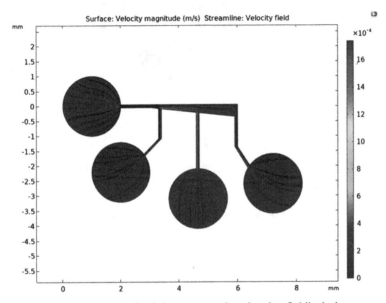

Fig. 6. Streamline for 6-degree tapered angle microfluidic design

The taper angle in a tapered microfluidic system is essential because it regulates the sedimentation rate (sedimentation velocity). As a result, it was discovered that sedimentation accelerates with increasing taper angles. As a result, it was discovered that sedimentation accelerates with increasing taper angle. A mixture of PS microbeads with different sizes and densities did not separate at taper angles of 6 and 12 degrees, as shown in Fig. 8. This is a result of the insufficient tapering area for sedimentation to occur.

Additionally, a group of particles begins to disperse as the taper angle reaches 20 degrees and can do so up to a 65-degree angle. As a result, we can assume that separation can happen at an angle of taper between 20 and 89 degrees. But the amount of particles gathered at outputs 1 and 3 gradually decreases as the taper's angle gets closer to 30 degrees. With 102 particles collected at both outputs 1 and 2, the ideal taper angle for cell separation is 27 degrees.

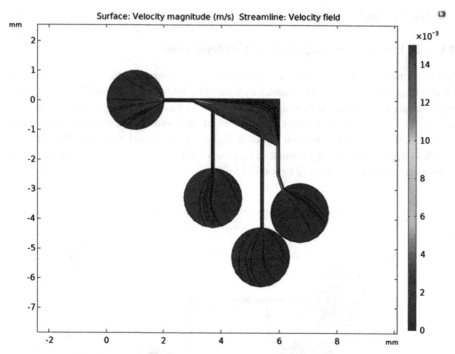

Fig. 7. Streamline for 27-degree tapered angle microfluidic design

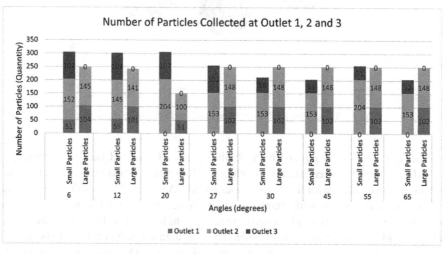

Fig. 8. Particles collected on each outlet for each tapered angle

4.2   Flow Rate's Impact on Particle Separation

The flow rate employed in this investigation ranges from 0.5 μl/min to 3.0 μl/min. These devices are particularly sensitive to inlet flow rates since they rely on hydrodynamic principles. Due to the rapid sedimentation that lower flow rates cause, particle separation may not be effective.

On the other hand, slower sedimentation due to decreased flow rates may result in larger particles being stuck at the farthest outflow. Therefore, figuring out the best flow rates permits particles to take the appropriate path and be caught at the right outlets. The impact of flow rates on particle separation at the ideal taper angle (27 degrees) is shown in Fig. 9.

According to Fig. 9, at 0.8333 x 10^{-11} m^3/s, no outlets are filled with big particles. This is because the huge particle cannot pass through the focusing region and taper area since the velocity is not high enough, causing the majority of the particles to stop at the inlet and in the focusing region. Next, a number of tiny particles are caught at outlet 1 at 1.5 to 2 x 10^{-11} m^3/s. This is because some small particles, which may have collided at the focusing zone, have velocities that are insufficient for continued movement.

At outlets 1 and 2, a combination of particles began to totally separate with 100% sample purity at a flow rate of 2.5 to 5 x 10^{-11} m^3/s. In addition, it appears that the amount of larger particles trapped at outlet 1 rises with the flow rate until it reaches 2.5 x 10^{-11} m^3/s, at which point it falls as a result of a late sedimentation process brought on by the quicker velocity of some larger particles. According to the number of particles trapped at each outlet, the appropriate flow rate for all larger particles captured at outlet 1 is 2.5 x 10^{-11} m^3/s, whereas small particles are trapped at outlets 2 and 3.

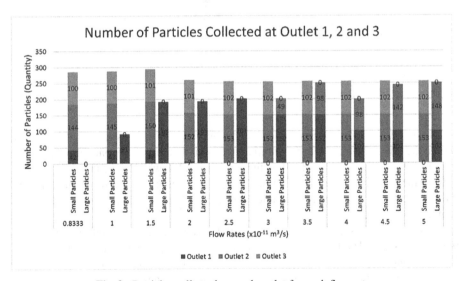

Fig. 9. Particles collected on each outlet for each flow rate

4.3 The Designs Comparison

Three designs have been developed in this investigation. The first design includes one inlet, a focusing region, a 27-degree tapered area, and three outlets, whereas the second design is identical to the first except that outlet two is connected to the tapered area. The third design is identical to the first, with the exception that it features two outlets and the length of the tapering section and height of the focusing region are each increased by 1 mm. These designs are shown in Figs. 10, 11 and 12.

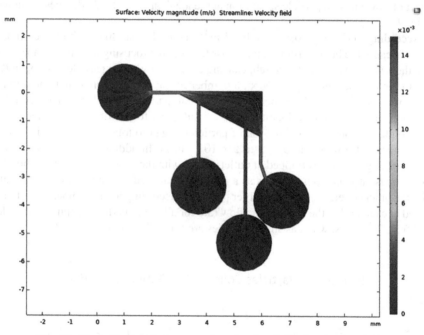

Fig. 10. Streamline inside the tapered angle microfluidic for the first design

Figures 10, 11 and 12 depict the flow within microfluidic devices subjected to laminar flow at the inlet. The trajectory is essential because it determines the path of the particle. Figures 10, 11 and 12 depict the flow through each inlet, which indicates that particles will accumulate at each outlet. Figure 11 depicts, in contrast, the streamline passing through outlets 1 and 3, but not outlet 2 (feedback), resulting in no particles being captured at outlet 2.

Figure 13 depicts a bar graph of the particles collected at each outflow for each of the three designs. The first design demonstrates that sample purity at outlets 1, 2, and 3 is one hundred percent, with large particles at outlet 1 and small particles confined at outlets 2 and 3. This is due to the optimal flow rates and angles used. In contrast to the other two designs, however, the quantity of collected particles is insufficient.

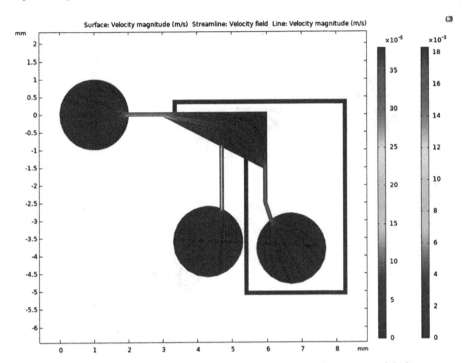

Fig. 11. Streamline inside the tapered angle microfluidic for the second design

The second design demonstrates that the sample purity of large particles at the first outflow is 70%, while that of small particles at the second outlet is 100%. The second design uses feedback by connecting outlet 2 to the tapered area in order to remove the particles trapped in outlet 2. Additionally, this design is capable of capturing more particles at each outlet than the original design.

Recent designs indicate that the sample purity of large particulates at the first outlet is 75%, and at the second outlet, it is 100%. By increasing the height of the concentrating region by 1 mm, the collision between particles, especially small particles, is reduced, resulting in a higher level of sample purity at the first outflow. My design has an advantage over the second one. During the initial discharge, particulates are captured. It may be due to the particles' decreasing velocity, which causes sedimentation to occur unexpectedly quickly. In addition, by increasing the length of the taper region, the third outlet captures a greater number of microscopic particles than the other two designs.

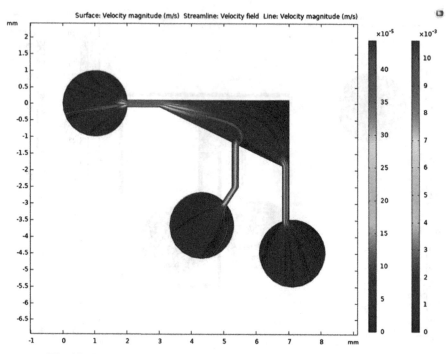

Fig. 12. Streamline inside the tapered angle microfluidic for the third design

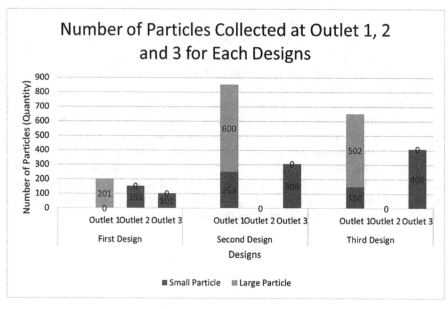

Fig. 13. Particles collected on each outlet for each design

5 Conclusion

Based on the hydrodynamic principle, this study demonstrates the optimal designs of tapered angle microfluidics for particulate separation. Several tapering angles and flow rates were utilized to create the optimal design. This produced favorable outcomes. The device operates using hydrodynamic principles and sedimentation processes. The first design is capable of separating a mélange of PS microbeads of varying sizes and densities with 100 percent sample purity at each outlet. To increase the number of particles captured at each outlet of the initial designs, two new designs were created. The designs can obtain 100 percent sample purity of small particles at outlet 3, but at outlet 1, only 70 and 75 percent sample purity of large particles can be obtained for the second and third designs, respectively. Nonetheless, these devices are able to surmount the limitations of previous cell separation devices based on hydrodynamic principles and have been shown to be capable of separating a mixture of particles using finite element simulation. In addition, the improved sample purity suggests that the proposed design could be used to construct inexpensive and user-friendly microfluidic cell separation devices.

Acknowledgment. This project is supported by the Ministry of Higher Education of Malaysia and Universiti Teknologi Malaysia under the Professional Development Research University Grant (no. Q.J130000.21A2.06E67) in the project "Development of Microfluidic Systems in PDMS for Microalgae Detection and Separation for Renewal Energy Application".

References

1. Wittekind, C., Neid, M.: Cancer invasion and metastasis. Oncology **69**(Suppl 1), 14–16 (2005)
2. Kim, M.-Y., et al.: Tumor self-seeding by circulating cancer cells. Cell **139**(7), 1315–1326 (2009)
3. de Bono, J.S., et al.: Circulating tumor cells predict survival benefit from treatment in metastatic castration-resistant prostate cancer. Clin. Cancer Res. **14**(19), 6302–6309 (2008)
4. Cohen, S.J., et al.: Relationship of circulating tumor cells to tumor response, progression-free survival, and overall survival in patients with metastatic colorectal cancer. J. Clin. Oncol. **26**(19), 3213–3221 (2008)
5. Parkinson, D.R., et al.: Considerations in the development of circulating tumor cell technology for clinical use. J. Transl. Med. **10**(1), 138 (2012)
6. Alix-Panabières, C., Pantel, K.: Clinical applications of circulating tumor cells and circulating tumor DNA as liquid biopsy. Cancer Discov. **6**(5), 479–491 (2016)
7. Allard, W.J., Terstappen, L.W.M.M.: CCR 20th anniversary commentary: paving the way for circulating tumor cells. Clin. Cancer Res. **21**(13), 2883–2885 (2015)
8. Cristofanilli, M., et al.: Circulating tumor cells, disease progression, and survival in metastatic breast cancer. N. Engl. J. Med. **351**(8), 781–791 (2004)
9. Bidard, F.-C., et al.: Clinical validity of circulating tumour cells in patients with metastatic breast cancer: a pooled analysis of individual patient data. Lancet Oncol. **15**(4), 406–414 (2014)
10. How does flow Cytometry work?. Nanocellect (2020). https://nanocellect.com/blog/how-does-flow-cytometrywork/. Accessed: 1 Jan 2022

11. Allan, A.L., Keeney, M.: Circulating tumor cell analysis: technical and statistical considerations for application to the clinic. J. Oncol. **2010**, 426218 (2010)
12. Wikipedia contributors, "Microfluidics," Wikipedia, The Encyclopedia (2021). https://en.wikipedia.org/w/php?title=Microfluidics&oldid=1055354635. Accessed Dec 2021
13. Taniguchi, K., Kajiyama, T., Kambara, H.: Quantitative analysis of gene expression in a single cell by qPCR. Nat. Methods **6**(7), 503–506 (2009)
14. Whitesides, G.M.: The origins and the future of microfluidics. Nature **442**(7101), 368–373 (2006)
15. Rahmanian, N., Bozorgmehr, M., Torabi, M., Akbari, A., Zarnani, A.-H.: Cell separation: potentials and pitfalls. Prep. Biochem. Biotechnol. **47**(1), 38–51 (2017)
16. Yamada, M., Nakashima, M., Seki, M.: Pinched flow fractionation: continuous size separation of particles utilizing a laminar flow profile in a pinched microchannel. Anal. Chem. **76**(18), 5465–5471 (2004)
17. Yamada, M., Seki, M.: Hydrodynamic filtration for on chip particle concentration and classification utilizing microfluidics. Lab Chip **5**(11), 1233–1239 (2005)
18. Manohar, S.M., Shah, P., Nair, A.: Flow cytometry: principles, applications and recent advances. Bioanalysis **13**(3), 181–198 (2021)
19. Zhang, J., Li, W., Alici, G.: "Inertial Microfluidics: Mechanisms and Applications", in Advanced Mechatronics and MEMS Devices II, pp. 563–593. Springer International Publishing, Cham (2017)
20. Chen, J., Li, J., Sun, Y.: Microfluidic approaches for cancer cell detection, characterization, and separation. Lab Chip **12**(10), 1753–1767 (2012)
21. "Sedimentation," BYJUS (2016). https://byjus.com/chemistry/sedimentation/. Accessed 1 Jan 2022
22. Ahmad, I.L., Ahmad, M.R., Takeuchi, M., Nakajima, M., Hasegawa, Y.: Tapered microfluidic for continuous micro object separation based on hydrodynamic principle. IEEE Trans. Biomed. Circuits Syst. **11**(6), 1413–1421 (2017)

Solving a Mobile Robot Scheduling Problem using Metaheuristic Approaches

Erlianasha Samsuria$^{(\boxtimes)}$, Mohd Saiful Azimi Mahmud, and Norhaliza Abdul Wahab

Department of Control and Mechatronics, Faculty of Electrical Engineering, Universiti Teknologi Malaysia Skudai, Johor, Malaysia
nenasha2@graduate.utm.my, azimi@utm.my

Abstract. Flexible Manufacturing Systems (FMS) comprised a number of machine tools alongside other material handling devices that forms a synergistic combination of productivity-efficiency transport and flexibility. In FMS, mobile robots are commonly deployed in the material handling system for the purpose of increasing the manufacturing process' productivity and efficiency. Due to the necessity of navigating from one location to another, it is crucial to properly designed the AMR's schedule in accordance to the real-time situation prior to planning its path. A reliable, efficient, and optimally scheduling is the most important in such manufacturing system. This paper presents the metaheuristic approaches i.e., Genetic Algorithm (GA) and Particle Swarm Optimization (PSO) to deal with the NP-hard problem of scheduling mobile robot in Job-Shop FMS environment. These algorithms are developed to find the feasible solution for the integrated problem with the goal to obtain a minimum completion time of all tasks (or makespan). The results indicated that the performance of GA provided the better solution quality in terms of minimal makespan, while PSO exhibited better convergence times.

Keywords: Genetic Algorithm · Scheduling · Particle Swarm Optimization · Flexible Manufacturing System

1 Introduction

Over the recent years, the market demands have become ever more volatile and abrupt in its changes that promises higher levels of flexibility, versatility and scalability based on reconfiguration of plants to meet the individual requirements of product and overall process [1]. As flexibility gains more importance continuously, flexible manufacturing systems (FMS) will be vital to the manufacturing industry today and into the future, especially in the era of technology that is characterized by the concept of the "smart factory" in the Industry 4.0 initiative [2, 3]. Basically, FMS can be defined as a computer-controlled system that is composed of a set of programmable, numerically controlled machines, interconnected with flexible tools, automated material handling system, loading and unloading stations, and real-time algorithms for scheduling and control [4, 5].

Through material handling system, the intention of flexibility in manufacturing can be achieved [6]. Material handling involves the transportation of raw materials, partially

© The Author(s), under exclusive license to Springer Nature Singapore Pte Ltd. 2024
F. Hassan et al. (Eds.): AsiaSim 2023, CCIS 1912, pp. 29–39, 2024.
https://doi.org/10.1007/978-981-99-7243-2_3

manufactured products, and goods from one location to another within manufacturing systems. It is usually carried out by mobile robots, due to their flexibility and adaptability [7, 8]. In order to optimally utilize them in an effective manner, the proper and effective scheduling of the mobile robot is one of the most important aspects for improving the performance of manufacturing system and current trends towards the use FMS [9, 10]. The problem of scheduling in FMS refers to a process in which mobile robots are assigned to a specific task, while considering cost and required operations' time constraints [11].

As part of the scheduling process, starting and ending times for the individual operations are determined in order to optimize a specified performance measure. Typically, minimizing the completion time of all tasks (or makespan) has been frequently used as the primary objective since it has a direct impact on resource efficiency [12]. The scheduling problem has been dealt before by using various optimization techniques. Focusing on metaheuristic approaches, several approximation algorithms have been applied to the scheduling problem that is generally considered as NP-hard combinatorial problem, such as tabu search [13], ant colony algorithm [14], genetic algorithm (GA) [15–17], artificial bee colony [18] and particle swarm optimization (PSO) [19, 20].

In the field of optimization, GA and PSO represent the most well-known metaheuristic techniques. GA is one of the intelligent and directed random search algorithms that have been broadly used for searching optimal or near-optimal solutions to a range of complex NP-hard problems. Rather than searching for an entire solution, the algorithm focuses on a specific part of it, and thus a near-optimal solution is expected to be attainable much faster [21]. On top of that, GA is the most frequency approaches that have been successfully applied to scheduling problems especially on solving for searching and optimization issues such as job shop scheduling (JSS) and travelling salesman problem (TSP). On the other hand, the main strength of PSO is its fast convergence [22]. It is favourable for its simplicity and ease of use, aside from the fact that it can be applied with a stable and fast convergence speed that can be guaranteed. In light of these characteristics, an extensive range of optimization problems can be solved with PSO.

This paper presents the application of GA and PSO in solving the integrated scheduling problem of mobile robot in Job-shop FMS environment. The issue is featured with both job-shop and mobile robot scheduling problem that is similar to the jobs' transportation between the machines in a given FMS environment where the such problems belong to NP-hard combinational problem. Minimizing the makespan is specified as a performance criterion. Briefly, this work primarily aims to determine the optimal schedule of the order of tasks' execution to be assigned to a single mobile robot for jobs transportation.

2 Methodology

2.1 Problem Formulation

In brief, the FMS entails variety of operations of assorted tasks (or jobs) processed on several machines and a single mobile robot. There are basically three functional areas in the FMS design layout: four different machines (or workstations), a loading/unloading (L/U) station and a charging station, with a single mobile robot to perform related transportation tasks.

Each job should be processed by machines which may require several machines or one machine in each of the stages. Consequently, jobs may be processing in one of the machines during each stage of production and may visit one or more stages of production throughout the course of the process, resulting in different routes of the process for each job. Once the job is started, each operation must be completed without interruption, i.e., no preemption.

Distribution/collection centers are located at L/U stations to manage components. In the production process, a mobile robot is employed for transportation tasks from one machine to another. The robot only able to carry one job at a time, and it is always located at the L/U station as a starting point of the production process. At any given time, the mobile robot can execute the following actions;

(i) Empty trip; the mobile robot travels to the machine to pick up a job once the current operation of the job is completed.

(ii) Loaded trip; the mobile robot carries the job and moves to the machine that perform the next operation.

The mobile robot can travel around the workplace for transporting components between the machines processing operations with a prescribed travelling time. The goal of the described scheduling problem for a particular FMS environment with machine placement in common FMS layouts is to simultaneously optimize an efficient jobs processing and transportation cost of mobile robot. In other words, the reasonable scheduling strategy is to obtain a sequence of jobs or operations on machines while considering the transportation times of mobile robot in the respective environment with the aim to shorten the makespan i.e., the total time spent for the completion of all jobs (as can see in Eq. 1).

In order to illustrate the application of GA and PSO algorithms, an instance of job set 1 from [23, 24] that comprises of 5 jobs, 13 operations and 4 machines is generated as shown in Table 1. Taken from the same reference, the layouts of machine locations which shows the travelling times of mobile robots between the machines are deployed for the computer experiments in this paper. Table 2 presents the machine-to-machine distances or mobile robot travel time matrix from machine-to-machine for layout 1 and 2, while the demonstration of machine locations and mobile robot flow paths in FMS layout can refer to Fig. 1.

Table 1. Job reference for an example FMS scheduling problem.

Ref. Data	Ref. No												
Jobs/Tasks	1			2			3			4		5	
Operations	1	2	3	1	2	3	1	2	3	1	2	1	2
Machines	1	2	4	1	3	2	3	4	1	4	2	3	1
Processing Times	8	16	12	20	10	18	12	8	15	14	18	10	15
Job No Representations	1	2	3	4	5	6	7	8	9	10	11	12	13

Table 2. Travel time matrices

FMS Layout 1 (L1)		L/U	M1	M2	M3	M4
	L/U	0	6	8	10	12
	M1	12	0	6	8	10
	M2	10	6	0	6	8
	M3	8	8	6	0	6
	M4	6	10	8	6	0
FMS Layout 2 (L2)		L/U	M1	M2	M3	M4
	L/U	0	4	6	8	6
	M1	6	0	2	4	2
	M2	8	12	0	2	4
	M3	6	10	12	0	2
	M4	4	8	10	12	0

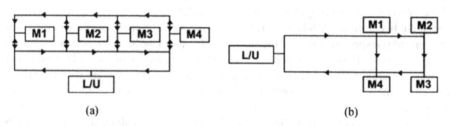

(a)　　　　　　　　　　　　　　　(b)

Fig. 1. Layout of FMS (a) Layout 1 (L1), (b) Layout 2 (L2)

2.2 Objective Function

The scheduling of job shop (or machine) and mobile robot relies on a mathematical model. Here, the model for the optimization of the following objective (fitness) functions is demonstrated mathematically by the equations as follows,

Objective Function,

$$f = \min C_{max} \tag{1}$$

The optimization process is intended to determine a feasible schedule that minimize the makespan, $C_{max} = \text{Max}(C_1, C_2, C_3, \ldots Cn)$, based on the total operations completion time which is defined by Eq. (2–4),

Operation completion time,

$$T_{ij} = t_{mm'} + t'_{mm'} \tag{2}$$

$$O_{ij} = T_{ij} + P_{ij} \tag{3}$$

Total completion time,

$$C_i = \sum O_{ij} \tag{4}$$

where $i = $ job, $j = $ operation, $T_{ij} = $ transportation/traveling times of mobile robot and P_{ij} = processing times. Considering scheduling is a combinatorial problem, it is necessary to choose an appropriate method to optimize the problem.

In this paper, the considered scheduling problem is focused on the flexible manufacturing environment that utilize mobile robot platform for material transport system. Particularly, the operation-based sequence is derived by the developed algorithms that determine how jobs will be scheduled. The assignment of mobile robot for each operation is independent which not include in the initial representation of optimization algorithms and called during the fitness evaluation module.

The problem of job operation-mobile robot scheduling is considerably similar to pick-up and drop-of jobs problem in a given environment and, certainly, it has been classified as NP-hard problems, for which a polynomial-time solution is highly unlikely expected to exist [25]. Hence, this paper attempts to employ GA and PSO for optimizing the feasible schedule with minimal makespan and the performance of their optimization is evaluated and compared.

2.3 Genetic Algorithm

The application of GA to the field of computer science was formerly invented by John Holland in the 1970s [21]. It requires only an appropriate encoding as well as a fitness (objective) function for quantitatively assessing an individual's quality within a set of encoded genes, which called as chromosome. This algorithm consists of three different phases in the search mechanism: the initiation of suitable chromosomes (or solutions) representation, fitness assessment, and genetic operators, namely as selection, crossover, and mutation. The related steps of the genetic algorithm in solving the described scheduling problem can be summarized as follows,

1. Create an initial chromosomes population that represents the operation sequence, each operation indicates a specific task carried out on a specific machine, which can be expressed by the job representation integer number begins with number 1 (as in Table 1). Individual chromosomes have a length equivalent to the total operations' number performed on all jobs.
2. Evaluate the chromosomes' fitness (or objective) value of makespan based on Eq. 1–4, in accordance to the given GA operation sequence.
3. Select the best individuals from the current population by which every individual is paired at random selection method and the fitter individual of the pair with a higher fitness value will be selected as the parent.
4. During crossover operation, copy and exchange the selected job-operation numbers located on parent chromosomes to produce the new offspring
5. Complete the remaining positions of the offspring with the unselected job-operations numbers, in the same order as on the parent's chromosomes.
6. After crossover, some offspring undergo operation shift mutation by inverting the substring between two randomly selected positions within a chromosome.

7. Repeat steps 4–7 until the order of and population size (N_p), crossover rate (P_c), or mutation rate (P_m) are completed.
8. Replace the population of chromosomes from the old one to the new one.
9. Sort the combined parent and child chromosomes based on fitness function cost and select the best of them.
10. Return to step 2 and repeat until the termination criteria is satisfied (i.e., the number of generations reach its maximum, G_{max}).

This developed GA coding structure utilizes the operations-based coding which is comparable to be used in the flexible job shop scheduling [26, 27], with no information of transportation included in the chromosome representation of GA except in the fitness evaluation.

2.4 Particle Swarm Optimization

The second algorithm used in this paper is PSO algorithm, once introduced by James Kennedy and Russell C. Eberhart in 1995 [28]. As the name implies, PSO algorithm works as multiple agents or a group of agents known as particles, which is based on the behavior of birds, fishes and bees that randomly swarm around large areas of places in search of food or even resources for survival purposed. By interacting with other individuals in a population of particles, the PSO algorithm aims to locate an optimal area in a multidimensional complex search space. To solve for the integrated scheduling problem, a PSO implementation steps are described briefly as follows,

1. Create an initial particle swarm by a group of particles, S_i. In this case, each possible sequence of jobs-operations is regarded as a particle, with its dimension representing the operation itself. A swarm of particles with velocities and positions are initialized at random.
2. The encoding of a particle involved with implementing the smallest position value (SPV) rule and assigning the dimensions' codes to the particles based on job representation numbers from Table 1.
3. Once the swarm is initiated, the fitness function of makespan (Eq. 1–4) is evaluated for each particle using the current particle position.
4. Determine and compare the performance of each particle to its personal best (P_{best}), i.e.,

$$\text{If } f(S_i(t)) < f(P_{best}); \ f(P_{best}) = f(S_i(t)); \ P_{best} = S_i(t)$$

5. Determine and compare the performance of each particle to its global best particle (P_{gbest}), i.e.,

$$\text{If } f(S_i(t)) < f(P_{gbest}); \ f(P_{gbest}) = f(S_i(t)); \ P_{gbest} = S_i(t)$$

6. Update the velocity and position from P_{best} and P_{gbest} values of particles, based on the following equations, respectively,

$$v_i(t+1) = w \, v_i(t) + c_1 \text{rand}_i (P_{ibest} - S_i(t)) + c_2 \text{rand}_i (P_{igbest} - S_i(t)) \qquad (5)$$

$$S_i(t+1) = v_i(t+1) \qquad (6)$$

7. Return to step 3 and repeat until the termination criteria is satisfied (i.e., reach the maximum iteration number, I_{max}), then the global best particles return as the best solution.

The setting parameters for implementing PSO including w represents as inertia weight where the inertia weight will decrease linearly during the searching process as described in equation as follow,

$$w = w_{max} - ((w_{max} - -w_{min})/t_{max}) * t \tag{7}$$

Besides, c_1 and c_2 indicate cognitive and social coefficients, respectively, which both coefficients have been tested between the value of 1 to 2. The $rand_i$ represents as the random numbers that to be assigned between [0,1].

3 Results and Discussion

In this section, the experimental results are examined which were aimed at comparing the implementation of GA and PSO algorithms to solve a single mobile robot – job shop scheduling problem in a given FMS environment as described in Fig. 1. The travelling time of mobile robot in the corresponding machine location layout is shown in Table 1 which include time taken of loading and unloading activities of each job.

3.1 Optimization Algorithms Parameter Setup

GA and PSO algorithms for solving the described scheduling problem was compiled with MATLAB 2022b and run under Intel Core i7, 16 GB RAM and Windows 11 operating system. The best selected on parameters used in computer experimental calculations for both GA and PSO are described as in Table 3.

Table 3. Parameter used in computer experiments

	Parameter	Value
Genetic Algorithm	Population size	100
	Maximum genetic generation	100
	Crossover probability	0.6
	Mutation probability	0.2
PSO Algorithm	Particles Number	50
	Maximum iteration	100
	Cognitive coefficient	1
	Social coefficient	1
	Weight of inertia (minimum)	0.4
	Weight of inertia (maximum)	0.6

According to Table 3, the parameters were selected by executing the algorithms repeatedly to solve the described problem until the results obtained were satisfied. Based on GA PSO in this paper, the running times for the algorithms only took a few seconds per execution.

In this section, a performance comparison between the application of PSO and GA was conducted based on convergence time and solution quality. The response for the total completion time of all tasks (or makespan) changes in every iteration for GA and PSO in corresponds to FMS layout 1 and 2 are presented in Fig. 2 and 3, respectively.

As in Fig. 2, the GA provided better solution quality with a total completion time for all tasks (makespan) of 150 as compared to PSO that had a total completion time of 158. Same goes to Fig. 3, the optimized makespan of GA is 116 while PSO makes a makespan value of 119. On the other hand, in terms of convergence performance, PSO performed better on the iterative process with convergence time at the 20th iteration for layout 1 and 13th iteration for layout 2, as compared to GA that had settled down at 50th iteration for layout 1 and 35th iteration. A graphical result in Fig. 4 embodies the GA and PSO optimization scheduling solution owing to the combination of mobile robot assignment particularly for layout 1 can be illustrated in the Gannt Chart representation. From Fig. 4 and 5, the arrow line displays the assigned tasks from machine to another machine locations that need to be visited sequentially by mobile robot, *LT* and *UL* indicate the loading and unloading time of each job, respectively, and the jobs-operation number is denoted as Ji ($i = 1 - 13$).

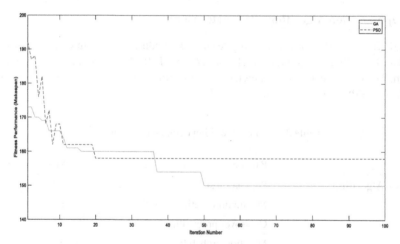

Fig. 2. Fitness cost in each iteration for FMS Layout 1

For FMS layout 1, the best scheduling solution obtained from GA optimization is given as $12 - 8 - 7 - 6 - 3 - 11 - 10 - 9 - 5 - 13 - 4 - 2 - 1$, while the best scheduling solution obtained from PSO optimization is given as $10 - 12 - 4 - 2 - 11 - 3 - 9 - 5 - 13 - 7 - 6 - 8 - 1$. On the other hand, for FMS layout 2, the best scheduling solution obtained from GA optimization is given as $10 - 9 - 2 - 6 - 11 - 3 - 4 - 7 - 13 - 5 - 8 - 12 - 1$, while the best scheduling solution obtained from PSO optimization is given

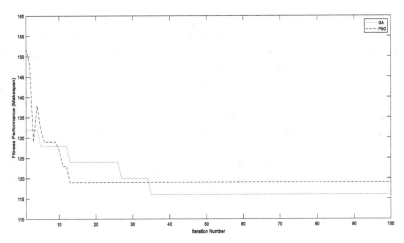

Fig. 3. Fitness cost in each iteration for FMS Layout 2

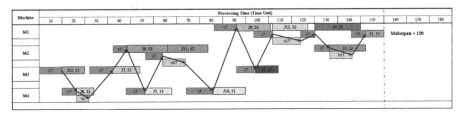

Fig. 4. The scheduling results for GA optimization

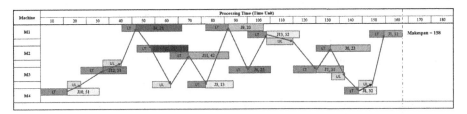

Fig. 5. The scheduling results for PSO optimization

as $4 - 2 - 11 - 8 - 10 - 9 - 1 - 12 - 6 - 3 - 7 - 13 - 5$. The given jobs sequences are presented in terms of jobs representation numbers as shown in Table 1.

4   Conclusion

This paper studies a combinatorial NP-hard single mobile robot scheduling problem in job-shop FMS environment which has been solved by a genetic algorithm and particle swarm optimization. The performance comparison between these two algorithms has been conducted in terms of convergence ability and solution quality of fitness function.

A single objective to obtain a minimum time spent to complete all the assigned tasks (or makespan) is selected as performance criterion in the optimization. In comparison with GA, the experimental results shows that the convergence of PSO over the iterations is significantly fast. However, the GA performed better in finding the best solutions for an optimized makespan value which is much lower than in PSO searching process. Use of GA ensure a quality solution and promising optimization results within an acceptable time frame. Hence, this approach had significant value as an effective method for solving in this kind of integrated scheduling problem.

Acknowledgement. The authors are grateful to the Universiti Teknologi Malaysia and the Ministry of Higher Education (MOHE), for their partial financial support through their research funds, Vote No. Q.J130000.3023.04M12 & R.J130000.7623.4B796.

References

1. Baumann, D., Mager, F., Wetzker, U., Thiele, L., Zimmerling, M., Trimpe, S.: Wireless control for smart manufacturing: recent approaches and open challenges. Proc. IEEE **109**(4), 441–467 (2021)
2. Gania, I.P., Stachowiak, A., Oleśków-Szłapka, J.: Flexible manufacturing systems: industry 4.0 solution. 24th International Conference Prod. Res. ICPR 2017, pp. 57–62 (2017)
3. Mehrabi, M.G., Ulsoy, A.G., Koren, Y.: Reconfigurable manufacturing systems: key to future manufacturing. J. Intell. Manuf. **11**(4), 403–419 (2000)
4. Chang, G.A.: Modeling and analysis of flexible manufacturing systems : a simulation study modeling and analysis of flexible manufacturing systems : a simulation study. In: 122nd ASEE Annual Conference & Exposition, p. 12726 (2015)
5. Mejthab, Y.M., Al-aubidy, K.M.: Design and implementation of real-time scheduling algorithms for flexible manufacturing systems. Int. J. Adv. Mechatron. Syst. **7**(4), 202–212 (2017)
6. Fragapane, G., Ivanov, D., Peron, M., Sgarbossa, F., Strandhagen, J.O.: Increasing flexibility and productivity in Industry 4.0 production networks with autonomous mobile robots and smart intralogistics. Ann. Oper. Res. **308**(1–2), 125–143 (2022)
7. Kumar, S., Manjrekar, V., Singh, V., Kumar, B.: Integrated yet distributed operations planning approach: a next generation manufacturing planning system. J. Manuf. Syst. **54**(June 2019), 103–122 (2020)
8. Reddy, B.S.P., Rao, B.S.P.: Flexible manufacturing systems modelling and performance evaluation using automod. Int. J. Simul. Model **10**, 78–90 (2011)
9. Dang, Q.V., Nguyen, L.: A heuristic approach to schedule mobile robots in flexible manufacturing environments. Procedia CIRP **40**, 390–395 (2016)
10. Petrović, M., Miljković, Z., Jokić, A.: A novel methodology for optimal single mobile robot scheduling using whale optimization algorithm. Appl. Soft Comput. J. **81**, 105520 (2019)
11. Mousavi, M., Yap, H.J., Musa, S.N.: A fuzzy hybrid ga-pso algorithm for multi- objective agv scheduling in fms. Int. J. Simul. Model. **16**, 58–71 (2017)
12. Abdelmaguid, T.F., Nassef, A.O., Badawia, A.: A hybrid GA / heuristic approach to the simultaneous scheduling of machines and automated guided vehicles. Int. J. Prod. Res. **42**(2), 267–281 (2004)
13. Shen, L., Dauzère-Pérès, S., Neufeld, J.S.: Solving the flexible job shop scheduling problem with sequence-dependent setup times. Eur. J. Oper. Res. **265**(2), 503–516 (2018)

14. Rashmi, M., Bansal, S.: Task scheduling of automated guided vehicle in flexible manufacturing system using ant colony optimization. Int. J. Latest Trends Eng. Technol. **4**(1), 177–181 (2014)
15. Sharma, D., Singh, V., Sharma, C.: GA based scheduling of FMS using roulette wheel selection process. Adv. Intell. Soft Comput. **131**(2), 931–940 (2012)
16. Dang, Q.V., Nielsen, I.: Simultaneous scheduling of machines and mobile robots. Commun. Comput. Inf. Sci. **365**, 118–128 (2013)
17. Sanches, D.S., da Silva Rocha, J., Castoldi, M.F., Morandin, O., Kato, E.R.R.: An adaptive genetic algorithm for production scheduling on manufacturing systems with simultaneous use of machines and AGVs. J. Control. Autom. Electr. Syst. **26**(3), 225–234 (2015)
18. Sharma, N., Sharma, H., Sharma, A.: Beer froth artificial bee colony algorithm for job-shop scheduling problem. Appl. Soft Comput. J. **68**, 507–524 (2018)
19. Petrović, M., Vuković, N., Mitić, M., Miljković, Z.: Integration of process planning and scheduling using chaotic particle swarm optimization algorithm. Expert Syst. Appl. **64**, 569–588 (2016)
20. Sreekara Reddy, M.B.S., Ratnam, C., Rajyalakshmi, G., Manupati, V.K.: An effective hybrid multi objective evolutionary algorithm for solving real time event in flexible job shop scheduling problem. Meas. J. Int. Meas. Confed. **114**, 78–90 (2018)
21. Candan, G., Yazgan, H.R.: Genetic algorithm parameter optimisation using Taguchi method for a flexible manufacturing system scheduling problem. Int. J. Prod. Res. **53**(3), 897–915 (2015)
22. Liu, H., Wang, Y., Tu, L., Ding, G., Hu, Y.: A modified particle swarm optimization for large-scale numerical optimizations and engineering design problems. J. Intell. Manuf. **30**(6), 2407–2433 (2019)
23. Bilge, U., Ulusoy, G.: Time window approach to simultaneous scheduling of machines and material handling system in an FMS. Oper. Res. **43**(6), 1058–1070 (1995)
24. Chakrabarti, S., Saha, H.N.: Simultaneous scheduling of machines and automated guided vehicles utilizing heuristic search algorithm. In: 2018 IEEE 8th Annual Computing and Communication Workshop and Conference (CCWC), pp. 54–59 (2018)
25. Optimization, D.C.: A deterministic single exponential time algorithm for most lattice problems based on voronoi cell computations. SIAM J. Comput. **23**(4), 1364–1391 (2013)
26. Wan, M., Xu, X., Nan, J.: Flexible job-shop scheduling with integrated genetic algorithm. In: Proceedings of 4th International Workshop on Advanced Computational Intelligence, IWACI 2011, pp. 13–16 (2011)
27. Kumar, M.V.S., Janardhana, R., Rao, C.S.P.: Simultaneous scheduling of machines and vehicles in an FMS environment with alternative routing. Int. J. Adv. Manuf. Technol. **53**, 339–351 (2011)
28. Kennedy, J., Eberhart, R.: Particle swarm optimization. Neural Networks, 1995. Proceedings., IEEE Int. Conf., **4**, pp. 1942–1948 (1995)

Research on General Automatic Test Platform for Onboard Automatic Train Protection

Zhang Guozhen[✉], Jiang Yunwei, and Shi Miao

Beijing National Railway Research and Design Institute of Signal and Communication Group
Co., Ltd., Beijing 100070, China
guozhen_cn@163.com

Abstract. Traditional train control systems experience problems such as low efficiency, tedious operation, high labor costs, poor visibility, and difficulty in designing and maintaining test cases during manual functional testing of onboard automatic train protection (ATP) equipment in a semi-physical testing environment. In the study, we conducted in-depth research on the functions of various ATP products and designed and implemented a general automatic test platform for ATP. Additionally, we developed a general script language for onboard ATP testing that can be converted into a set of specific script instructions based on train type. For the hardware, an extensible interface platform was designed and developed and combined with a robotic arm control unit and driver–machine interface image recognition unit to simulate ATP external input signals and driver operation. For the software, we designed a general communication protocol between devices; developed a series of toolsets including script parsing and execution, communication protocol conversion, and test process management; and realized the automatic control of the test process. An actual engineering application indicates that this ATP general automatic test platform can be used to reduce development costs and improve test production efficiency, thereby showing high practical value.

Keywords: Onboard ATP · Automatic Test Platform · Test Script Language · Test Process Management

1 Introduction

Onboard automatic train protection (ATP) equipment monitors the safe operation of trains based on line data and train information wayside equipment and is the key core equipment of railroad train operation control systems. In-depth and comprehensive functional testing of the ATP equipment is important to ensure its reliability. In recent years, researchers have extensively studied the design of onboard ATP test platforms, mainly focusing on semi-physical simulation testing, by building a simulation test interface platform to simulate wayside conditions (including track circuit and balise information) as well as external input signals of ATP (including simulated train cabin, driver operation, output feedback, speed, and distance) to provide the external input information required for testing ATP, which is used to simulate a real operating environment and to control the test process by manually operating the simulation software [1–4]. Some researchers

F. Hassan et al. (Eds.): AsiaSim 2023, CCIS 1912, pp. 40–49, 2024.
https://doi.org/10.1007/978-981-99-7243-2_4

have also automated the test process by designing and writing test information files or test scripts in a specific format to describe the test cases; however, many of these test scripts are in a complex format that is difficult to edit and understand, have no automatic control of the entire test process [5–7], or require expensive test script execution engines [8].

Based on the above analysis, this study designed a general automatic test platform for onboard ATP comprising a common ATP testing script language and script execution parsing engine, an extensible hardware interface platform, a robotic arm and driver–machine interface (DMI) image recognition unit to simulate external input and output signals, and the driver operation of ATP. The platform includes a communication protocol between various devices, toolsets for test script editing, and test process management to realize test process automation control.

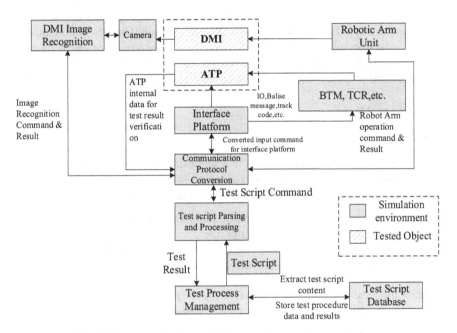

Fig. 1. Structure of onboard ATP general automatic test platform.

The structural diagram of the onboard ATP general automatic test platform is shown in Fig. 1. The ATP and DMI devices are the objects under test. The extensible hardware interface platform simulates the external interface signals of the ATP device, providing signals of both digital and analog quantities such as pressure, speed transmission pulse signals, serial port, and controller area network (CAN). It also receives various bus and input/output (I/O) device outputs of the ATP device, calculates the train travel distance, etc., and forwards this information to the protocol conversion software. The robotic arm unit receives control commands, simulates the train driver operation of the DMI, and feeds back the key results. The image recognition unit collects and recognizes key information of the DMI in real time. The protocol conversion software acts as a data transmission and conversion channel for the interface platform, robotic arm, DMI image recognition and other test peripherals, ATP status record data, and script analysis software. It converts the received script command data into the corresponding data format according to the communication protocol with the test equipment and converts the received test equipment and train location information into the format required by the script parsing software. The test script parsing software parses and executes the test instructions according to the onboard ATP test script specification and sends the instruction execution status to the test process management software in real time. The test process management software realizes the configuration and automatic execution of test case scripts, real-time judgment and display of test status, automatic switching of multiple test cases after execution, etc., to automatically control and manage the test process.

The overall test is based on a script-driven approach. The test platform provides input signals for the ATP according to the driver instructions, operates the DMI keys, collects the bus and I/O status information of the ATP, and sends it to the communication protocol conversion software, which then forwards it to the script parsing software as the basis for judging whether the test execution is correct, and finally displays the script execution results in the test flow management software to complete the test verification.

2   Software Design

2.1   Generic Scripting Language Design

The ATP automatic test platform is built in a script-driven manner. Test scripts are the basis of automatic testing, and the design of scripts should maximize the realization of ATP functional testing. The design of the test script language consists of three aspects:

1. Study different standards of ATP test cases, summarize the common behaviors of test scenarios, test methods, and design script rules;
2. Analyze the working mode of ATP peripheral devices, adopt keyword technology to abstract the communication between the script simulation device and ATP, and formulate script instructions;
3. Specify the message description format of the script simulation device according to the communication protocol and message composition of the ATP and peripheral devices.

- **Script Rule Design**

  Script rules define the structure of script composition, including script comments and script directives; the script comment uses "//" to indicate a single-line comment. The script instruction contains the train position, instruction name, and instruction field key-value pair content. When the train position meets the instruction execution condition (forward: train position > instruction trigger position; reverse: train position < instruction trigger position), the script instruction is executed.
- **Script Command Abstraction**

  The test script must simulate the operation of ATP peripheral devices (balise data transmission unit, track circuit receiver, radio block center (RBC), etc.) and summarize the common behaviors of the devices. The interaction between ATP peripheral devices and the ATP generally falls within one of the following two cases:

1. ATP peripheral devices only send data and the ATP receives them, such as balise, track circuit receiver devices, etc.;
2. ATP peripheral devices interact with the ATP for data sending and receiving, and they reply to the corresponding messages according to the content of the messages sent by the ATP, such as RBC devices, etc. Accordingly, two types of script instructions are designed: message sending and message receiving.

Typical instructions of the message-sending class:

1. SEND_MSG: send a message with the specified content once;
2. START_PERIOD_SEND_MSG: Send the contents of the message defined in the script at the specified period;
3. START_AUTO_ANSWER_MSG: A cycle check automatically sends a message specified in the script after receiving the specified message.

Typical instructions of the message-reception class:

1. WAIT_MSG: Wait for a message to be sent by the ATP and continue with subsequent script instructions if the current message matches the content;
2. EXPECT_MSG: Determine whether the content of the received message is correct and whether the test result meets the expectation.

This is supplemented by special keywords (direction of travel, test maximum time), automatic configuration of message reply content, and other instructions to increase the usability and scalability of the script language.

- **Message Content Configuration**
  The message description format and configuration file structure of the script are specified according to the communication protocol and the transmitted message composition between the ATP and peripheral devices, such as balise transmission module(BTM), radio block center (RBC), and track circuit receivers. The message configuration is written in XML format and contains information such as message length, packet number, field type, placeholder, and packet nesting. For transmitting commands, the script parsing software converts the field key-value pairs into a byte stream format according to the message configuration XML file and sends them (e.g., transponder messages, RBC messages, orbital circuit low-frequency carrier information). For receiving commands, the script parsing software parses the received byte stream information into message field key-value pairs according to the XML configuration file and compares the fields individually to determine whether they meet the expectations.

2.2 Test Script Management

Based on the test script structure, a relational database is used to deconstruct and store the script content, and a tree data table structure is designed from the project to the test command fields according to the top-down dependency relationship, as shown in Fig. 2.

A specialized script editing tool is developed to support testers and developers to write test cases according to scripting rules by graphical means; store real-time script editing information in the corresponding form; support structured online editing of scripts;

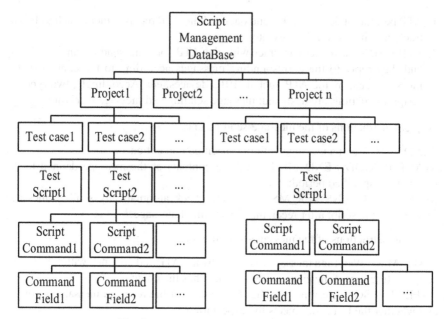

Fig. 2. Test script database storage structure.

realize script instruction copy, insert, modify, delete, parse, and other operations; and support previewing script content and importing existing scripts to effectively improve test scripting efficiency and accuracy.

Through the script editing tool, the test scripts written by the tester are classified according to the project. When testing, the tester can directly select the sequential test, or select several test scripts for testing separately, and a variety of test categories is supported.

2.3 Test Process Management

● **Test Process Management Architecture Design**

Based on the design concept of the cloud platform, the test process management software adopts a browser/server (B/S) architecture with front-end and back-end separation in which the test business logic, protocol conversion, and network communication are realized on the server side; the browser is used as the human–computer interaction interface, with only business data interaction between the front and back ends, which can be developed and deployed separately, thereby enabling less software coupling, high scalability, high flexibility, and easy maintenance.

The management software architecture, shown in Fig. 3, comprises four layers: (i) the human–computer interaction layer provides a friendly user interface, including script editing and management, test execution control, and historical data browsing; (ii) the business logic layer realizes specific business functions, including data storage, script generation, and automated testing; (iii) the access layer realizes the interaction between

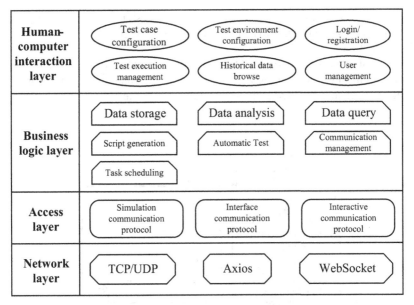

Fig. 3. Test process management software architecture diagram.

software platforms based on different communication protocols; (iv) the network layer defines the Ethernet communication mode and supports transmission control protocol (TCP), user datagram protocol (UDP), WebSocket, and other communication methods.

Test process management software is required to perform the following functions:

1. Test environment configuration to achieve test item selection, simulation unit configuration, and test sequence generation;
2. Test execution management to enable the test-dependent software to start in an orderly manner, graphical status monitoring, and support for automatic succession testing of multiple test cases;
3. Historical data management to achieve orderly and centralized storage of test history data and multi-mode browsing and querying;
4. Test report generation to support customized test report templates and automatic report generation after test completion;
5. User management to support user addition, deletion, password reset, and permission management.

- **Testing Process**
  When the tester uses automatic test platform for testing, the whole testing process is divided into three stages: test preparation, test execution and test result analysis.

1. Test Preparation
   Testers first need to configure the test environment, including test mode, startup software, and track code configuration, and then complete the test sequence configuration for the product functional test requirements, including test case selection and sequence

adjustment. The configuration information is textually stored in the JavaScript Object Notation (JSON) data format to facilitate test process parsing and configuration reuse.

2. Test Execution

The test management software then implements automated testing and real-time status monitoring based on the test configuration information. As shown in Fig. 4, the test sequence is selected at the front end to start the test.

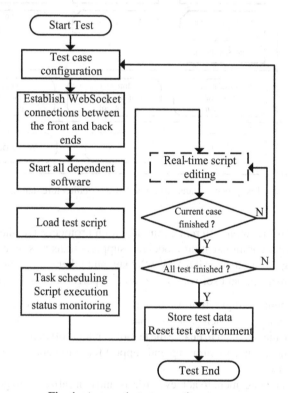

Fig. 4. Automatic test execution process.

Step 1: The back end parses the test configuration to obtain the test sequence, including the test case execution order, script information contained in the test case, and test resource configuration.

Step 2: Unlike with traditional asynchronous communication, WebSocket allows the back end to actively send data to the front-end page, changing the passive reply and improving the system sensitivity.

Step 3: The test mode is determined. The software logic test sequentially starts the simulation simulator; the semi-physical simulation test powers the device and checks the device status.

Step 4: The script database information is extracted to generate test scripts.

Step 5: According to the interaction scenario, multi-script task scheduling and execution is conducted, parsing test scripts individually and sending them to peripheral test devices for execution by protocol conversion. The back end receives test status data in real time and sends them to the front end for display. Simultaneously, it supports real-time editing of unexecuted test script statements during the testing process.

Step 6: The test progress is continuously monitored. If the execution of all test scripts of the current test case is completed, the current test case is considered to have completed the test and proceeds to the next step.

Step 7: The test data are stored, the test environment is reset, and the test is completed.

3. Test Result Analysis

The test result analysis consists of report generation and exception analysis. The test report generation module supports customized report template selection (Word/Excel format), report path storage setting, and report emailing, including execution case results, execution error instruction content, and actual test data compared with expectations. Test exception analysis displays test process data in the form of graphs and charts, including information such as received test history data and test-dependent software interaction status. Testers can quickly locate test problems based on reports and historical data.

3   Hardware Design

The hardware of the ATP general test platform includes the interface platform, robotic arm unit, and DMI image recognition unit. Among them, the robotic arm unit simulates the driver's key operation on the DMI according to the script control command, supports multiple operation modes, and contains the execution feedback function. The DMI image recognition unit is used to collect and recognize the display image of the DMI and forward the recognized key information as one of the bases of the test judgment results. The interface platform is the core of the semi-physical simulation and receives the control commands sent by the communication protocol conversion software, parses and converts them into the output of the corresponding channel, and provides feedback on the operation status of the ATP.

The structure of the interface platform, shown in Fig. 5, consists of four parts: the control unit, interface adapter box, interface cable, and power supply. The interface adapter box is used to directly connect the board resources of the control unit with the ATP through a cable and contains a speed adapter board to realize the simulation of the speed transmission pulse signal. The interface cable is used to connect the control unit, interface adapter box, controllable transponder, and ATP. The power supply is used to provide the ATP logic working power and the input and output interface power. The control unit consists of a CompactPCI chassis, central processing unit (CPU) card, I/O card, communication card, and controllable transponder. Among them, the CPU card runs the interface platform program to control the board signal output and acquisition and interacts with the test management software in real time. The I/O card completes the signals, including digital input (DI), digital output (DO), analog input (AI), and analog output (AO) signal acquisition and output. Specifically, the DO card considers the flow limitation of the output node of the board and adds relays on the channel to achieve electrical isolation to avoid damaging the board drive circuit. The communication card supports the multifunction vehicle bus (MVB), CAN bus, and serial port communication.

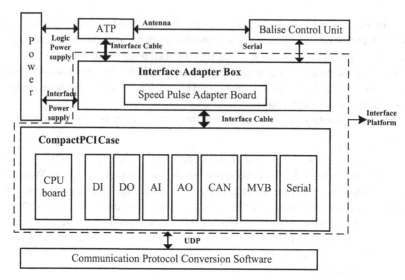

Fig. 5. Structure of hardware interface platform.

The controllable transponder controls the transmission of transponder messages such that the transponder antenna of the tested equipment can receive the message data in time. The interface unit communicates with the communication protocol conversion software via Ethernet, receives the converted script command data, and forwards the ATP output, train travel distance, and other information calculated according to the speed pulse.

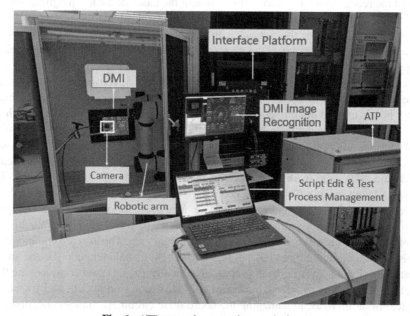

Fig. 6. ATP general automatic test platform.

4   Conclusion

To solve the existing problems of low time efficiency, cumbersome operation, and high labor cost in the semi-physical test environment of ATP, this study researched and designed a general automatic test platform for ATP, as shown in Fig. 6.

By designing a general scripting language for testing, test cases are described in the form of script instructions to provide a basis for automatic testing. The hardware layout of the test platform was optimized, with an expandable vehicle interface platform, robotic arm unit, and DMI image recognition unit to realize unmanned automatic testing of the platform and reduce testing labor costs. Additionally, test process management software was developed to realize test case configuration, automatic execution, and report generation. The entire test process is controlled automatically. Through an actual engineering application, the onboard ATP general automatic test platform was proven to be fully functional, interactive and friendly, and highly automated, which effectively reduces the work intensity of testers and significantly improves the efficiency of ATP product testing.

References

1. Ao, Q., Xie, Z.: On-board ATP test platform technology based on PXI general bus technology. Railway Communication and Signal Eng. Technol. **17**(11), 13–15 (2020)
2. Dai, M.: Design of on-board ATP test system. Railway Communication and Signal Eng. Technol. **12**(01), 15–16 (2015)
3. Wei, L.: On board ATP system simulation test platform. Railway Communication and Signal Eng. Technol. **10**(04), 22–24 (2013)
4. Yang, K., Ke, C., Xue, R.: Design and implementation of interface unit in on-board ATP simulation test environment. Control and Information Technol. **03**, 52–55 (2019)
5. Luo, F., Li, Y.: Design and implementation of automatic simulation test platform for CTCS-2 train control on-board equipment. Railway Communication Signal **54**(04), 75–79 (2018)
6. He, X., Wang, Y., Zhou, Z.: Research on key technologies of train control on-board software automatic test system. Locomotive Electric Drive **05**, 89–91 (2014)
7. Hou, Q., Li, K., Zhang, Y.: Research on test method of ATP on-board equipment based on script drive. Jiangsu Science and Technol. Inf. **36**(22), 46–48 (2019)
8. Li, Z., Xu, X., Ou, G.: Design and implementation of automatic test scheme of train control on-board system based on TestStand. Railway Communication Signal **57**(04), 5–9 (2021)

Research and Implementation of Simulation Testing Technology for Train Control System Based on Satellite Positioning

Yang Chen[1], Xiaoyu Cao[1], and Xingbang Han[2(✉)]

[1] China Energy Xinshuo Zhunchi Railway (Shanxi) Co., Ltd., Shanxi 038500, China
[2] China Railway Signal and Communication Research and Design, Institute Group Co., Ltd., Beijing 100070, China
han_xingbang@aliyun.com

Abstract. In this study on satellite-based train control systems, the data structure of electronic maps was analyzed. Moreover, a semi-physical simulation system for laboratory system integration testing was designed and implemented. By parsing and computing electronic maps and combining them with train control engineering data, the proposed design swiftly constructs a system integration testing environment, improves the real-time calculation efficiency of satellite simulation data, supports the testing and verification of the functions and data of satellite-based train control systems, and provides agile and powerful technical support for equipment development and laboratory testing. The design's viability was confirmed via system deployment and laboratory tests in an engineering project, fulfilling the laboratory testing criteria.

Keywords: Train control system · Simulation test · Satellite positioning · Electronic map

1 Overview

Satellite-based train control systems are emerging as next-generation train control systems. The application of satellite positioning technology can enhance existing train speed and distance measurement methods while reducing the deployment of ground-based balises and other wayside devices used for assisting train positioning, thus saving system construction and maintenance costs [1]. With the promotion and application of the BeiDou Navigation Satellite System (BDS) 3.0 [2], various types of satellite-based train control systems have been developed [3].

Train control systems are important for ensuring safe train operation and improving transportation efficiency [4]. Therefore, sufficient system testing and verification are required during system development and engineering implementation to determine whether a train control system meets the specification requirements. For satellite-based train control systems, a simulation testing system is required to support laboratory system functional and data testing [5, 6]. Currently, most testing systems used in domestic

F. Hassan et al. (Eds.): AsiaSim 2023, CCIS 1912, pp. 50–59, 2024.
https://doi.org/10.1007/978-981-99-7243-2_5

applications for satellite-based train control systems are designed only for onboard equipment. There is a lack of mature and systematic design solutions for semi-physical system integration testing environments that can simultaneously integrate interlocking, wayside train control, and onboard equipment for satellite-based train control systems. This hinders testing and verification of satellite positioning-related equipment and functions of train control systems within a large system.

All in all, in this study, a complete simulation testing scheme for satellite-based train control systems was designed and implemented based on the design architecture of existing train control systems, including a full set of solutions such as generation, use, and monitoring of satellite simulation data. The system can quickly generate satellite simulation data and complete the testing environment construction based on engineering data and electronic maps, thus satisfying the laboratory system integration testing requirements of satellite-based train control systems.

2 Architectural Design

The integration testing system architecture of existing train control systems is relatively mature. Taking the CTCS(China Train Control System)-3-level train control system as an example, a semi-physical system integration testing environment is constructed by providing train, wayside equipment, and train-ground communication channel simulations, and integrating on-board, wayside train control, and station interlocking equipment. A human-machine interface is provided through train and wayside equipment simulations used to simulate normal system operation and abnormal scenarios, thereby testing and verifying the system functions as well as data of the integrated physical equipment and entire system.

Based on the existing train control system integration testing architecture, the basic functions of train, wayside equipment, and train-ground communication channel simulations were maintained, and satellite data simulation-related modules were added. On-board, wayside train control, and station interlocking equipment for satellite-based train control systems was integrated to construct a semi-physical simulation testing system, as shown in Fig. 1.

The integration testing simulation system has a unified interface among modules, and communication is performed using Ethernet to reduce system construction costs. The physical interface layer module has an actual interface with the object under test, including Ethernet, and digital and analog signals.

An additional satellite simulation data generation interface (including the processing module) was added to calculate and generate a table including satellite positioning data for all track sections in advance. This table is used to provide the simulation host with real-time conversion of train position satellite data, thereby reducing the satellite simulation calculation during the operation of the simulation system.

The simulation host is the core module of existing train-control integration testing systems. It reads the line data from the database, calculates the train's simulated position information in real time, returns the wayside equipment information (track circuits, balises, etc.) of the train's location, and interacts with the track section occupancy. In this design, the existing calculation logic for train positioning in the simulation testing

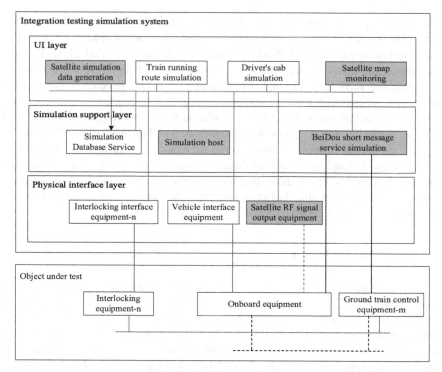

Fig. 1. Architectural design.

system was reused. Based on the position of the train in the track section, real-time conversion of the satellite simulation data of the current position point was added. This is used to generate satellite RF(Radio Frequency) signals for the onboard equipment. These signals are displayed on the satellite map monitoring interface.

The satellite RF signal output equipment is the key device for onboard equipment access. This equipment is used to convert satellite simulation data into RF signals consistent with the output specifications of real satellite systems, realizing the timing and positioning of satellite receiving terminals and the satellite positioning receiving antenna interface of onboard equipment. The satellite receiving and processing units of the onboard equipment were included in the measurement range. The system ensures that the onboard equipment receives only simulated satellite positioning data through laboratory signal shielding conditions and power modulation of the satellite signal output equipment, thereby achieving train positioning and operation at any engineering line position in the laboratory.

The satellite map monitoring interface uses GIS (Geographic Information System) and other technologies to display the real-time position and running trajectory of a train, providing users with an observation of the train's running status. The operation of the satellite positioning simulation was monitored by comparing the position of the train on the topological track at the simulation interface with the running line.

The BeiDou short message service simulation module is suitable for simulation testing environments based on the BeiDou navigation positioning system. It supports the interface of train and ground equipment and uses the BeiDou short message service for train-to-ground interactions. It provides a simulation channel for the BeiDou short message service to construct a simulation testing environment based on the full function "Timing + Positioning + Short message" of the BeiDou navigation system.

3 Key Technology

3.1 Calculation of Satellite Simulation Data

Key for this simulation system is the calculation of satellite simulation data for the current position of the train model with real-time changing position. It is necessary to comprehensively con-sider the accuracy, calculation efficiency, and environmental construction cost of the real-time calculation output expressed in terms of longitude, latitude, and altitude. A solution is proposed to pre-calculate the satellite simulation data and simplify the calculation using a table lookup during the real-time calculation process, as shown in Fig. 2. Electronic maps and engineering data tables are used as input files.

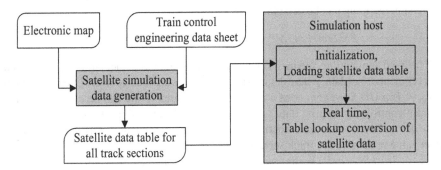

Fig. 2. Calculation process of satellite simulation data.

Electronic map data form the basis for implementing satellite positioning [7] and include the geo-graphic information files of the track and fixed application data. The track geographic information file contains the position information (longitude, latitude, and altitude) of the track segment division and track segment record points surveyed for the line. Using the electronic map structure of a new type of train control system as an example, the track geographic information data structure is illustrated in Fig. 3.

The principle of dividing the track segment in an electronic map is that the over-all geometric shape of the track conforms to the actual situation of a straight-line or curved-line segment. The curved-line segment is approximated by multiple straight-line segments (with a fixed heading angle) [8, 9]. The electronic map contains the longitude, latitude, and altitude values of the recorded points at both ends of each track segment, defined as $P_S < Lon,Lat,Alt >$. Combined with $P_S < Lon,Lat,Alt >$ of the recorded

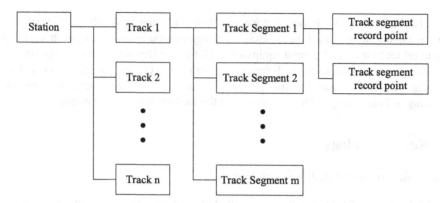

Fig. 3. Structure of track geography information data of electronic map.

points at both ends of the track segment, the position information $P_S < Lon,Lat,Alt >$ of the target point can be calculated using linear interpolation by searching for the track, track segment, and offset value of the target position point. The combination of the track, track segment, and offset value of the target position point is defined as $P_{eMap} < TrackID,SegID,Sm >$, where $TrackID$ is the track number, $SegID$ is the track segment number, and Sm is the offset value of the target position point on the track segment. Conversion mapping can be established as follows:

$$PS\langle Lon, Lat, Alt \rangle \leftrightarrow PeMap\langle TrackID, SegID, Sm \rangle \tag{1}$$

The fixed application data in the electronic map contain the track, track segment, and track segment offset data, where devices such as balises are located in the line from which $P_{eMap} < TrackID,SegID,Sm >$ of a certain balise position point can be obtained. Using the train-control engineering data table, the position relationship of the corresponding balise device based on the track section in the line can be calculated and expressed by the offset value relative to the track section endpoint, defined as $P_{line} < TrackSectionID,St >$, where $TrackSectionID$ is the track section number, and St is the offset value of the target position point on the track section. Therefore, by representing the position of a certain balise in both the track-line and electronic-map systems, conversion mapping can be established as follows:

$$PeMap\langle TrackID, SegID, Sm \rangle \leftrightarrow Pline\langle TrackSectionID, St \rangle \tag{2}$$

By using Eqs. (1) and (2), Eq. (3) from the target point based on the track section and offset value to the satellite positioning data can be obtained:

$$PS\langle Lon, Lat, Alt \rangle \leftrightarrow Pline\langle TrackSectionID, St \rangle \tag{3}$$

Equation (2) can be applied using a full-line track-section mapping table. During the real-time calculation process of the simulation host, $P_{eMap} < TrackID,SegID,Sm >$ can be obtained by looking up the table with $P_{line} < TrackSectionID,St >$ of the current position (target point) of the train; thus, the satellite positioning data $P_S < Lon,Lat,Alt >$ of the target point can be obtained by a linear calculation.

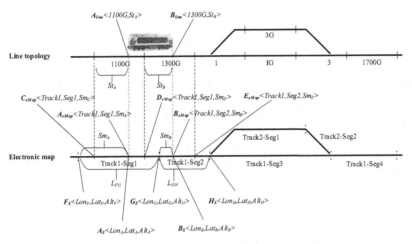

Fig. 4. Example of satellite simulation data calculation.

Taking Fig. 4 as an example, the positions of target points A and B at the head and tail of the train in the topological position of the line are known (A_{line} and B_{line} in Fig. 4). The process of calculating the longitude, latitude, and altitude of positioning points A and B (A_S and B_S in Fig. 4) is as follows:

1. Using Eq. (2), the track and track segment of target points A and B in the electronic map can be obtained by looking up the table; simultaneously, the specific position of the endpoints of the track section where target points A and B are located in the topological position of the line can be obtained in the electronic map (C_{eMap} and D_{eMap} in Fig. 4).

2. The offset Sm of target points A and B in the track segment of the electronic map can be calculated using Eqs. (4) or (5), and the specific positions of points A and B in the electronic map can be obtained (A_{eMap} and B_{eMap}); L_{FG} in Eq. (5) is the length of the track segment.

$$SmA = SmC + StA \tag{4}$$

$$SmB = StB - (LFG - SmD) \tag{5}$$

3. According to the track segment where A_{eMap} and B_{eMap} are located, the longitude, latitude, and altitude of the corresponding endpoints of the track segment can be obtained by looking up the table (F_S, G_S, and H_S in Fig. 4).

4. Using Eqs. (6)–(8), the longitude, latitude, and altitude of positioning point A (A_S in Fig. 4) can be calculated. By replacing A with B in the formulas, the longitude, latitude, and altitude of positioning point B (B_S in Fig. 4) can be obtained.

$$\frac{(LonA - LonF)}{(LonG - LonF)} = \frac{SmA}{LFG} \tag{6}$$

$$\frac{(LatA - LatF)}{(LatG - LatF)} = \frac{SmA}{LFG} \tag{7}$$

$$\frac{(AltA - AltF)}{(AltG - AltF)} = \frac{SmA}{LFG} \tag{8}$$

Based on the above, the accuracy of the satellite simulation data is ensured by fully utilizing electronic map data and the train control engineering data table. By pre-calculating and moving part of the conversion logic to the system initialization, the real-time calculation of satellite simulation data is reduced and its efficiency is improved. Standardization of the electronic map and train control engineering data tables is utilized to automatically parse data resulting from the application of Eq. (2) through computer analysis, thereby achieving rapid construction of satellite simulation data and reducing environmental construction costs.

3.2 Satellite RF Signal Generation

Generating satellite RF signals at any position along the line in a closed laboratory environment and transmitting them to the satellite receiver antenna of onboard equipment constitutes a key technology for the inclusion of the satellite receiving and processing units of the onboard equipment within the test scope. This improves the authenticity of the semi-physical system and expands the boundaries of system testing.

By selecting an independent satellite RF signal output device and receiving control commands from the simulation host to generate RF signals, the calculation of the satellite simulation data and generation of RF signals are decoupled, making the system modular.

3.3 System Implementation

Incremental development of the satellite simulation data generation module, real-time calculation module for satellite data on the simulation host, and satellite map monitoring interface was added to the existing train control system simulation test system. A mature third-party device was used as the satellite signal output equipment to achieve a simulation test system suitable for satellite-based train control systems.

The system is capable of generating pre-calculated satellite data tables with a single click while performing data correctness verification on electronic map files (Fig. 5).

The real-time calculation provided by the semi-physical simulation system meets the requirement of maintaining stable satellite and fusion positioning for onboard equipment while conducting tests on a train running at a speed of 350 km/h in an indoor laboratory environment.

The train running route simulation interface (Fig. 6), satellite map monitoring interface (Fig. 7), and interface of the third-party satellite RF signal output device (Fig. 8) allow for real-time observation of the current positioning data and support the construction of abnormal satellite signal scenarios (such as signal obstruction and positioning anomalies), effectively supporting laboratory system integration testing.

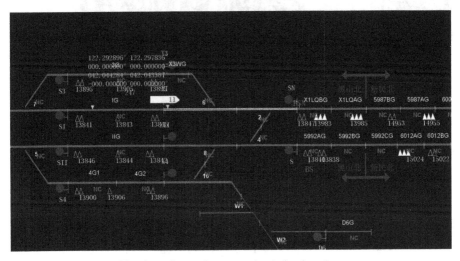

Fig. 5. Satellite simulation data generation interface.

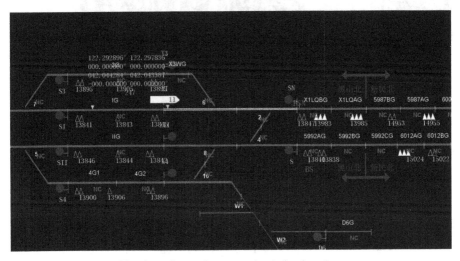

Fig. 6. Train running route simulation interface.

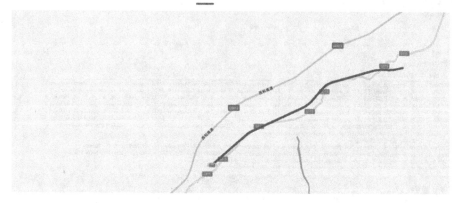

Fig. 7. Satellite map monitoring interface.

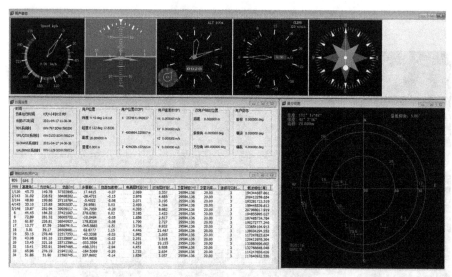

Fig. 8. Interface of the third-party satellite RF signal output device.

4 Conclusions

In this study, based on the analysis and utilization of electronic map data for satellite-based train control systems, a simulation test system suitable for satellite-based train control systems was designed. Pre-calculation of satellite simulation data using electronic maps ensures data accuracy and improves calculation efficiency, enabling rapid generation of satellite simulation data for testing environments. Additionally, real-time monitoring of the satellite simulation status through numerical and GIS methods supports laboratory integration testing of satellite-based train control systems. The feasibility of this design was demonstrated through system implementation and laboratory testing of an engineering project, which met the requirements of laboratory testing.

In the future, system optimization will focus on enriching the interpolation algorithm of the satellite simulation data calculation module to improve calculation accuracy. The electronic map-parsing module can be expanded to adapt to different electronic map formats in various train control systems, thereby enhancing the system's usability. Additionally, further optimization and improvement of the construction function for abnormal satellite positioning system scenarios will be performed to enrich the executable cases for laboratory testing.

References

1. Yingying, L., Hao, Z.: Research on train control technology based on electronic map and satellite positioning. Railway Signal and Communication Eng. Technol. (RSCE) **19**(12), 23–27 (2022). https://doi.org/10.3969/j.issn.1673-4440.2022.12.005
2. Ting, Y., Guangwu, C., Xiaochun, Z.: Train communication system based on beidou short message. Electronics World **18**, 54–55 (2016). https://doi.org/10.3969/j.issn.1003-0522.2016.18.038
3. Guo, H., Jia, Y., Cui, J.: Application of satellite positioning technology in lightweight train control system. Proceedings of the 10th China Satellite Navigation Academic Annual Conference, pp. 1–4 (2019)
4. Zhison, M., Hongfei, A.: Research on comprehensive autonomous positioning technology of new train control system. J. China Railway Society **44**(1), 56–64 (2022). https://doi.org/10.3969/j.issn.1001-8360.2022.01.008
5. Jiajia, X.: Research on satellite navigation simulation based on localized lTCS. Railway Signalling & Communication **56**(9), 56–59 (2020). https://doi.org/10.13879/j.issn1000-7458.2020-09.19393
6. Zhang, Y.: Train track simulation platform based on satellite navigation. Railway Signalling & Communication, **57**(7), 21–24 (2021). https://doi.org/10.13879/j.issn.1000-7458.2021-07.20368
7. Kang, X., Xiaoyu, Z., Zhang, Y.: Research on automatic generation methed of electronic map based on engineering line data. Railway Signalling & Communication **58**(7), 28–33 (2022). https://doi.org/10.13879/j.issn.1000-7458.2022-07.22139
8. China Railway Co., Ltd.: TJ/DW238–2020 Provisional Technical Requirements for Electronic Maps of New Train Control Systems. China Railway Co., Ltd., Beijing (2020)
9. Yaozhen, W.: Discussion on transmission scheme of TSR in digital map for new train control system. Railway Signal and Communication Engineering Technology (RSCE) **18**(11), 38–41 (2021). https://doi.org/10.3969/j.issn.1673-4440.2021.11.008

A Study on Correspondence Point Search Algorithm Using Bayesian Estimation and Its Application to a Self-position Estimation for Lunar Explorer

Hayato Miyazaki, Hiroki Tanji, and Hiroyuki Kamata[✉]

School of Science and Technology, Meiji University, Tokyo, Japan
kamata@meiji.ac.jp

Abstract. The purpose of this research is to improve the accuracy of the lunar surface high-precision landing technology for small unmanned probes in the SLIM (Smart Lander for Investigating Moon) project [1] of JAXA. As a method for realizing high precision landing technology, matching by image collation using the images taken by the space probe and the lunar map image is being studied. However, shooting in outer space tends to cause positional errors due to low resolution caused by disturbances. Conventionally, position estimation has been performed by matching based on k-NN Matching and Ratio Test. However, the accuracy of the estimated position depends on the selected correspondence points, and no error correction is performed in this method. Therefore, we propose a matching model that minimizes the error of position estimation by using Bayesian estimation. Bayesian estimation is an estimation approach that uses the likelihood calculated from observed data and the prior distribution of parameters. In the proposed method, the estimation problem based on the feature values of local features is formulated as a constrained optimization problem, and a Bayesian model is constructed by designing the likelihood and prior distribution. In this paper, we realize a matching algorithm with the function of conventional self-localization by Bayesian estimation.

Keywords: self-location estimation · Bayesian estimation · image matching

1 Introduction

The purpose of this research is to improve the accuracy of the landing technique of a small unmanned space probe on the Moon in the SLIM (Smart Lander for Investigating Moon) project [1] of the Institute of Space and Astronautical Science (JAXA/ISAS), Japan Aerospace Exploration Agency (JAXA). The SLIM project aims to shift from conventional "landing where it is easy to land" to "landing where you want to land" [2]. One of the main objectives of the SLIM project is to achieve a "pinpoint landing" within a landing error of 100 m from the target landing site. To achieve this goal, the project employs a method of estimating its own position based on image information from the space probe's onboard camera [3, 4].

F. Hassan et al. (Eds.): AsiaSim 2023, CCIS 1912, pp. 60–70, 2024.
https://doi.org/10.1007/978-981-99-7243-2_6

The position estimation in this project is assumed to be a two-stage process in order to provide redundancy. The first stage is the detection of craters by principal component analysis on the space probe and position estimation using them as feature points. At this stage, the method chosen considers the grade of computing resources available on the space probe. The second stage is the position estimation performed by the ground backup process in case the position estimation on the space probe fails. In the ground backup process, images of the lunar surface taken by the space probe are compressed to approximately 8 Kbytes in JPEG format and used to estimate the space probe's self-position by matching them with local features extracted from the high-resolution map image prepared in advance. An overview of location estimation in the SLIM project is shown in Fig. 1.

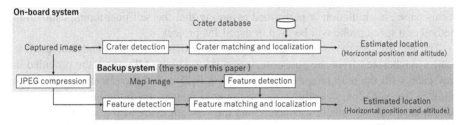

Fig. 1. Overview of the self-localization system used in the SLIM project [5].

The estimation method based on the crater detection is already in the practical stage. Currently, the method based on feature matching is a backup process in the event that position estimation fails on the space probe's process, but in the near future we aim to implement it as an onboard process on the space probe. Therefore, an algorithm for high-precision position estimation by matching images of the lunar surface taken by the space probe and map images of the lunar surface in the database is examined, and improvements are made to address the problems.

In the ground backup process, it is assumed that the captured images of the lunar surface with various disturbances are greatly compressed to reduce communication costs. Since this is likely to cause matching errors, it is necessary to consider matching that is less likely to affect self-position estimation.

Therefore, this study proposes an image matching method based on Bayesian estimation. Bayesian estimation is an approach that estimates the posterior distribution of parameters using the likelihood calculated from observed data and the prior distribution of parameters. To derive a Bayesian method for the image matching, we formulate the image matching problem as a constrained optimization problem. Furthermore, by designing the likelihood and prior distribution based on the formulation, a Bayesian model for image matching is constructed. The posterior distribution of the Bayesian model is estimated using a variational Bayesian method. In this paper, we simulate image matching using the Bayesian estimation model. The proposed matching method is compared with the conventional method through the space probe's self-position estimation.

In conventional correspondence point selection, the validity of correspondence points is determined based on the distance of local features and used for self-location estimation.

This study examines a method to achieve highly accurate location detection by simultaneously narrowing down correspondence points and weighting correspondence points by judging validity based on probabilities derived from the optimization of correspondence point detection.

2 Self-position Estimation of Lunar Explorer

This section provides an overview of the datasets and conventional estimation methods used to evaluate lunar space probe's self-position estimation [4].

2.1 Dataset

In this paper, a simulation is performed assuming that the self-positioning estimation method of a space probe will be used in the SLIM project.

The data used for the evaluation are map images provided by JAXA and photographic images synthesized from the past observation information. All images are provided in grayscale, with the map image being 1883px × 1615px and the photographed image being 512px × 512px. In addition to the captured images, the horizontal position, the altitude at which the image was taken, and the yaw angle are also given as odometry information of the space probe. Using the image and odometry information, the position is estimated by matching the map image with the captured image. In this paper, this odometry information is called GNC (guidance, navigation, and control) data. The map image and captured image used in the simulation of this study are shown in Fig. 2.

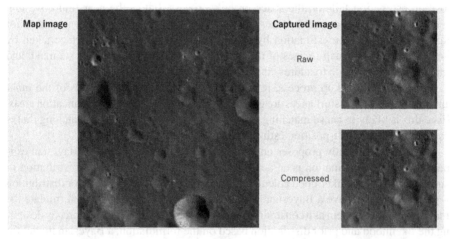

Fig. 2. Map images and Captured images used for self-localization estimation.

Note that 1000 of these captured images are available for verification. In this paper, to verify the effectiveness of image matching using Bayesian estimation, we used 1000 nominal images with no artificially added disturbance.

2.2 Conventional Method

The success of the space probe's self-position estimation is judged based on whether or not the "estimated value" and the "true value" obtained from the GNC data and image information are within 3 px of each other. This 3px corresponds to an actual distance of 100m, and is assumed to be within the 100 m landing error in the SLIM project.

In the literature, the estimation process is as follows. The flow of this method is shown in Fig. 3.

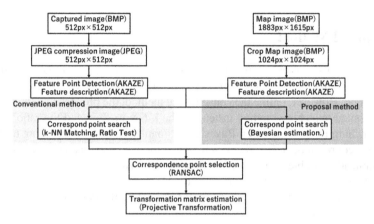

Fig. 3. Flowchart for self-localization estimation in the SLIM project.

(1) **Determination of search area**

Crop a 1024 px × 1024 px area from the map image based on the approximate position by the GNC data.

(2) **Feature point detection and feature description**

AKAZE features are used for feature point detection and feature description in captured and map images. It is tolerant of image scale and rotation [6]. Based on previous research, we believe that this is an effective feature point detection and feature description method for lunar images with disturbance [4].

(3) **Correspondence point search**

k-NN (k-Nearest Neighbor) method is used to search for candidate correspondence points from feature points in the captured image and the map image [7]. The distance between feature vectors of feature points in both images is calculated by brute force; in Match Ratio, a combination of correspondence points with high confidence is selected as correspondence points using the Ratio Test [8]. In this case, let α be $0 < \alpha < 1$.

$$\text{(Closest distance feature)} < \text{(Second nearest distance feature)} * \alpha \qquad (1)$$

(4) **Correspondence point selection and outlier exclusion**

Candidate estimates of the horizontal position are obtained from the coordinates of local features in the captured image and the corresponding local features in the map image. Outliers in the candidate estimates are removed by random sample consensus (RANSAC) [9].

(5) **Transformation matrix estimation**

The singular value decomposition creates a transformation matrix from the selected corresponding points. This allows the coordinates of the captured image to be transformed onto the map image to estimate self-location. The transformation matrix uses a projective transformation that allows estimation after considering the rotation of the captured image, changes in scale, and the inclination of the space probe.

3   Proposed Method

We propose a self-location estimation method using an image matching method based on Bayesian estimation. The process for (1), (2), (4), and (5) is the same as the conventional method. However, unlike the conventional method, the proposed method uses the Bayesian image matching method instead of the k-NN method and the ratio test. The image matching method based on Bayesian estimation is derived according to the following procedure. In Sect. 3.1, the matching method for image matching is formulated as a minimization problem of a cost function. On the basis of this formulation, Sect. 3.2 sets the likelihood and prior distribution for Bayesian estimation. In Sect. 3.3, We derive a variational Bayesian inference algorithm for estimating the posterior distribution.

3.1   Formularization

The matching in the image is computed by the feature vectors detected by AKAZE and therefore follows the following

$$\sum_{m=1}^{M} \left\| x_m - \sum_{n=1}^{N} z_{mn} y_n \right\|^2 \tag{2}$$

where $x_m (m = 1, \ldots, M)$ and $y_n (n = 1, \ldots, N)$ are the feature descriptors of local features extracted from the captured and map images, respectively. They are observed data acquired by AKAZE. $z_{mn} = [z_{m1}, \ldots, z_{mN}]$ is a one-hot vector satisfying the following Eq. (3), which works as follows when the feature point in the k_{th} captured image corresponds to the $\hat{n}_k$ th feature point in the map image.

$$z_{mn} = \begin{cases} 1 \ (n = \hat{n}_k) \\ 0 \ (otherwise) \end{cases} \tag{3}$$

In this study, Eq. (2) is regarded as the error magnitude to be minimized. Therefore, the cost function is derived as

$$\mathfrak{I}_1 \sum_{m=1}^{M} \| x_m - Y z_m \|^2 \tag{4}$$

We express $y_{(\hat{n}_m)} = \sum_{n=1}^{N} z_{mn} y_n = Y z_m$ by setting $Y = \{y_1, y_2, \ldots, y_N\}$.

Assuming that some local features of the captured image do not correspond to any feature of the map image due to disturbances and compression, we rewrite the cost function using positive weights $w_m(w_1, \ldots, w_M)$. Thus, the cost function for image matching [5] can be formulated as

$$\mathfrak{I}_2 = \sum_{m=1}^{M} w_m \|x_m - Yz_m\|^2 \tag{5}$$

The reason for introducing w_m as a constraint condition is to examine the correspondence of local features while reducing the influence of unnecessary correspondence points. In the minimization problem of Eq. (5), w_m is constrained to $w_m > 0$, for all m.

3.2 Bayesian Estimation

(1) **Likelihood**

Feature point matching by Bayesian estimation requires setting the likelihood of observed data and updating the posterior distribution by the set likelihood and prior distribution.

Therefore, using the observed data $Ð = \{x_1, x_2, \ldots, x_M, y_1, y_2, \ldots, y_N\}$, and the parameter $\theta = \{w_1, w_2, \ldots, w_M, z_1, z_2, \ldots, z_M\}$, the likelihood is

$$p(Ð|\theta) = \prod_{m=1}^{M} p(x_m, Y|w_m, z_m)$$
$$= \left(\frac{1}{(2\pi)^{\frac{MD}{2}}}\right)\left(\prod_{m=1}^{M}|w_m|^{\frac{D}{2}}\right) \exp\left(-\frac{1}{2}\sum_{m=1}^{M} w_m\|x_m - Yz_m\|^2\right) \tag{6}$$

The likelihood $p(Ð|\theta)$ in this paper can be expressed as a normal distribution (D is the dimensionality of the local feature).

(2) **Prior distribution**

Next, we set the prior distribution in Bayesian estimation. Since the posterior distribution is derived using $p(\theta|Ð) = p(Ð|\theta)p(\theta)/p(Ð)$, the prior distribution is preferably conjugate to the likelihood in the inference of the posterior distribution. The parameters to be estimated in the correspondence point search are w_m and z_m. For the prior distribution of z_m we introduce the parameter π_m. The prior distribution $p(\theta)$ is

$$p(\theta) = p(w, z, \pi) = \prod_{m=1}^{M} p(w_m)p(z_m) = \prod_{m=1}^{M} p(w_m)p(z_m|\pi_m)p(\pi_m) \tag{7}$$

(2)-1 $p(w_m)$

We assume that is drawn from the gamma distribution $p(w_m)$ defined as

$$p(w_m) = \text{Gamma}(w_m; \eta_w, \beta_w) = \frac{\beta_w^{\eta_w}}{\Gamma(\eta_w)} w_m^{\eta_w - 1} \exp(-\beta_w w_m) \tag{8}$$

where η_w and β_w are the shape and scale parameters, respectively, and $\Gamma(\eta_w)$ is the gamma function. The gamma distribution is conjugate to the likelihood.

(2)-2 $p(z_m|\pi_m)$

     We use the categorical distribution for $p(z_m|\pi_m)$ considering the one-hot vector property. Categorical distribution is based on a vector of probabilities that each element of z_m is 1, satisfies $\sum_{n=1}^{N} \pi_{mn} = 1$. and can be expressed by

$$p(z_m|\pi_m) = \text{Categorical}(z_m; \pi_m) = \prod_{n=1}^{N} \pi_{mn}^{z_{mn}} \tag{9}$$

(2)-3 $p(\pi_m)$

     $p(\pi_m)$ is the Dirichlet distribution, which is the conjugate of the categorical distribution. The Dirichlet distribution has a parameter $\alpha_{mn} > 0$ and can be expressed by

$$p(\pi_m) = \text{Dir}(\pi_m; \alpha_m) \propto \prod_{n=1}^{N} \pi_{mn}^{\alpha_{mn}-1} \tag{10}$$

3.3 Bayesian Estimation Algorithm

To estimate the posterior distribution $p(\theta|Đ)$, we derive the inference algorithm from the likelihood and prior distribution defined in Sect. 3.2. In our algorithm, a variational Bayesian method is introduced. In this method, the variational posterior distribution $q(\theta)$ is introduced, and $q(\theta)$ is obtained using an iterative method so that the following functional $\mathcal{L}[q]$ is maximized [5, 10].

     The functional $\mathcal{L}[q]$ to be maximized with respect to $q(\theta)$ is

$$\mathcal{L}[q] = \int q(\theta) \log \frac{p(Đ,\theta)}{q(\theta)} d\theta =$$
$$\sum_{m=1}^{M} \int q(w_m)q(z_m)q(\pi_m) \log \frac{p(x_m, Y|w_m, z_m)p(w_m)p(z_m|\pi_m)p(\pi_m)}{q(w_m)q(z_m)q(\pi_m)} d(w_m, z_m, \pi_m) \tag{11}$$

For convenience, we approximate $q(\theta)$ as

$$q(\theta) = \prod_{m=1}^{M} q(w_m)q(z_m)q(\pi_m) \tag{12}$$

By solving $\frac{\partial \mathcal{L}[q]}{\partial q(\theta)} = 0$ with respect to $q(\theta)$ in the above equation, we obtain the variational posterior distribution as follows:

$$q(w_m) = \text{Gamma}\left(w_m; \hat{\eta}_{w_m}, \hat{\beta}_{w_m}\right) = \frac{\hat{\beta}_{w_m}^{\hat{\eta}_{w_m}}}{\Gamma\left(\hat{\eta}_{w_m}\right)} w_m^{\hat{\eta}_{w_m}-1} \exp\left(-\hat{\beta}_{w_m} w_m\right) \tag{13}$$

$$q(z_m) = \text{Categorical}\left(z_m; \hat{\pi}_m\right) = \prod_{n=1}^{N} \hat{\pi}_{mn}^{z_{mn}} \tag{14}$$

$$q(\pi_m) = \text{Dir}\left(\pi_m; \hat{\alpha}_m\right) = \prod_{n=1}^{N} \pi_{mn}^{\hat{\alpha}_m-1} \tag{15}$$

where

$$\hat{\eta}_{w_m} = \eta_w + \frac{D}{2} - 1 \tag{16}$$

$$\hat{\beta}_{w_m} = \frac{1}{2}\left(x_m^T x_m - 2\sum_{n=1}^{N} x_m^T y_n E_{q(z_m)}[z_{mn}] + \sum_{n=1}^{N} y_n^T y_n E_{q(z_m)}[z_{mn}]\right) + \beta_w \quad (17)$$

$$\hat{\alpha}_{mn} = \alpha_{mn} + E_{q(z_m)}[z_{mn}] \quad (18)$$

$$\log \hat{\pi}_{mn} = E_{q(w_m)}[w_m]\left(x_m^T y_n - \frac{1}{2}y_n^T y_n\right) + E_{q(\pi_m)}[\log \pi_{mn}] \quad (19)$$

The $E_{q(\theta)}[\theta] = \int^\theta q(\theta)d\theta$ is the expectation of θ.

In this study, the above parameters are updated and the respective expected values $E_{q(w_m)}[w_m]$, $E_{q(z_m)}[z_{mn}]$, and $E_{q(\pi_m)}[\log \pi_{mn}]$ are updated by an iterative method. As an estimate of the probability that the result $z_{mn} = 1$, the MAP estimate of the value of π_{mn} can be calculated to perform a correspondence point search.

The initial value of $\hat{\pi}_{mn}$ is drawn from a random process and then normalized by

$$\hat{\pi}_{mn} \leftarrow \frac{\hat{\pi}_{mn}}{\sum_{n=1}^{N} \hat{\pi}_{mn}} \quad (20)$$

Algorithm 1

Set initial values.

$\alpha_{mn}, \beta_w, \eta_w, \log \hat{\pi}_{mn} \leftarrow$ random number

/* Update expected values and parameters by iterative method. */

for do

$E_{q(w_m)}[w_m] \leftarrow \frac{\hat{\eta}_{w_m}}{\hat{\beta}_{w_m}}$

$E_{q(\pi_m)}[\log \pi_{mn}] \leftarrow \Psi(\hat{\alpha}_{mn}) - \Psi\left(\sum_{n=1}^{N} \hat{\alpha}_{mn'}\right)$

$\log \hat{\pi}_{mn} \leftarrow E_{q(w_m)}[w_m]\left(x_m^T y_n - \frac{1}{2}y_n^T y_n\right) + E_{q(\pi_m)}[\log \pi_{mn}]$

$\hat{\pi}_{mn} \leftarrow \text{softmax}[\log \hat{\pi}_{mn}]$

$\hat{\alpha}_{mn} \leftarrow \alpha_{mn} + E_{q(z_m)}[z_{mn}]$

$\hat{\beta}_{w_m} \leftarrow \frac{1}{2}\left(x_m^T x_m - 2\sum_{n=1}^{N} x_m^T y_n E_{q(z_m)}[z_m] + \sum_{n=1}^{N} y_n^T y_n E_{q(z_m)}[z_m]\right) + \beta_w$

end for

In Algorithm 1, Ψ is the digamma function defined by $\Psi(x) = \frac{d}{dx}\log\Gamma(x)$, and softmax$[x_{mn}] = \frac{\exp(x_{mn})}{\sum_{n=1}^{N}\exp(x_{mn})}$ is the softmax function.

4 Verification

In this study, verification will be conducted in two stages.

4.1 Image Matching Method Based on Bayesian Estimation

In the first, correspondence point search is performed using the photographed image for simulation and the map image. The operation of the refinement search was verified by using π_{mn} as an estimate of the probability that $z_{mn} = 1$, and removing those with small values from the corresponding points in order to minimize the position error. This refinement is called "Probability refinement".

4.2 Self-position Estimation

Second, we simulated self-location estimation using the matching method in Bayesian estimation described above. The effectiveness of this study was verified by comparing the results with methods using k-NN Matching and Ratio test.

5 Experimental Result

Verifications are shown in Sect. 5.

5.1 Image Matching Method based on Bayesian Estimation

The matching results according to the validation are shown in 5.1 and compared with the matching by k-NN method. The results of the k-NN matching method according to Section 2.2 (3) are shown in Fig. 4.

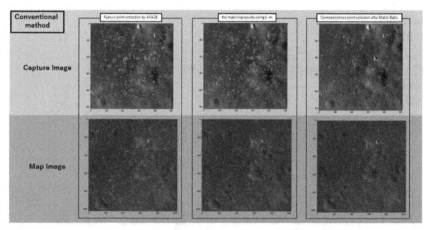

Fig. 4. Results of correspondence point selection using the conventional method. (correspondence point search: k-NN matching, correspondence point selection: Ratio test)

In contrast, the matching results using Bayesian estimation of the proposed method are shown in Fig. 5.

These results show that the proposed method is able to select correspondence points as well as the conventional method. In addition, "Probability refinement" significantly reduces the number of correspondence points that can be assumed to be incorrectly matched.

5.2 Self-position Estimation

Based on the results in Sect. 5.1, a self-position estimation of the space probe using Bayesian estimation is simulated. The self-position estimation is performed according

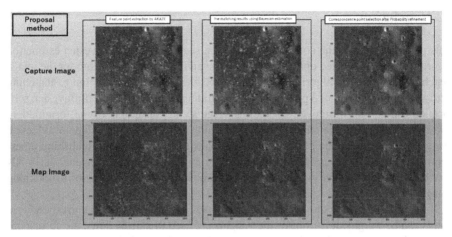

Fig. 5. Results of correspondence point selection by Proposal method. (correspondence point search: Bayesian estimation, Correspondence point selection: Probability refinement)

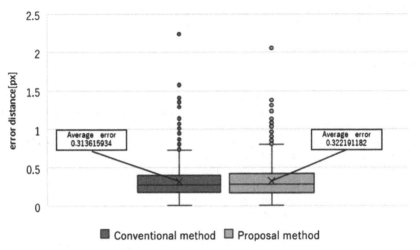

Fig. 6. Error in self-localization estimation by conventional and proposed methods. (Verified with 1000 Capture Images)

to Sect. 2.2. Estimation results for the conventional and proposed methods are shown in Sect. 5.2. This result is a simulation of self-position estimation using 1000 images from the dataset in Sect. 2.1, and the error is shown by a box-and-whisker diagram.

The validation of the method with 1000 Capture Images shows that both the conventional and the proposed methods can meet the SLIM project's goal of an error within 3[px]. Compared to the conventional method, there is no variation in the mean value or error, indicating that they are able to estimate the same level.

6 Conclusion

This paper formulates the correspondence point search using the Bayesian estimation algorithm as a constrained optimization problem and derives an algorithm for the image matching method. The algorithm satisfies the SLIM project's goal of image matching with an estimation error of 3 px or less, demonstrating the effectiveness of Bayesian self-location estimation. Moreover, by adding weights and constraints to the formulation, correspondence point search using k-NN matching and correspondence point selection using the ratio test can be performed by Bayesian estimation, thus establishing a new approach for the SLIM project. This algorithm is superior to the conventional k-NN matching in that it can give weights to each correspondence point, in addition to obtaining comparable results.

In the future, we would like to take advantage of the feature that correspondence points can be weighted. Currently, the weights are used to narrow down the correspondence points, but we will make efforts to develop a self-position estimation algorithm that uses the weights to reduce the self-position estimation error. We will also evaluate the performance of the Bayesian self-location estimation method on a dataset with various disturbances added to the captured images to demonstrate the effectiveness of Bayesian self-location estimation on space probes.

References

1. Yoshikawa, S., et al.: Conceptual study on the guidance, navigation and control system of the smart landing for investigating moon (SLIM). In: Proceedings of Global Lunar Conference 2010, pp. 59–62 (2010)
2. Ishida, T., et al.: Crater-based optical navigation technologies for lunar precision landing in SLIM project. In: Proceedings of the 16th IPPW-2019 (2019)
3. Komatsubara Y., et al.: Crater detection and backup processing based on the feature extraction method. In: Proceedings of 62th Space Sciences and Technology Conference, 1D10 (2018)
4. Ohara, S., et al.: Study on high precision matching between captured image with disturbance and map image. In: Proceedings of 63th Space Sciences and Technology Conference, 1B08 (2019)
5. Tanji, H., et al.: Study on image navigation for the unmanned lunar lander using high precision image matching based on a variational Bayesian approach. In: Proceedings of 65th Space Sciences and Technology Conference, 65th ROMBUNNO.3S11 (2021)
6. Alcantarilla, P.F., Nuevo J., Bartoli A.: Fast explicit diffusion for accelerated features in nonlinear scale spaces. In: British Machine Vision Conference (BMVC) (2013)
7. Dasarathy, B.V.: Nearest Neighbor (NN) Norms: NN Pattern Classification Techniques. IEEE Computer Society Press; IEEE Computer Society Press Tutorial (1991)
8. Lowe, D.: Distinctive image features from scale-invariant keypoints, cascade filtering approach. Int. J. Comput. Vision 60, 91–110 (2004)
9. Raguram, R., Chum, O., Pollefeys, M., Matas, J., Frahm, J.: USAC: a universal framework for random sample consensus. IEEE Trans. Pattern Anal. Mach. Intell. 35, 2022–2038 (2013)
10. Qu, H.-B.: Robust point set matching under variational Bayesian framework. In: IAPR 22nd International Conference on Pattern Recognition, pp. 58–63 (2014)

Calculation of Stress Intensity Factors Under Mode-I Using Total Lagrangian Smoothed Particle Hydrodynamics

Shen Pan[1,2], Khai Ching Ng[1(✉)], and Jee-Hou Ho[1]

[1] University of Nottingham Malaysia, Jalan Broga, 43500 Semenyih, Malaysia
edxsp1@nottingham.edu.my
[2] Dongguan University of Technology, Dongguan 523808, China

Abstract. This paper presents a study on calculating Stress Intensity Factors (SIFs) under mixed mode using a simulation method based on Smoothed Particle Hydrodynamics (SPH). By modeling crack behaviors using the total Lagrangian SPH (TLSPH) method, we can obtain displacement and stress fields at the crack tip and further calculate the SIFs. This study used the TLSPH method to simulate cracks with mode-I loading to obtain displacement and stress field data. Using these data, we calculated SIFs under mode-I using different methods. Better results can be achieved using the extrapolation method with a relatively short calculation time. This study obtained displacement and stress field data for calculating SIFs under mode-I through SPH simulation. The simulation results show that the extrapolation stress method delivers better accuracy and time-efficiency with errors of 0.20%, 1.97%, and 4.39% for particle diameters of 0.125 mm, 0.25 mm, and 0.5 mm, respectively. These findings are essential for the study and practical engineering applications of SIFs under mixed mode.

Keywords: TLSPH · Stress Intensity Factor · Fatigue Crack

1 Introduction

This article investigates various methods for calculating SIFs under mode crack conditions and compares their accuracy and computation time. Mixed mode cracking refers to a situation where the crack tip is subjected to loading of both Mode I and Mode II stresses, a common problem in engineering applications.

The methods used in this study were maximum principal stress, extrapolation displacement, extrapolation stress, and J-integral methods. The J-integral method was first used. It is an energy-based method for calculating SIFs at the crack tip. The J-integral method has a high level of generality and is widely used for various geometries and loading conditions. Secondly, extrapolation displacement and extrapolation stress methods were considered. These methods are based on the displacement and stress fields to estimate SIFs by extrapolating either displacement or stress. Finally, the maximum stress method was simplified, which assumes that the SIF is proportional to the maximum

© The Author(s), under exclusive license to Springer Nature Singapore Pte Ltd. 2024
F. Hassan et al. (Eds.): AsiaSim 2023, CCIS 1912, pp. 71–85, 2024.
https://doi.org/10.1007/978-981-99-7243-2_7

stress at the crack tip. The accuracy of different methods was compared by analyzing the results.

Our study aims to find a method for calculating SIFs in mode-I cracking conditions, which is both highly accurate and yields results quickly. Through this research, we hope to provide a valuable reference for calculating SIFs in mode-I cracking in engineering applications. This will help engineers choose the best method for specific problems while considering accuracy and computation efficiency.

2 Literature Review

J-integral, extrapolation displacement, extrapolation stress, and maximum stress methods are common methods for calculating SIFs, but they differ in principles and applications.

The J-integral serves as an energy-based approach to compute Stress Intensity Factors (SIFs) at the crack tip. Rooted in the elastic strain energy and the power within the stress field, this method assesses SIFs by evaluating an integral over a closed contour encircling the crack tip. The J-integral method [1] applies to complex geometries and loading conditions and is suitable for linear and nonlinear elastic materials [2].

On the other hand, the extrapolation displacement method [3] is based on the displacement field for calculating SIFs at the crack tip. This method estimates the SIFs by extrapolating the displacement field at the crack tip [4]. The prerequisite of this method is the information on the crack path, and it assumes that the displacement field near the crack tip is linear. It is particularly well-suited for linear elastic materials.

Another approach, equally applicable to linear elastic materials, is the extrapolation stress method [5]. It is a stress field-based method for calculating SIFs at the crack tip. This method estimates the SIFs by extrapolating the stress field at the crack tip. It also requires information about the particles in the direction of crack extension in advance and assumes that the stress field near the crack tip is linear.

The maximum principal stress method [6] is simplified for estimating SIFs at the crack tip. This method assumes that the SIF is proportional to the principal stress at the crack tip. The maximum principal stress method also requires knowing the crack size in advance and has the same assumption as the extrapolation stress method. This method is relatively simple but only applies to specific geometries, loading conditions, and relatively small crack sizes.

In conclusion, the J-integral method is a general and comprehensive method suitable for complex situations. However, it is computationally more complex. The extrapolation displacement method, extrapolation stress method, and maximum stress method are relatively simplified methods suitable for particular conditions, but their accuracy may be lower. The choice of the appropriate method should be evaluated based on specific requirements and feasibility.

3 Methodology

The SIFs were introduced in 1957 by George R Irwin. It stands as one of the most essential and valuable parameters within fracture mechanics. It characterizes the stress condition at the crack's endpoints, establishing a significant connection with the rate of crack propagation. Furthermore, this parameter establishes the criteria for the failure caused by the crack.

Figure 1 illustrates three types of SIFs: Mode I (opening), Mode II (shearing) and Mode III (tearing).

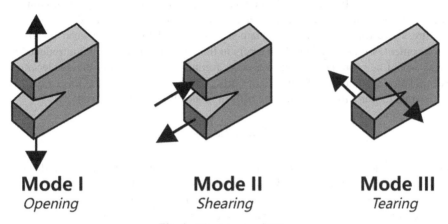

Mode I   **Mode II**   **Mode III**
Opening   *Shearing*   *Tearing*

Fig. 1. Three types of SIFs.

For the different modes, Irwin's formula is usually written as follows:

$$K_I = Y\sigma_{xx}(\pi a)^{1/2} \tag{1}$$

$$K_{II} = Y\sigma_{yy}(\pi a)^{1/2} \tag{2}$$

$$K_{III} = Y\tau_{xy}(\pi a)^{1/2} \tag{3}$$

where, Y is a dimensionless shape factor dependent on the crack geometry, σ_{xx}, σ_{yy}, τ_{xy} is the stress on the crack face and a is crack length.

The different methods for calculating the SIFs are described below.

3.1 Maximum Principal Stress

The SIF K is a fundamental quantity in fracture mechanics, and it can be derived by the Irwin formula to measure the strength of the local stress field near the crack tip [7].

In the fatigue crack propagation simulation by SPH method, K values for all crack tip particles must be calculated. In this study, K values are straightforwardly Computed using Eqs. 1, 2 and 3.

Notice that all three equations have a parameter 'a', which determines the dependence of stress with distance from the crack tip and reflects the singularity when $a = 0$.

3.2 Extrapolation Method Based on Nodal Displacement

In traditional Finite Element Methods, more accurate results of displacements can be derived than stresses because they are the primary solution variable. Like the finite element method, the displacement extrapolation method is also applicable to the TLSPH method. In the TLSPH method, the behavior of matter is modelled by the motion and interaction of particles, where displacement is the main solution variable.

Stress is a quadratic variable obtained from the relationship between strain and displacement, which may introduce errors in the calculation. The errors introduced by stress would be more significant in a weak formulation. Therefore, the displacement extrapolation method is preferable to the stress extrapolation method in this case.

Irwin's formula for describing the SIFs in modes I, II and III is given:

$$K_I = \sqrt{2\pi}\frac{F_1(r)}{\sqrt{r}} \tag{4}$$

where, $F_1(r)$ denotes the stress distribution functions in modes I and r is the distance from current particle to the crack tip. The displacement field near the crack can be expressed as: $u = \Delta u(r)$. Here, Δu represents the displacements in the x direction.

According to the theory of linear elasticity, the relationship between stress and displacement is expressed as

$$\sigma = G\varepsilon \tag{5}$$

where, G is the shear modulus, ε is the strain, σ is the stress. The strain can be obtained from the displacement using the strain-displacement relation.

Given that the stress is derived from displacement, the stress distribution function in Irwin's formula is replaced by the corresponding displacement.

The mode-I SIFs on the material point of the displacement crack surface can be obtained through displacement [3]:

$$K_I = \sqrt{2\pi}\frac{G}{1+\kappa}\frac{|\Delta v|}{\sqrt{r}} \tag{6}$$

Here, the term $(1 + \kappa)$ appears because of plane strain conditions. The exact derivation of the equations is complicated and depends on additional assumptions such as plane strain or plane stress conditions, isotropic or anisotropic material behaviors, etc.

As shown in the Fig. 2, Δv, Δu, and Δw are the relative displacements of the crack surface with respect to another crack surface.

Since the Eq. 6 retains the singularity of r, the accurate result can be only obtained when $r \to 0$. However, when r is too close to the crack, conventional elements cannot correctly reflect the crack tip's singularity, which affects the simulation result. By plotting the curve of the data pair (r_i, K_{Ii}) and using the extrapolation method to fit the data points, the intercept of the fitted line is the SIF.

$$K_I \approx \frac{\sum r_i \sum r_i K_I - \sum r_i^2 \sum K_{Ii}}{\left(\sum r_i\right)^2 - N \sum r_i^2} \tag{7}$$

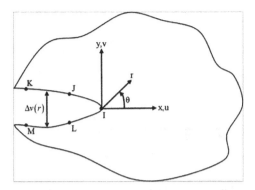

Fig. 2. Crack tip displacement [3].

3.3 Extrapolation Method Based on Particle Stress

The equation for calculating SIFs by stress extrapolation is the same as the maximum principal stress method. However, this method does not calculate the crack for the crack tip particle to ensure that the singularity of the crack tip particle is not affected during the search for SIFs,

The most direct method of calculating the SIF is based on stress extrapolation. From Eq. 7, it can be seen that K_1 corresponds to the value when $r \to 0$ at the crack tip, which cannot be attained by direct numerical calculation. Therefore, the conceptual method is shown as follows.

It is easy to read the stress value σ_y directly, and corresponding coordinate value, r, at the integration point in the crack front element using Finite Element Analysis, which can be directly achieved in commercial finite element software. If the relationship between σ_y and r is plotted, the stress distribution curve shown in Fig. 3 can be obtained. With the refinement of the element, the stress value tends to infinity, i.e., stress singularity. Although numerical methods cannot directly calculate the value of K_I at the crack tip, the non-singular stress values at the crack front are known [7]. For every $r_i > 0$, there is a non-singular stress value σ_{yi} and the corresponding K_{Ii} [5].

$$K_{Ii} = \sigma_{yi}\sqrt{2\pi r_i} \tag{8}$$

3.4 J-integral

A closed path around the crack tip evaluates the J-integral. Its value is calculated by:

$$J = \int_\Gamma \left(W dy - T \frac{\partial u}{\partial x} ds \right) \tag{9}$$

where Γ is a curve surrounding the notch tip, W is the strain energy density. The integral evaluation is performed in an anticlockwise direction, commencing from the lower flat notch surface and the integration continues along the path Γ to the upper flat surface. T

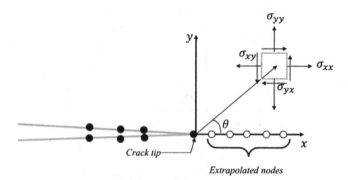

Fig. 3. Line of points near the crack tip [5].

is the traction vector, which can be defined as the outward normal the path Γ, u is the displacement vector, and ds is an element of the arc length along Γ.

In practice, Eq. 9 is unsuitable for numerical calculations because it is not feasible to calculate stresses and strains along the integration path. In addition, the results obtained are only sometimes consistent when integration loop is very close to the crack tip.

To address these issues, Moran [8] and Shivakumar [9] proposed the equivalent integral domain method for numerical computation of the J-integral. Using the divergence theorem, the J-integral is calculated by replacing the integration loop with a finite area near the crack tip.

Based on this approach, Eq. 10 can be transformed into the following:

$$J = \int_A \left(\sigma_{ij} \frac{\partial u_j}{\partial x_1} - w\delta_{1i} \right) \frac{\partial q}{\partial x_i} dA \tag{10}$$

For 2-D problems, it can be specifically expanded as follows:

$$J = \int_A \left[\left(\sigma_{xx} \frac{\partial u}{\partial x} + \tau_{xy} \frac{\partial v}{\partial x} - w \right) \frac{\partial q}{\partial x} + \left(\tau_{xy} \frac{\partial u}{\partial x} + \sigma_{yy} \frac{\partial v}{\partial x} \right) \frac{\partial q}{\partial y} \right] dA \tag{11}$$

The function q [10] is any smooth function with a value of zero at the outer boundary and unity at the inner boundary. A particular area is chosen around the crack tip. The selected domain could have any shape, such as rectangular or circular.

The distribution of the q is shown as follows:

$$q = \begin{cases} 1, \, \textit{at the inner boundary} \\ 0, \, \textit{at the outer boundary} \end{cases} \tag{12}$$

The variables in the element can be approximated by the nodal values using interpolation of shape functions.

After obtaining the strain, the stress can be obtained through the constitutive relation:

$$\sigma = C\varepsilon \tag{13}$$

For plane stress problems:

$$C = \frac{E}{1 - v^2} \begin{pmatrix} 1 & v & 0 \\ v & 1 & 0 \\ 0 & 0 & \frac{1-v}{2} \end{pmatrix} \tag{14}$$

For plane strain problems:

$$C = \frac{E}{(1 - 2v)(1 + v)} \begin{pmatrix} 1 - v & v & 0 \\ v & 1 - v & 0 \\ 0 & 0 & \frac{1-2v}{2} \end{pmatrix} \tag{15}$$

After obtaining the stress and strain, the strain energy density factor can be obtained from the following formula:

$$w = \frac{1}{2}\sigma\varepsilon^{\mathrm{T}} \tag{16}$$

Finally, the J-integral over a cell can be computed as follows:

$$\overline{J} = \int_{-1}^{1}\int_{-1}^{1} I(r, s)\mathrm{d}r\mathrm{d}s \tag{17}$$

where: $I(r, s) = \left[\left(\sigma_{xx}\frac{\partial u}{\partial x} + \tau_{xy}\frac{\partial v}{\partial x} - w\right)\frac{\partial q}{\partial x} + \left(\tau_{xy}\frac{\partial u}{\partial x} + \sigma_{yy}\frac{\partial v}{\partial x}\right)\frac{\partial q}{\partial y}\right]det(J^e)$

using Gaussian integration, the integral of Eq. 17 can be approximated by summation over the integration points:

$$\overline{J} \approx I(r_1, s_1) + I(r_2, s_2) + I(r_3, s_3) + I(r_4, s_4) \tag{18}$$

where: $r = \left\{-\frac{1}{\sqrt{3}} \frac{1}{\sqrt{3}} \frac{1}{\sqrt{3}} - \frac{1}{\sqrt{3}}\right\}$ and $s = \left\{-\frac{1}{\sqrt{3}} - \frac{1}{\sqrt{3}} \frac{1}{\sqrt{3}} \frac{1}{\sqrt{3}}\right\}$

The above shows the process of calculating the J-integral with area.

4   Result

This section demonstrates that TLSPH can accurately determine Mode-I SIF. Three scenarios were considered in total, which are (i) a finite plate with a center crack.

Table 1 lists the various parameters used in this study to analyze Case 1. Here, E represents the Young's Modulus, G stands for the Shear Modulus, ρ_0 denotes the density, B is the bulk modulus, Δt is the time step and $h/\Delta x_0$ is h spacing ratio.

The first case involves a plate with dimensions of $2b \times 2t = 30.50$ mm $\times$ 30.625 mm, an externally applied stress $\sigma_0 = 1.0$MPa, and a crack of length $2a$ (6.0 mm) is positioned at the center of the plate (Fig. 4).

In mode-I SIFs of this case, the SIF can be obtained by solving the crack array problem in an infinite periodic medium. Typical Analytical Solution [11]:

$$K_{th1} = \sigma_0\sqrt{\pi a}\left[\frac{1 - \frac{a}{2b} + 0.326\left(\frac{a}{b}\right)^2}{\sqrt{1 - a/b}}\right] \tag{19}$$

Table 1. The different properties in the center crack simulations.

Parameter	Value
E	71.4 GPa
G	28.6 GPa
ρ_0	2700 kg/m^3
B	47.6 GPa
Δt	10^{-9} s
$h/\Delta x_0$	1.6

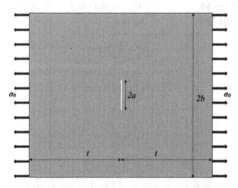

Fig. 4. A plate of finite size containing a central crack [11].

When particle spacing Δx_0 =0.125 mm, the plate is divided into a matrix of 61 rows × 61 columns; For particle spacing Δx_0 =0.25 mm, the plate is divided into a matrix of 121 rows × 121 columns; For particle spacing = $\Delta x_0$0.5 mm, the plate is divided into a matrix of 241 rows × 241 columns.

4.1 Maximum Principal Stress

The following figures illustrate the zoomed-in view around center crack with various particle spacings, with Δx_0 equals to 0.125 mm, 0.25 mm and 0.5 mm, respectively (Fig. 5).

As the particle spacing becomes smaller, the magnitude of principal stress at the crack-tip also increases. According to the theory of Linear Elastic Fracture Mechanics (LEFM), the smaller particle spacing decreases the effective distance between the crack tip and the crack-front particle, which lead to an increase in stress with the proximity of the crack tip. In this simulation, the SIFs at the crack front are calculated by the maximum principal stress method.

The mode-I $K_{numerical}$ at the crack front is calculated for three different particle spacings and compared with its corresponding analytical solution K_{th}. The following table and figure display the computation results of various particle spacings.

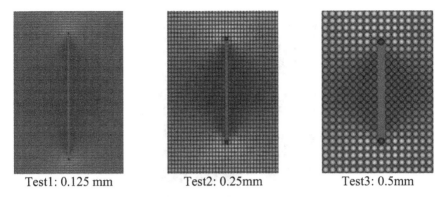

Test1: 0.125 mm   Test2: 0.25mm   Test3: 0.5mm

Fig. 5. The magnified view focusing on the central crack for the three scenarios.

Table 2. The effect of discretization.

Parameter	Δx_1	Δx_2	Δx_3
$2a/mm$	6.125	6.25	6.5
$\Delta x_0/m$	0.125	0.25	0.5
$\sigma_{p,max}$	6.57031	4.86839	3.72343
$K_{numerical}$	0.1302	0.1364	0.1476
K_{th}	0.1002	0.1013	0.1035
Error	29.94%	34.69%	42.64%

where, $\sigma_{p,max}$ is the maximum principal stress, $K_{numerical}$ is the numerical solution of SIF, K_{th} is the theoretical solution of SIF. Although finer discretization can reduce the error between the numerical and analytical SIFs, it also results in higher computational costs.

The results are presented in Table 2. The effect of discretization., where the maximum principal stress method has difficulties in calculating the SIF for mode I. The errors are 29.94%, 34.69%, and 42.64% for particle diameters of 0.125 mm, 0.25 mm, and 0.5 mm, respectively. This is because the particle diameter at the tip of the fracture is not as high as it should be. The high stress values caused by the singularity of the particles at the crack tip requires the setting of very small spacings to obtain relatively accurate results, which is significantly computationally expensive. Therefore, Mode I SIFs was calculated to avoid the use of crack tip values.

4.2 Extrapolation method based on particle stress.

Figure 6 shows the distribution of particles required for the stress extrapolation method. All particles extending ahead of the crack tip particles are selected. However, since the particle at the crack tip have a significant impact on the results, it needs to be discarded when analyzing the fracture problem.

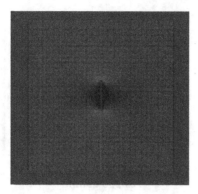

Fig. 6. Line of points near the crack tip.

The following table displays the simulation results with various particle spacings, Δx_0 equals to 0.125 mm, 0.25 mm and 0.5 mm respectively.

Table 3. The effect of discretization.

Parameter	Δx_1	Δx_2	Δx_3
$2a/mm$	6.125	6.25	6.5
$\Delta x_0/mm$	0. 125	0.25	0.5
$K_{numerical}$	0.1000	0.0993	0.1080
K_{th}	0.1002	0.1013	0.1035
$Error$	0.20%	1.97%	4.39%

The advantages of using stress extrapolation to calculate the SIF are its relatively small error and convenience.

The results are presented in Table 3, with errors of 0.20%, 1.97%, and 4.39% for particle diameters of 0.125 mm, 0.25 mm, and 0.5 mm respectively. This shows that for a specific crack size (crack length of approximately 6.25 mm), stress extrapolation provides an accurate and reliable way to calculate the SIF.

Another advantage of this method is its convenience. The stress extrapolation method can determine the location of line particles near the crack tip efficiently, which is essential for accurate estimation of the SIFs. This indicates that we can find and deal with those particles that influence the crack tip stress and crack extension behaviors most more easily.

Finally, it is worth noting that the SIF for Mode I can be accurately represented using TLSPH simulations, with a maximum error of less than 4.39% between numerical and the analytically SIF. This further demonstrates the feasibility and accuracy of the stress extrapolation method in the numerical simulation of crack mechanics.

4.3 Extrapolation Method Based on Particle Displacement

Figure 7 shows the particle distribution required for the displacement extrapolation method. All particles in the vicinity of the crack are selected. This method does not require the calculation of K values at the crack tip, but these particles are not on the same line as the crack tip particles. Therefore, it is necessary to calculate the angle between the particles on the selected line and the crack tip particles with the corresponding correction formula to calculate the SIFs.

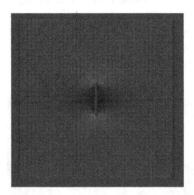

Fig. 7. Line of points near the crack tip.

The following table displays the simulation error of various particle spacings, Δx_0 equals to 0.125 mm, 0.25 mm and 0.5 mm respectively.

Table 4. The effect of discretization.

Parameter	Δx_1	Δx_2	Δx_3
$2a/mm$	6.125	6.25	6.5
$\Delta x_0/mm$	0. 125	0.25	0.5
$K_{numerical}$	0.0992	0.1029	0.1079
K_{th}	0.1002	0.1013	0.1035
Error	1.02%	1.59%	4.27%

The results are presented in Table 4, with errors of 1.02%, 1.59%, and 4.27% for particle diameters of 0.125 mm, 0.25 mm, and 0.5 mm respectively. This indicates that for a specific crack size (crack length of approximately 6.25 mm), stress extrapolation method provides an accurate and reliable result of the SIFs.

The use of displacement extrapolation can therefore improve the accuracy of the SIFs calculation. This method usually requires extremely fine particle delineation near the crack tip to obtain a more accurate singular stress field. Although this may increase the complexity and time of the calculation, such fine delineation is necessary to obtain more accurate results.

Overall, the displacement extrapolation method is seen as a more accurate method to calculate SIFs, both in the finite element method and in the TLSPH method, especially when dealing with fracture problems containing singular stress fields.

4.4 J-integral

Figure 8 shows the distribution of particles required for the displacement extrapolation method. All particles extending ahead of the crack tip particles are selected. This method does not require calculating the K at crack tip, but these particles are not on the same line as the crack tip particle. Therefore, the angle between the particles on the selected line with particle at the crack tip should be calculated.

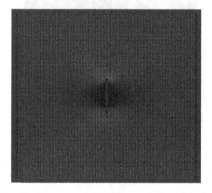

Fig. 8. Contour of points around the crack tip.

The table below shows the simulation results with various particle spacings, Δx_0 equals to 0.125 mm, 0.25 mm and 0.5 mm respectively.

Table 5. A comparison of mode-I SIFs obtained analytically K_{th} with those obtained through SPH simulations K_I.

Parameter	Δx_1	Δx_2	Δx_3
$2a/mm$	6.125	6.25	6.5
$\Delta x_0/mm$	0. 125	0.25	0.5
$K_{numerical}$	0.0991	0.1039	0.1087
K_{th}	0.1002	0.1013	0.1035
Error	1.12%	2.53%	5.02%

The results are shown in Table 5, with errors of 1.12%, 2.53% and 5.02% for particle diameters of 0.125 mm, 0.25 mm and 0.5 mm respectively. This clearly shows that for a specific crack size (crack length of approximately 6.25 mm), J-integral provides an accurate and reliable method of calculating the SIFs. Although the values obtained by J-integral are also relatively accurate, the process of finding the profile requires a greater computational cost. In addition, as the particle diameter increases, the error in the calculation increases accordingly. This may be because the increase in particle diameter leads to a reduction in the number of particles located at the crack tip, which in turn reduces the accuracy of the simulation. Therefore, for problems with large crack lengths, it may be necessary to use smaller particle diameters to reduce the computational error and obtain more accurate SIF values.

4.5 A Comparison of Mode-I SIFs Calculation Errors by Different Methods

The following graph shows a comparison of calculation errors using different methods for deriving SIFs values. Due to the large errors that occur when estimating the value of the principal stress at crack tip, maximum principal stress was excluded from the analysis.

It can be found that the SIFs directly calculating by using the maximum principal stress at crack tip have the largest error, while the results obtained through extrapolation displacement are most accurate. When the crack length is 11.75mm, the error between numerical and analytical is 3%. However, with the extrapolation stress method, the line of particle near the crack tip is easy to find, i.e., $\theta = 0$. There is no need to consider the issue of angles. Due to the large errors that occur when estimating the value of the principal stress at crack tip, the following graph show a comparison of different methods for calculating SIFs value. However, Maximum principal stress has been excluded from the analysis.

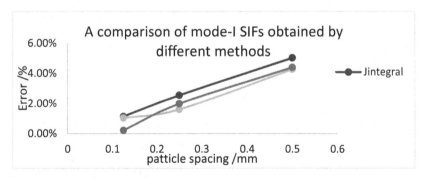

Fig. 9. A comparison of mode-I SIFs error by different methods.

Figure 9 compare three methods for calculating SIFs in terms of accuracy and computational efficiency. The three methods used are displacement extrapolation, stress extrapolation and the J-integral method. By comparing the calculation errors of the three methods for different particle diameters, we derive the following results.

The displacement extrapolation method gives the best results, with calculation errors of 1.02%, 1.59%, and 4.27% for particle diameters of 0.125 mm, 0.25 mm, and 0.5 mm respectively. Although this method may add complexity and time to the calculation, it provides an accurate and reliable way to calculate SIFs, especially when dealing with fracture problems containing singular stress fields.

The stress extrapolation method is a close second, with errors of 0.20%, 1.97%, and 4.39% for particle diameters of 0.125 mm, 0.25 mm, and 0.5 mm respectively. The main advantage of this method is its convenience in determining the location of particles near the crack tip efficiently and quickly, which is essential for accurate estimation of SIFs.

The J-integration method is slightly less effective, with errors of 1.12%, 2.53% and 5.02% for particle diameters of 0.125, 0.25 and 0.5 mm respectively. Although the J-integral method is less computationally efficient, it provides an accurate and reliable result of SIFs.

Overall, all three methods provide a relatively accurate and reliable result of calculating SIFs, but the displacement extrapolation method and the stress extrapolation method are more suitable for dealing with fracture problems containing singular stress fields than the J-integral method.

5   Conclusion

In this study we compared several methods used to calculate SIFs in mixed mode and evaluated their accuracy and calculation times. As a result of our analysis and comparison, we have come to the following conclusions:

Firstly, the J-integral method is suitable for complex geometries and loading conditions and delivers high accuracy. However, it has high computational complexity and may require long computation time.

Secondly, the displacement extrapolation method is a displacement field-based method for estimating the SIF under mode I. Although it can provide good results in some cases, it requires prior knowledge of cracking information and assumes that the displacement field is linear.

Finally, the stress extrapolation method showed relatively good results in our study, especially when calculating the SIF under mode I. This method estimates the SIF by applying extrapolated stresses without the necessity to know the crack size in advance and is suitable for linear elastic materials. There is no need to calculate the angle between the desired stress particle and the crack tip particle under mode I.

It is worth noting that our study found that the stress extrapolation method achieved good results and short calculation times when calculating the SIF under mode I. The stress extrapolation method estimates the SIF by applying extrapolated stresses without the prior knowledge of the crack size and is suitable for linear elastic materials.

Overall, the stress extrapolation method can be a good choice when calculating SIFs in mode I case based on our results. However, the J-integral method or the displacement extrapolation method may still be advantageous for specific problems and conditions. Therefore, the choice of method should be assessed and decided according to the specific requirements of the problem. In future work, accurately calculating stress intensity factors could lead to more accurate results in predicting the fatigue life of engineering devices, such as wave energy generators.

References

1. Wang, C., Wang, S., Chen, G., Yu, P., Peng, X.: Implementation of a J-integral based maximum circumferential tensile stress theory in DDA for simulating crack propagation. Eng. Fract. Mech. **246**, 107621 (2021). https://doi.org/10.1016/j.engfracmech.2021.107621
2. Zhu, Y., Zhang, C., Yu, Y., Hu, X.: A CAD-compatible body-fitted particle generator for arbitrarily complex geometry and its application to wave-structure interaction. J. Hydrodyn. **33**(2), 195–206 (2021). https://doi.org/10.1007/s42241-021-0031-y
3. Zhu, N., Oterkus, E.: Calculation of stress intensity factor using displacement extrapolation method in peridynamic framework. J. mech. **36**(2), 235–243 (2020). https://doi.org/10.1017/jmech.2019.62
4. Kuang, J.H., Chen, L.S.: Comparison of displacement extrapolation methods to "benchmark" surface crack problems. Eng. Fract. Mech. **47**(4), 473–477 (1994). https://doi.org/10.1016/0013-7944(94)90248-8
5. Elmhaia, O., Belaasilia, Y., Askour, O., Braikat, B., Damil, N.: An efficient mesh-free approach for the determination of stresses intensity factors. Eng. Anal. Bound. Elem. **133**, 49–60 (2021). https://doi.org/10.1016/j.enganabound.2021.08.001
6. Tazoe, K., Tanaka, H., Oka, M., Yagawa, G.: Analyses of fatigue crack propagation with smoothed particle hydrodynamics method. Eng. Fract. Mech. **228**, 106819 (2020). https://doi.org/10.1016/j.engfracmech.2019.106819
7. Long, Z., Xiaomin, Z., Jiyun, S., Hengwei, Z.: Thermal fracture parameter analysis of MEMS multilayer structures based on the generalized thermoelastic theory. Microelectron. Reliab. **98**, 106–111 (2019). https://doi.org/10.1016/j.microrel.2019.04.017
8. Moran, B., Shih, C.F.: A general treatment of crack tip contour integrals. Int. J. Fract. **35**(4), 295–310 (1987). https://doi.org/10.1007/BF00276359
9. Shivakumar, K.N., Raju, I.S.: An equivalent domain integral method for three-dimensional mixed-mode fracture problems. Eng. Fract. Mech. **42**(6), 935–959 (1992). https://doi.org/10.1016/0013-7944(92)90134-Z
10. Gu, P., Dao, M., Asaro, R.J.: A simplified method for calculating the crack-tip field of functionally graded materials using the domain integral. J. Appl. Mech. **66**(1), 101–108 (1999). https://doi.org/10.1115/1.2789135
11. Ganesh, K.V., Islam, M.R.I., Patra, P.K., Travis, K.P.: A pseudo-spring based SPH framework for studying fatigue crack propagation. Int. J. Fatigue **162**, 106986 (2022). https://doi.org/10.1016/j.ijfatigue.2022.106986

Enriching On-orbit Servicing Mission Planning with Debris Recycling and In-space Additive Manufacturing

Bai Jie, Chao Tao$^{(\boxtimes)}$, Ma Ping, Yang Ming, and Wang Songyan

Control and Simulation Center, Harbin Institute of Technology, Harbin 150000, China
chaotao2000@163.com

Abstract. On-orbit servicing (OOS) is a crucial research area with vast application prospects. There is currently a lack of studies focusing on a novel type of OOS scenario. In this scenario, In-Space Additive Manufacturing (ISAM) and Debris Recycling (DR) are incorporated into OOS. Traditional OOS obtains the necessary products for failed satellites through recycling space debris and ISAM, thereby improving sustainability and efficiency in the servicing process. We construct an optimization model. In this model, the total cost of velocity is chosen as the optimization objective. And mission sequences, orbital transfer time, service position, and service time are considered as design variables. Compared to traditional OOS, the novel optimization model is more prone to trapping optimizers in local optima. Therefore, we employ the Whale Optimization Algorithm (WOA) due to its ability to escape local optima. Simulation results demonstrate the effectiveness of the optimization model we have developed in representing the novel problem accurately. Moreover, the solution obtained through the utilization of the WOA closely approximates the optimal solution.

Keywords: On-orbit Service (OOS) · Mission Planning · In-space additive manufacturing (ISAM) · Debris recycling (DR) · Whale Optimization Algorithm (WOA)

1 Introduction

The concept of on-orbit service (OOS) for spacecraft was first proposed in the 1960s [1, 2]. The cost of launch vehicles and satellites is exorbitant, and OOS offers significant benefits by extending the lifespan and enhancing the performance of space systems, while also reducing mission expenses to a large extent. Typical tasks performed in OOS include on-orbit repairs, active debris removal, refueling operations, and satellite operational status monitoring, et al. [3]. Mission planning plays a crucial role in OOS as it enables improved work efficiency and minimizes fuel consumption during missions. Scholars from around the world have conducted extensive research in this field, contributing to the advancement of the mission planning OOS.

Chen [4] has proposed a mixed strategy in which a mission planning problem including a depot, target satellites and servicing spacecraft (SSc) is solved. In this study, the

© The Author(s), under exclusive license to Springer Nature Singapore Pte Ltd. 2024
F. Hassan et al. (Eds.): AsiaSim 2023, CCIS 1912, pp. 86–100, 2024.
https://doi.org/10.1007/978-981-99-7243-2_8

economic cost and benefit models are established as design objectives. Han [5] investigated the on-orbit repairing mission planning of multiple geosynchronous orbit (GEO) satellites and proposed a novel algorithm, the large neighborhood search-adaptive genetic algorithm, to tackle the problem. Some active debris removal strategies have been studied in the past few years. Liu [6] developed a preliminary plan for a multi-nanosatellite active debris removal on LEO. In his study, a modified genetic algorithm is also proposed to solve the dynamic multi-objective TSP. Yu [7] employed a modified multi-objective particle swarm optimization to solve the mission planning of GEO debris removal with multiple SSc.

Meanwhile, In-Space Additive Manufacturing (ISAM), also known as 3D printing in space, has been actively explored and studied by countries worldwide. In 2014, the National Aeronautics and Space Administration (NASA) launched the first zero-gravity 3D printer to the International Space Station (ISS) and successfully printed the first-ever part manufactured in space [8]. Following suit, in 2015, the European Space Agency (ESA) commissioned an Italian company to develop its own in-space 3D printer, which was subsequently sent to the space station [9]. In recent years, the application of 3D printing technology in space has expanded extensively, with the aerospace industry becoming a new hotspot for its implementation. Some 3D-printed objects have already been deployed in orbit, and it is anticipated that the number of 3D-printed items in space will grow exponentially in the coming years [10].

As previously enumerated, while numerous scholars have conducted research on OOS, few studies have focused on a novel OOS mission scenario. In this mission scenario ISAM and Debris Recycling (DR) are incorporated into OOS mission planning. This novel mission scenario presents two challenging aspects that need to be addressed. The first challenge is how to construct an optimization model specifically tailored to this new mission scenario. The second challenge relates to the characteristics of mathematical models, which often pose difficulties for optimizers, leading to the problem of getting stuck in local optima. Therefore, it is necessary to explore a more efficient algorithm to solve this mathematical optimization model.

To address this novel mission scenario, we develop a mixed-integer optimization model. Unlike the models that describe existing OOS, the optimization model incorporating DR and ISAM is more susceptible to being trapped in local optima. Additionally, the Whale Optimization Algorithm (WOA) demonstrates a remarkable ability to balance the exploration and exploitation phases, making it highly competitive in escaping local optima [11]. Considering the challenges posed by this optimization model, we employ the WOA to tackle the problem effectively. In the solving procedure, we introduce the concepts of Swap Operators and Swap Sequences. These concepts are employed to handle discrete variables, and enhances the suitability of the optimization model for the WOA.

The remaining sections of the paper are organized as follows. Section 2 provides a detailed description of the novel mission scenario. In Sect. 3, we present the optimization model specifically designed for this novel mission scenario. Subsequently, in Sect. 4, we introduce the optimization algorithm utilized to solve the problem. The computational experiments conducted are presented in Sect. 5, followed by the concluding remarks in Sect. 6.

2 Mission Scenario Description

In this mission scenario ISAM and DR are incorporated into OOS mission planning. Specifically, the mission scenario involves a Space Station acting as the mothership, multiple failed satellites, SSc, and a considerable quantity of debris. The mothership serves dual purposes as a warehouse and a manufacturing facility, utilizing feedstock from Earth and space debris to produce 3D-printed spare parts. The SSc is responsible for recycling space debris and shuttling between the mothership and target satellites with the spares.

The service scenario follows the following process: First, the SSc transfers from one selected debris to the next. Second, SSc brings the captured space debris to the Space Station and retrieves the printed components. Third, SSc, carrying the printed components, transfers from the Space Station to the failed satellites one by one. An example of the mission is shown as Fig. 1.

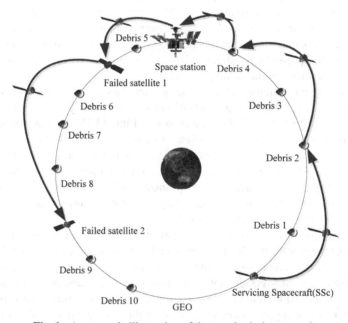

Fig. 1. An example illustration of the novel mission scenario

Besides that, three constraints should be mentioned as follows:

- SSc retrieves components from the Space Station to repair failed satellites, so SSc needs to rendezvous with the Space Station first before rendezvousing with the failed satellites.
- A number of debris are selected from a large number of space debris. After recycling the space debris, they need to rendezvous with the Space Station.
- SSc provides different maintenance times for different targets. It takes longer to maintain failed satellites and less time to recycle space debris.

The goal of this mission is to find the optimal rendezvous sequence, rendezvous time and servicing time to ensure optimal benefits.

3 Mathematical Optimization Model

3.1 Dynamics Model

The two-body model offers a reasonable assessment of fundamental principles and long-term performance in select conditions. For on-orbit servicing (OOS) missions, classical mechanics-based simplified two-body model typically fulfills most requirements. The simplified two-body model is usually expressed as

$$
\begin{aligned}
\ddot{\vec{r}} &= -\tfrac{\mu}{r^3}\vec{r} \\
\dot{\vec{r}} &= \vec{v}
\end{aligned}
\tag{1}
$$

Let $(\vec{r}_0, \vec{v}_0)$ be the initial state of the object, where r_0 and v_0 respectively denote the radius vector and velocity vector. The end state of object $(\vec{r}_{\Delta t}, \vec{v}_{\Delta t})$ after a time of Δt is calculated by using the Lagrange coefficients f, g and its' derivatives $\dot{f}$, $\dot{g}$[12].

$$
\begin{aligned}
\vec{r}_{\Delta t} &= f\,\vec{r}_0 + g\,\vec{v}_0 \\
\vec{v}_{\Delta t} &= \dot{f}\,\vec{r}_0 + \dot{g}\,\vec{v}_0
\end{aligned}
\tag{2}
$$

where $f, g, \dot{f}, \dot{g}$ are calculated by

$$
\begin{aligned}
f &= 1 - \chi^2 C(\alpha\chi^2)/r_0 \\
g &= \Delta t - \chi^2 S(\alpha\chi^2)/\sqrt{\mu} \\
\dot{f} &= \sqrt{\mu}[\alpha\chi^3 S(\alpha\chi^2) - \chi]/rr_0 \\
\dot{g} &= 1 - \chi^2 C(\alpha\chi^2)/r
\end{aligned}
\tag{3}
$$

In the above equations, μ denote the gravitational coefficient and its value is 398,600 km^3/s^2. α is the reciprocal of the semi major axis. $C(x)$ and $S(x)$ are Stumpff functions. χ is the universal variable. Details of the χ can be found in Ref [12].

3.2 Lambert Problem

The relation between the Lagrange coefficients $f, g, \dot{f}, \dot{g}$ and the change of true anomaly $\Delta\theta$ can be described as below

$$
\begin{aligned}
f &= 1 - \tfrac{\mu r_2}{h^2}(1 - \cos\Delta\theta) \\
g &= \tfrac{r_1 r_2}{h}\sin\Delta\theta \\
\dot{f} &= \tfrac{\mu}{h}\tfrac{1-\cos\Delta\theta}{\sin\Delta\theta}\left[\tfrac{\mu}{h^2}(1-\cos\Delta\theta) - \tfrac{1}{r_1} - \tfrac{1}{r_2}\right] \\
\dot{g} &= 1 - \tfrac{\mu r_1}{h^2}(1 - \cos\Delta\theta)
\end{aligned}
\tag{4}
$$

with the help of Eq. (2), (3), we can get the following Eq. (4), Newton iterative method is used to gain the root of the equation.

$$
\sqrt{\mu}\Delta t = \left[y(z)/C(z)\right]^{\frac{3}{2}} S(z) + A\sqrt{y(z)}
\tag{5}
$$

where $y(z)$, A, z can be calculated by

$$y(z) = r_1 + r_2 + A[zS(z) - 1]/\sqrt{C(z)} \tag{6}$$

$$A = \sin\Delta\theta\sqrt{r_1 r_2/(1 - \cos\Delta\theta)} \tag{7}$$

$$z = \alpha\chi^2. \tag{8}$$

With the help of the Lagrangian coefficient, the Lambert transfer orbit can be obtained by formulas (1) and (2). Given the initial orbital velocity V_{in} and the terminal orbital velocity V_{out} V_{out}

$$\Delta V = \Delta V_{in} + \Delta V_{out} \tag{9}$$

4   Optimization Algorithm

4.1   Design Variables

During the first stage of the mission, SSc will select three out of ten debris and move towards them in a sequential manner to rendezvous and recycle. The design variables for this optimization process include the visiting sequence Q, the duration of orbital transfer (DUR), and the service time required for each rendezvous mission (SER). The maneuver time and impulses within each rendezvous:

$$Q = (q_i, q_j, q_k), i, j, k \in (1, 2, ...10) \tag{10}$$

$$DUR = (dur_1, ...dur_d), d = 4 \tag{11}$$

$$SER = (ser_1, ..., ser_s), s = 3 \tag{12}$$

Once all the targeted debris has been recycled, the SSc will then proceed towards the space station. As a result, there will be more instances of DUR compared to SER for the entire mission.

4.2   Objective Function

The objective function is a combination of the total velocity increment and the mean mission completion time.

$$\min f = \sum_{i=1}^{n} \Delta v_{in_i} + \Delta v_{out_i} \tag{13}$$

4.3 Constraints

In the practical mission, there are time constraints for each transfer time dur_i and each service time ser_i. For space fragments with large quality and high complexity, more time needs to be allocated to complete the mission. The continuous design variables can be described as $(dur_1, dur_2, dur_3, dur_4, ser_1, ser_2, ser_3)$.

The upper constraints of the continuous design variables can be expressed as

$$UB = (ub_1, ub_2, ..., ub_7) \tag{14}$$

The lower constraints of the continuous design variables can be expressed as

$$LB = (lb_1, lb_2, ..., lb_7) \tag{15}$$

4.4 The Whale Optimization Algorithm

Whale Optimization Algorithm (WOA) is a population-based algorithm proposed by Mirjalili et al. [11]. WOA can be divided into two phases: exploitation and exploration. The exploration phase is used to randomize the search for prey by making movements as much as possible. During the exploitation phase, whales use bubble-net attacking and encircling techniques to capture their prey, while investigating the promising areas of the search space. This phase can be defined as the process of detailed investigation of the search space's promising areas. To simulate the movements of a whale encircling its prey during the exploitation phase, the mathematical model can be described as follows

$$\vec{D} = \left| \vec{C}\vec{X}^*(t) - \vec{X}(t) \right|$$
$$\vec{X}(t+1) = \vec{X}^*(t) - \vec{A}\vec{D} \tag{16}$$

where $\vec{X}(t)$ indicates the position vector at time step t, $\vec{X}^*(t)$ represents the best solution. $||$ is the absolute value. The vectors $\vec{A}$ and $\vec{C}$ are expressed as follows, respectively.

$$\vec{A} = 2\vec{a}\vec{r} - \vec{a}$$
$$\vec{C} = 2\vec{r} \tag{17}$$

where $\vec{r}$ is random vector. $\vec{a}$ decreases from 2 to 0 with the course of iterations. a can be expressed as below:

$$a = 2 - 2t/MaxIter \tag{18}$$

where $MaxIter$ is the max number and t represents the time step. The mathematical model of spiral movement of bubble-net can be described as below

$$\vec{X}(t+1) = D' \cdot e^{bl} \cdot \cos(2\pi l) + \vec{X}^*(t) \tag{19}$$

where $D' = \left| \vec{X}^*(t) - \vec{X}(t) \right|$, b is a constant for defining the shape of the logarithmic spiral, and l is a random number from -1 to 1. To simulate the two movements, shrinking encircling prey and the spiral-shaped path, the probability of switching between two movements is 50%.

$$\begin{cases} shrinking\ encircling\ prey\ (p < 0.5) \\ the\ spirl - shaped\ path\ \ \ (p > 0.5) \end{cases} \tag{20}$$

In the exploration phase, a random whale is created to maximize the randomness of movements. This mathematical model can be represented by Eq. (20).

$$\vec{D} = \left| \vec{C}\vec{X}_{rand}(t) - \vec{X}(t) \right|$$
$$\vec{X}(t+1) = \vec{X}_{rand}(t) - \vec{A}\vec{D} \tag{21}$$

where $\vec{X}_{rand}(t)$ is a random whale and the variable $\vec{A}$ has the same meaning as $\vec{A}$ in Eq. (16).

To provide a better understanding of the WOA, the following pseudo-code is presented.

Algorithm 1: Whale Optimization Algorithm (WOA)

Initialize the whale population X_i ($i = 1,2,3,\ldots,n$)

Calculate the fitness of each search agent

$X^* =$ the best search agent

1. **while** ($t <$ maximum number of iterations)
2.   **for** each search agent
3.   Update a, A, C, I and p
4.   **if 1** ($p < 0.5$)
5.   **if 2** ($|A| < 1$)
6.   Update the position of the current search agent by the Eq. (16)
7.   **else if 2** ($|A| >= 1$)
8.   Select a random search agent (X_{rand})
9.   Update the position of the current search agent by the Eq. (21)
10.   **end if 2**
11.   **else if 1** ($p >= 0.5$)
12.   Update the position of the current search by the Eq. (19)
13.   **end if 1**
14.   **end for**
15.   Check if any search agent goes beyond the search space and amend it
16.   Calculate the fitness of each search agent
17.   Update X^* if there is a better solution
18. $t = t + 1$
19. **end while**
20. return X^*

4.5 A Method for Handling Discrete Variables

Since the *SER* and *DUR* variables are continuous, it is necessary to process the discrete variable, Q, to achieve optimal results in the solution. The concepts of swap operator and swap sequence [13] are described as below: The solution sequence of the N-node combinatorial optimization (CO) problem is $S = (a_i)$, $i = 1, \ldots, N$. The swap operator $SO(i_1, i_2)$ exchanges node a_{i_1} with a_{i_2} in solution S. The plus sign " $+$ " is utilized to signify this new meaning.

For instance, consider a CO problem with five nodes, where the following solution $S = (1, 3, 4, 2)$ is represented. The Swap Operator is $SO(1, 2)$, then

$$
\begin{aligned}
S' &= S + SO(1, 2) \\
&= (1, 3, 4, 2) + (1, 2) = (3, 1, 4, 2).
\end{aligned}
\tag{22}
$$

A Swap Sequence, *SS*, consists of one or more Swap Operators.

$$
SS = (SO_1, SO_2, SO_3 \ldots, SO_n)
\tag{23}
$$

where $SO_1, SO_2, SO_3 \ldots, SO_n$ are Swap Operators, and the order of these operators is significant. When a Swap Sequence acts on a solution, each Swap Operator within the sequence acts on the solution in a specific order. This relationship can be represented mathematically by the following formula:

$$
\begin{aligned}
S' &= S + SS \\
&= S + (SO_1, SO_2, SO_3 \ldots, SO_n) \\
&= ((S + SO_1) + SO_2) + \ldots + SO_n)
\end{aligned}
\tag{24}
$$

When different Swap Sequences act on the same solution, they may produce the same new solution. This set of equivalent Swap Sequences is referred to as the equivalent set. Among these sequences, the one with the fewest Swap Operators is called the Basic Swap Sequence (BSS).

Furthermore, by merging multiple Swap Sequences, we can define the operator $\oplus$ as a means of creating a new Swap Sequence. For example, suppose there are two Swap Sequences, *SS*1 and *SS*2, which act on solution S in a specific order (i.e., *SS*1 first and then *SS*2), resulting in a new solution S'. If there exists another Swap Sequence SS' that acts on the same solution S and results in the same solution S', we can represent this as follows:

$$
SS' = SS1 \oplus SS2
\tag{25}
$$

As a result, both SS' and $SS1 \oplus SS2$ belong to the same equivalent set. To construct a Basic Swap Sequence *SS* that transforms solution *B* into solution *A*, we define $SS = A - B$ (where the minus sign has a new meaning). By swapping the nodes in *B* according to the order in *A* from left to right, we can obtain *SS*. Therefore, the equation $A = B + SS$ holds.

Regarding the Whale Algorithm, we can reconstruct its calculation formula as follows. For Formula (16), we will adopt the following strategies

$$
\begin{aligned}
\vec{D} &= ROUND(\vec{C} * \left| \vec{X}^*(t) \ominus \vec{X}(t) \right|) \\
\vec{X}(t+1) &= \vec{X}^*(t) \ominus \vec{D}
\end{aligned}
\tag{26}
$$

For formula (19) we take the following strategy to reconstruct the formula:

$$\vec{X}(t+1) = D' \cdot e^{bl} \cdot \cos(2\pi l) \oplus \vec{X}^*(t) \tag{27}$$

For formula (21) we take the following strategy to reconstruct the formula:

$$\vec{D} = ROUND(\vec{C} * \left| \vec{X}_{rand}(t) \ominus \vec{X}(t) \right|)$$
$$\vec{X}(t+1) = \vec{X}_{rand}(t) \ominus \vec{D} \tag{28}$$

According to Eq. (26), (27) and (28), the combination optimization variables are calculated by the WOA.

4.6   Mixed-Integer Optimization Model

The algorithm for the dynamic model of mission planning is illustrated in Fig. 2, while Fig. 3 presents the optimization model used to solve the special OOS problem in this study. The overall algorithm begins by processing discrete variables using Eqs. (22), (23) and (24), after which the WOA searches for new positions of whales representing the visiting sequence Q, transfer time DUR, and service time SER. These new positions of whales and the initial state of each spacecraft (represented by six orbital elements) are then used as inputs for the dynamical model. The dynamical model calculates the cost of the total speed, which is also referred to as the fitness of whales, by continuously incrementing its value with time. After computing the number of six orbits following T time, the algorithm moves on to the next iteration. The optimal solution is obtained once the maximum number of iterations is reached.

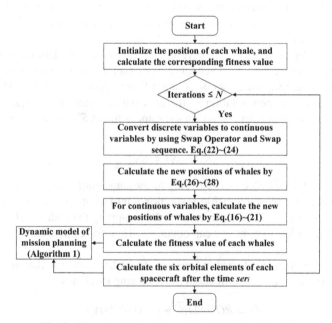

Fig. 2. Flow chat of the dynamic model (Algorithm 1)

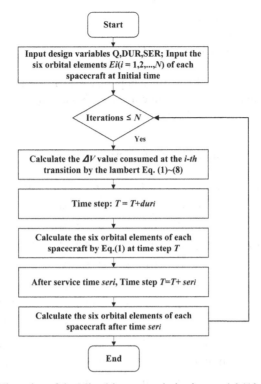

Fig. 3. Flow chat of the Mixed-integer optimization model (Algorithm 2)

5 Computation Experiments

In this section, we apply the method proposed above to a practical rendezvous mission, which involves 10 space debris, 2 failed satellites, a space station, and a SSc from Ref. [14]. These spacecrafts operate in a 30-day repeating-ground-track orbit, with their initial orbital elements specified as semi-major axis, eccentricity, inclination, RAAN, argument of periapsis, and true anomaly (the abbreviation was used in the table). The initial orbital elements of all debris are listed in Table 1, while the initial state of the failed satellites, SSc, and space station are given in Table 1.

By applying the algorithm outlined in reference [12] and utilizing the specific orbital element parameters provided by Table 1 and Table 2, we obtained the state vectors $\vec{r}$ and $\vec{v}$ for all spacecraft in the geocentric equatorial frame, as shown in Table 3 and Table 4. Subsequently, we utilized Algorithm 1 and Algorithm 2 to calculate the optimal task sequence, as well as the corresponding optimal service time and transfer time based on the obtained state vectors. In the computation process, we employed the WOA algorithm with 30 search agents, optimizing for 1500 iterations. The optimization involved 11 continuous design variables, with constraints defined as presented in Table 5.

Using a personal computer with a CPU frequency of 3.0 GHz, we perform six trials under identical computational conditions, each requiring approximately 2 min to complete. Finally, we obtain five sets of optimized service sequences, and identify the optimal sequence with the best fitness value as Experiment Number 1. This optimal sequence is presented in Table 6.

The outcome is presented as follows: The optimized service sequence is Debri2 → Debri5 → Debri7 → Failed satellite1 → Failed satellite 2 → Space station, with a total velocity cost of 58.26m/s. An alternative suboptimal sequence is Debri2 → Debri5 → Debri4 → Failed satellite1 → Failed satellite 2 → Space station, resulting in a total velocity increment ΔV_{total} of 78.60m/s.

Table 1. The orbital elements of all debris

Number	SMA(m)	ECC	INC (°)	RAAN (°)	ARG (°)	TA (°)
Debri 1	4.2165e+07	0.0001940	0.0179	118.3294	220.4078	17.2256
Debri 2	4.2165e+07	0.0002402	0.0152	116.2403	205.9543	75.2179
Debri 3	4.2165e+07	0.0002054	0.0171	108.8864	265.7792	120.5719
Debri 4	4.2165e+07	0.0001203	0.0151	115.9406	274.3255	157.4630
Debri 5	4.2165e+07	0.0000483	0.0173	114.2483	239.2561	174.9399
Debri 6	4.2165e+07	0.0003282	0.0145	113.6512	224.7644	193.3435
Debri 7	4.2165e+07	0.0001852	0.0180	112.6962	200.2765	235.9404
Debri 8	4.2164e+07	0.0003039	0.0164	108.3989	239.1021	293.5546
Debri 9	4.2164e+07	0.0001587	0.0116	118.2457	217.6796	314.9439
Debri 10	4.2164e+07	0.0003111	0.0180	113.8442	224.2112	358.0863

Table 2. The orbital elements of all Failed satellite, SSc and Space station

Name	SMA (m)	ECC	INC (°)	RAAN (°)	ARG (°)	TA (°)
Space STA	4.2165e+07	0.0001840	0.0179	118.3294	200.4078	300.2256
Failed SAT 1	4.2165e+07	0.0002402	0.0152	116.2403	240.9543	320.2165
Failed SAT 2	4.2165e+07	0.0002402	0.0152	116.2403	240.9543	290.2101
SSc	4.2165e+07	0.0002402	0.0152	116.2403	205.9543	30.1351

Following the experiment results, we performed an evaluation of the algorithm and its outcomes. To illustrate the convergence plot of the algorithms utilized in this paper, Fig. 4 was generated. The plot indicates a relatively strong convergence property of the algorithms. To assess the effectiveness of the algorithms presented in this paper, we verify the service sequence of SSc → Debri2 → Debri5. Starting from the initial time point, we plot the correlation between the transfer time and total velocity increment for SSc and Debri2, as shown in Fig. 5. After a specific interval of $dur_1 + ser_1$, SSc begins the transfer from Debri2 to Debri5. In dur_2 intervals, SSc meets Debri5 through a Lambert transfer orbit. The relationship between the cost of velocity and dur_2 is illustrated in Fig. 6. The two figures demonstrate that a strong correlation exists between

Table 3. The initial state vectors of debris in J2000 coordinate systems

Number	X (m)	Y (m)	Z (m)	V_X (m/s)	V_Y (m/s)	V_Z (m/s)
Debri 1	42052574.26	−2968041.30	−11133.06	216.68	3067.55	−0.51451
Debri 2	33488848.82	25615737.47	−10981.25	−1867.53	2442.70	0.15817
Debri 3	−29941615.39	29694414.33	5591.57	−2165.22	−2182.47	0.8230
Debri 4	−41786573.05	−5671378.37	10565.89	413.31	−3046.37	0.2530
Debri 5	−41312319.51	8446919.73	10334.32	−615.89	−3012.16	0.5430
Debri 6	−41742935.78	6045669.21	9071.44	−440.33	−3041.94	0.4111
Debri 7	−41660152.94	−6533564.39	12875.41	476.78	−3037.11	0.2299
Debri 8	8084448.66	−41376473.04	1542.15	3017.80	590.51	−0.8740
Debri 9	8085109.92	−41379857.96	1090.81	3017.67	590.07	−0.6181
Debri 10	38549949.49	−17049431.85	−8919.37	1243.97	2812.78	−0.7150

Table 4. The initial state vectors of failed satellites, SSc and space station vectors in J2000 coordinate systems

Name	$X(m)$	$Y(m)$	$Z(m)$	$V_X(m/s)$	$V_Y(m/s)$	$V_Z(m/s)$
Space STA	−8071583.70	−41381242.82	8357.73	3018.13	−588.20	−0.7435
Failed SAT 1	31034094.76	−28532715.66	−4043.37	2080.99	2264.13	−0.7613
Failed SAT 2	12611238.27	−40231190.19	1716.82	2933.90	920.41	−0.8067
SSc	41779037.16	−5626771.33	−9289.96	410.84	3047.70	−0.4554

Table 5. Lower constraints and upper constraints of continuous design variables

Variable	dur_1	dur_2	dur_3	dur_4	dur_5	dur_6	ser_1	ser_2	ser_3	ser_4	ser_5
$LB(s)$	18000	18000	18000	18000	18000	18000	7200	7200	7200	180000	180000
$UB(s)$	180000	180000	180000	180000	180000	180000	—	—	—	360000	360000

Table 6. The optimization results: optimal sequence, service time and transfer time

No	dur_1	dur_2	dur_3	dur_4	dur_5	dur_6	ser_1	ser_2	ser_3	ser_4	ser_5	ΔV_{total}
The service sequence: Debri2 → Debri5 → Debri7 → Failed satellite1 → Failed satellite 2 → Space station												
1	107100	87600	112800	88200	38700	37800	9300	15430	18600	260380	247830	58.26
2	107458	87591	125060	87264	39800	38500	15050	15679	13620	180240	247830	63.75
3	97446	87591	156800	78200	39800	70630	18604	12679	15630	340256	187490	96.58
4	107445	87583	124806	88200	39800	68500	16385	14689	15263	250260	247830	88.47
The service sequence: Debri2 → Debri5 → Debri4 → Failed satellite1 → Failed satellite 2 → Space station												
5	107100	87600	93400	86380	38800	35794	9320	15430	16580	264320	246845	78.60

phase difference as well as the cost of velocity, which is closely related to the initial positions of the spacecraft. Given that all spacecraft operate on orbits that are close to near-circular, the cost of velocity doesn't show significant variation with respect to the device phase. These two verification experiments effectively confirm the validity and effectiveness of the algorithm proposed in this paper.

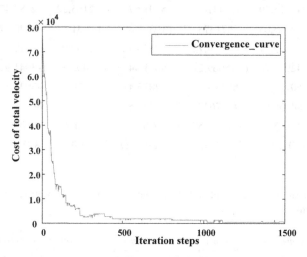

Fig. 4. Convergence curve of the mixed-integer optimization model solving by WOA

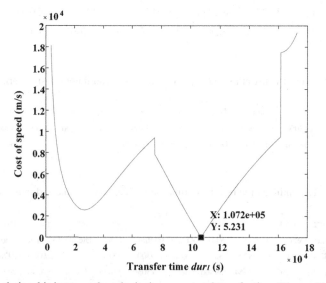

Fig. 5. The relationship between the velocity increment and transfer time (The position coordinate of the lowest point is (107200, 5.231), with the optimal transfer time being 107200 s. This result is consistent with the optimization outcomes obtained previously.)

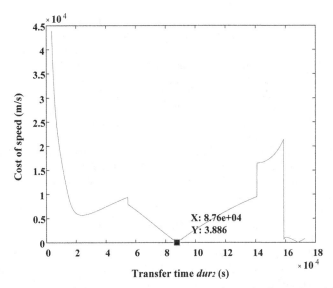

Fig. 6. The relationship between the velocity increment and transfer time (The minimum cost of velocity is 3.886 m/s, with a corresponding transfer time of dur_2 equal to 87600 s. This is almost equivalent to the optimized output value of 87583 s.)

6 Conclusion

In this paper, a comprehensive mission planning method is proposed for a novel mission scenario that combines In-Space Additive Manufacturing (ISAM) and Debris Recycling (DR) in OOS. Firstly, we construct a mixed-integer optimization model specifically designed for this novel mission scenario. Then, considering the inherent challenge of mathematical optimization models, which often lead to the issue of local optima, we employ the WOA algorithm in conjunction with the swap operator and swap sequence to solve it. The experimental results validate the convergence and effectiveness of our algorithm. This method holds significant value for operational rendezvous mission planning.

References

1. Li, W.J., Cheng, D.Y., Liu, X.G., et al.: On-orbit service (OOS) of spacecraft: a review of engineering developments. Prog. Aerosp. Sci. **108**, 32–120 (2019)
2. NASA, Ā.: On-orbit satellite servicing study project report. In-Space Nondestructive Inspection Technology Workshop (2010)
3. Cui, N.G., Wang, P., Guo, J.F., et al.: A review of On-orbit servicing. J. Astronaut. **28**(4), 805–811 (2007)
4. Chen, X.Q., Yu, J.: Optimal mission planning of GEO on-orbit refueling in mixed strategy. Acta Astronaut. **133**, 63–72 (2017)
5. Han, P., Guo, Y., Li, C., et al.: Multiple GEO satellites on-orbit repairing mission planning using large neighborhood search-adaptive genetic algorithm. Adv. Space Res. **70**(2), 286–302 (2022)

6. Liu, Y., Yang, J., Wang, Y., et al.: Multi-objective optimal preliminary planning of multi-debris active removal mission in LEO. Sci. China Inf. Sci. **60**(7), 1–10 (2017)
7. Jing, Y., Chen, X.Q., Chen, L.H.: Biobjective planning of GEO debris removal mission with multiple servicing spacecraft. Acta Astronaut. **105**(1), 311–320 (2014)
8. Prater, T., Werkheiser, N., Ledbetter, F., et al.: 3D Printing in Zero G Technology Demonstration Mission: complete experimental results and summary of related material modeling efforts. The Int. J. Adv. Manuf. Technol. **101**, 391–417 (2019)
9. Joshi, S.C., Sheikh, A.A.: 3D printing in aerospace and its long-term sustainability. Virtual Phys. Prototyping **10**(4), 175–185 (2015)
10. Peverini, O.A., Lumia, M., Addamo, G., et al.: How 3D-printing is changing RF front-end design for space applications. IEEE J. Microw. **3**(2), 800–814 (2023)
11. Mirjalili, S., Lewis, A.: The whale optimization algorithm. Adv. Eng. Softw. **95**, 51–67 (2016)
12. Curtis, H.D.: Orbital mechanics for engineering students, 3rd edn. Butterworth-Heinemann, Florida (2013)
13. Wang, K.P., Huang, L., Zhou, C.G., et al.: Particle swarm optimization for traveling salesman problem. In: Proceedings of the 2003 International Conference on Machine Learning and Cybernetics (IEEE cat. no. 03ex693), pp. 1583–1585. IEEE, Xi'an (2003)
14. NORAD GP Element Sets Current Data. http://celestrak.com/NORAD/elements/. Last accessed 28 May 2023

Attitude and Altitude Control of Quadrotor MAV using FOPID Controller

Aminurrashid Noordin[1,2], Mohd Ariffanan Mohd Basri[1(✉)], Zaharuddin Mohamed[1], and Izzuddin Mat Lazim[3]

[1] Faculty of Electrical Engineering, Universiti Teknologi Malaysia, UTM Johor Bahru, 81310 Johor, Malaysia
ariffanan@fke.utm.my
[2] Faculty of Electrical Technology and Engineering, Universiti Teknikal Malaysia Melaka, Hang Tuah Jaya, 76100 Durian Tunggal, Melaka, Malaysia
[3] Faculty of Engineering and Built Environment, Universiti Sains Islam Malaysia, 71800 Nilai, Negeri Sembilan, Malaysia

Abstract. This paper presents the implementation of a Fractional Order Proportional-Integral-Derivative (FOPID) controller for attitude and altitude stabilization of a quadrotor Micro Air Vehicle (MAV). FOPID extends the conventional Proportional-Integral-Derivative (PID) form by introducing fractional orders to enhance control. It maintains the basic structure of PID with the addition of λ for integration order and μ for differential order, allowing for the study of system performance and increased flexibility in controller design. The simulation in this study utilizes the Fractional-order Modeling and Control (FOMCON) toolbox in MATLAB. At the same time, the system's performance is assessed through the measurement of Integral Square Error (ISE) and Integral Square Control Input (ISCI). Regarding attitude control, the FOPID controller performs better transient response, and ISE is comparable to the PD controller. However, when considering ISCI, the Proportional-Derivative (PD) controller consumes less energy. Regarding altitude control, the FOPID controller performs better ISE but exhibits a 35% overshoot compared to the PD controller. However, the energy consumption of the FOPID controller is not too significant compared to the energy consumption of the PD controller.

Keywords: Quadrotor · PID · Fractional order PID · FAMCON

1 Introduction

In the past two decades, quadrotor research has gained significant popularity due to its diverse mapping, transportation, surveillance, and education applications. However, quadrotors pose challenges as they are highly under-actuated devices. To achieve successful outcomes in various missions, quadrotor controllers need to withstand perturbations. Quadrotors are nonlinear systems with multiple inputs and outputs, characterized by a high degree of dynamic coupling. Thus, reliable autonomous flight capabilities are

F. Hassan et al. (Eds.): AsiaSim 2023, CCIS 1912, pp. 101–112, 2024.
https://doi.org/10.1007/978-981-99-7243-2_9

crucial for accomplishing desired missions in terms of attitude and navigation control. In pursuit of this objective, a diverse range of control strategies has been extensively investigated. These strategies encompass Proportional-Integral-Derivative (PID), Linear Quadratic Regulator (LQR), Backstepping, Feedback Linearization Control (FLC), Sliding Mode Control (SMC), Model Predictive Control (MPC), Neural Network, H-infinity, Fuzzy Logic, and Adaptive Control.

The simplicity of the traditional PID controller offers advantages for both linear and nonlinear applications. Many researchers have described PID designs for quadrotor Unmanned Aerial Vehicles, (UAVs) [1–3]. The primary goal of the PID controller is to minimize the error by adjusting the control input, $u(t)$, over time. Simultaneously, the error value, $e(t)$, is continuously computed, and the PID controller applies a correction based on the proportional gain (k_p), integral gain (k_i), and derivative gain (k_d).

Recently, studies have indicated that the performance of PID controllers can be further enhanced by utilizing fractional-order PID (FOPID) controllers [4–10]. The FOPID controller is designed to improve the system's robustness. However, a drawback of using FOPID controllers is that they introduce additional complexity to the tuning procedures, as there are more parameters to be adjusted. Tuning methods for FOPID controllers can be broadly categorized into three main approaches: analytical, rule-based, and numerical tuning methods.

This paper aims to investigate the effectiveness of a Fractional Order Proportional-Derivative (FOPID) controller in stabilizing a small-scale quadrotor with a mass less than 0.1 kg. The primary focus of the study is on attitude and altitude control, as compared to the classical PD controller. The FOMCON toolbox in MATLAB is utilized for the simulation, and the measurement of Integral Square Error (ISE) and Integral Square Control Input (ISCI) is recorded for performance assessment.

2 Quadrotor Systems Modelling

Quadrotors are dynamic helicopters with four input forces and six output coordinates, requiring a dynamic underactuated system. They have four rotors arranged symmetrically, with diagonal motors (1 and 3) and (2 and 4) rotating in opposite directions. However, quadrotors are less stable than traditional helicopters, making them more difficult to control and increasing the risk of crashes. The MAV quadrotor is presented in this paper, with its dynamics described by two reference frames, the earth-fixed initial reference frame {E}, and the body-fixed reference frame {Q} as shown in Fig. 1.

Appropriate assumptions are made to simplify the controller design for the MAV quadrotor mathematical model. The underlying assumptions are as follows:

1. The quadcopter is assumed to be a rigid body.
2. The quadcopter's structure is assumed as symmetric with respect to the XY-axis.
3. The center of mass and the origin of the body fixed frame coincide.
4. The propellers are thought to be rigid because there is no blade flapping.
5. The four propellers always operate under the same circumstances, so their respective thrust and reaction torque coefficients are constant.

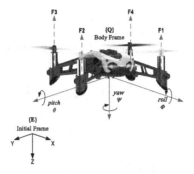

Fig. 1. Parrot Mambo Mini-drone MAV.

2.1 Quadrotor Dynamic Model

In this section, we present the mathematical model of the quadrotor. The dynamics of the quadrotor's translations and orientations are described using a 6 degree-of-freedom (DOF) model. This model is defined by two state vectors: $\xi = [x, y, z]^T$ and $\eta = [\phi, \theta, \psi]^T$, representing the quadrotor's positions and angles respectively.

As shown in Fig. 1, the translation from the initial fixed frame {E} to the body fixed frame {Q} requires a rotation matrix R, where in (1), C represents cosine and S represents sine.

$$R = \begin{bmatrix} C\theta C\psi & S\phi S\theta C\psi - C\phi S\psi & C\phi S\theta C\psi + S\phi S\psi \\ C\theta S\psi & S\phi S\theta S\psi + C\phi C\psi & C\phi S\theta S\psi - S\phi C\psi \\ -S\theta & S\phi C\theta & C\phi C\theta \end{bmatrix} \tag{1}$$

As the rotor generates the thrust force, where b is the thrust factor and Ω_i is the rotor speed, using Newton Euler second law of motion, we can derive a first set of differential equations that describe the quadrotor acceleration:

$$\ddot{\xi} = -g \cdot \begin{pmatrix} 0 \\ 0 \\ 1 \end{pmatrix} + R \cdot \frac{b}{m} \sum_{i=1}^{4} \Omega_i^2 \begin{pmatrix} 0 \\ 0 \\ 1 \end{pmatrix} \tag{2}$$

where, g is the gravitational coefficient, m is quadrotor mass and $i = 1, 2, 3, 4$. A second set of differential equations is generated by Newton's second law of rotation:

$$I\ddot{\eta} = -\dot{\eta} \times I\dot{\eta} - \sum_{i=1}^{4} J_r \left(\dot{\eta} \times \begin{pmatrix} 0 \\ 0 \\ 1 \end{pmatrix} \right) \Omega_i + \tau \tag{3}$$

Equations (2) and (3) describe a drone's translational and rotational motion, determining orientation and angular velocity, where, $I = diagonal[I_{xx}, I_{yy}, I_{zz}]^T$ is the inertia matrix, J_r is rotor the inertia and the vector τ expresses the torque. The vector τ is describes the as:

$$\tau = \begin{pmatrix} lb(\Omega_1^2 - \Omega_2^2 - \Omega_3^2 + \Omega_4^2) \\ lb(\Omega_1^2 + \Omega_2^2 - \Omega_3^2 - \Omega_4^2) \\ d(-\Omega_1^2 + \Omega_2^2 - \Omega_3^2 + \Omega_4^2) \end{pmatrix} \tag{4}$$

where d is the drag factor and l the length of the arm. The four rotational speeds of the rotors Ω_i, serve as the input variables for the real vehicle, and the model that was created is appropriate for the transformation of the inputs. Thus, the new artificial input variables can be formulated as:

$$u_1 = b\left(\Omega_1^2 + \Omega_2^2 + \Omega_3^2 + \Omega_4^2\right) \tag{5a}$$

$$u_2 = b\left(\Omega_1^2 - \Omega_2^2 - \Omega_3^2 + \Omega_4^2\right) \tag{5b}$$

$$u_3 = b\left(\Omega_1^2 + \Omega_2^2 - \Omega_3^2 - \Omega_4^2\right) \tag{5c}$$

$$u_4 = d\left(-\Omega_1^2 + \Omega_2^2 - \Omega_3^2 + \Omega_4^2\right) \tag{5d}$$

However, in (4), a new variable is discovered. The additional variable is similarly influenced by the rotors' speed rates. It must be regarded as addition artificial input as:

$$\Omega_d = -\Omega_1 + \Omega_2 - \Omega_3 + \Omega_4 \tag{6}$$

Assessment produces a dynamic model divided into underactuated and fully actuated subsystems, affecting flight and stability as follows:

$$under\ actuated \begin{cases} \ddot{x} = -(C\phi S\theta C\psi + S\phi S\psi)\frac{u_1}{m} \\ \ddot{y} = -(C\phi S\theta S\psi - S\theta C\psi)\frac{u_1}{m} \end{cases} \tag{7}$$

$$fully\ actuated \begin{cases} \ddot{z} = -g + (C\phi C\theta)\frac{u_1}{m} \\ \ddot{\phi} = \left(\frac{I_{yy}-I_{zz}}{I_{xx}}\right)\dot{\theta}\dot{\psi} - \frac{J_r\Omega_d}{I_{xx}}\dot{\theta} + \frac{l}{I_{xx}}u_2 \\ \ddot{\theta} = \left(\frac{I_{zz}-I_{xx}}{I_{yy}}\right)\dot{\phi}\dot{\psi} + \frac{J_r\Omega_d}{I_{yy}}\dot{\phi} + \frac{l}{I_{yy}}u_3 \\ \ddot{\psi} = \left(\frac{I_{xx}-I_{yy}}{I_{zz}}\right)\dot{\phi}\dot{\theta} + \frac{1}{I_{zz}}u_4 \end{cases} \tag{8}$$

2.2 Preliminary on Fractional Order Calculus

Fractional Calculus is a generalized extension of integer order calculus, offering flexibility in derivatives and integrals for real numbers. It is crucial for developing algorithms for Fractional Order Controllers (FOCs) and has extensive applications in modeling and control design. Fractional differentiation originated in the seventeenth century and underwent significant development in the nineteenth century. It includes a few well-known definitions, the most noteworthy of which are as follows:

i. The Riemann-Liouville (R-L) definition

$$D_t^\alpha f(t) = \frac{1}{\Gamma(\alpha)}\int_0^t \frac{f(\tau)}{(\tau - t)^{1-\alpha}}d\tau \tag{9}$$

where, α is fractional order with range of $0 \le \alpha < 1$

ii. The Caputo definition

$$D_t^\alpha f(t) = \frac{1}{\Gamma(n - \alpha)} \int_0^t \frac{f(\tau)}{(\tau - t)^{\alpha - n + 1}} d\tau \tag{10}$$

where the fractional order α has range of $n - 1 \le \alpha < n$, and $\Gamma(n - \alpha)$ is gamma function.

A fractional order PID controller (FOPID) in the form of PID was proposed in 1999 by [11], with the transfer function taking the following form:

$$G(s) = K_p + K_i \frac{1}{s^\lambda} + K_d s^\mu \tag{11}$$

where K_p stands for proportional gain, K_i for integral gain, K_d for differential gain, λ stands for integration order, and μ stands for differential order. The addition of λ and μ adds two configurable parameters, allowing for the study of system performance and increased flexibility in controller design. Additionally, FOPID can be used to generalize integral PID as

i. If $\lambda = 1$ and $\mu = 1$, a conventional PID is obtained as (12)
ii. If $\lambda = 1$ and $\mu = 0$, a PI is obtained.
iii. If $\lambda = 0$ and $\mu = 1$, a PD is obtained.
iv. If $\lambda = 0$ and $\mu = 0$, a P is obtained.

$$G(s) = K_p + K_i \frac{1}{s} + K_d s \tag{12}$$

Therefore, as illustrated in Fig. 2, the controller can be programmed to operate within the four possible FOPID control regions if and were adjusted to arbitrary values between 0 and 1. Although FOPID is more flexible, the tuning process may become more complex as a result. In this study, the modelling, control tuning, and control design processes for fractional-order systems are streamlined using the FOMCON MATLAB Toolbox, created by Aleksei et al. [12]. This toolbox is useful for streamlining the difficulties posed by fractional-order systems and improving control design procedures.

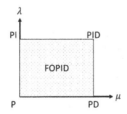

Fig. 2. Fractional-order control plane

2.3 Controller Design

In this paper, the FOPID controller as well as PD controller is designed and used as illustrated in general block diagram in Fig. 3 which comprises of attitude controller, and altitude controller with four control input u_1, u_2, u_3, and u_4.

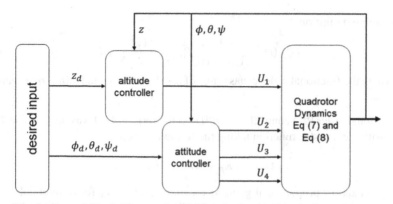

Fig. 3. General Block Diagram for MAV Attitude and Altitude Control System

2.3.1 Attitude Controller Design

The desired inputs for attitude are ϕ_d, θ_d and ψ_d, therefore, the tracking error for attitude angles are defined as:

$$e_\phi = \phi_d - \phi \tag{13a}$$

$$e_\theta = \theta_d - \theta \tag{13b}$$

$$e_\psi = \psi_d - \psi \tag{13c}$$

Hence the designed FOPID for attitude controller are as follows:

$$U_2 = k_p e_\phi + k_i \frac{1}{s^\lambda} e_\phi + k_d s^\mu e_\phi \tag{14a}$$

$$U_3 = k_p e_\theta + k_i \frac{1}{s^\lambda} e_\theta + k_d s^\mu e_\theta \tag{14b}$$

$$U_4 = k_p e_\psi + k_i \frac{1}{s^\lambda} e_\psi + k_d s^\mu e_\psi \tag{14c}$$

where k_p, k_d, and k_d are the proportional, integral, and differential coefficients of the FOPID controller while λ and μ are the integral order and differential order.

2.3.2 Altitude Control

The altitude tracking error for altitude z is set as:

$$e_z = z_d - z \tag{15}$$

Therefore, the designed FOPID control input is as follows:

$$U_1 = k_p e_z + k_i \frac{1}{s^\lambda} e_z + k_d s^\mu e_z \tag{16}$$

where k_p, k_d, and k_d are the proportional, integral, and differential coefficients of the FOPID controller while λ and μ are the integral order and differential order.

3 Simulation Results

The quadrotor MAV parameters used in this study are based on a Parrot Mambo Mini-drone are listed in Table 1 which are obtained from [13].

Table 1. Parrot Mambo MAV Model Physical Parameters

Specification	Parameter	Unit	Value
Quadrotor mass	m	kg	0.0630
Lateral moment arm	l	m	0.0624
Thrust coefficient	b	Ns^2	0.0107
Drag coefficient	d	Nms^2	0.7826400×10^{-3}
Rolling moment of inertia	I_{xx}	kgm^2	0.0582857×10^{-3}
Pitching moment of inertia	I_{yy}	kgm^2	0.0716914×10^{-3}
Yawing moment of inertia	I_{zz}	kgm^2	0.1000000×10^{-3}
Rotor moment of inertia	J_r	kgm^2	0.1021×10^{-6}

Two simulations are conducted to test the effectiveness of the FOPID controller in comparison to the PD controller. These simulations include an Attitude test and an Altitude test. The sampling period is set at 0.001 s, and the simulations run for a duration of 30 s. Table 2 displays the FOPID and PD gain parameters utilized in this simulation.

Table 2. FOPID and PD gains parameters

States	PD		FOPID				
	K_p	K_d	K_p	K_i	K_d	λ	μ
z	1.8	2.5	0.2	0.2	0.1	0.4	0.5
ϕ	0.02	0.01	0.2	-	0.1	0	0.75
θ	0.02	0.01	0.2	-	0.1	0	0.75
ψ	0.02	0.01	0.2	-	0.1	0	0.75

3.1 Attitude Simulation Results

In this simulation, two scenarios are conducted. In the first scenario, the initial attitude is set at 0.1 rad, with the desired attitude set to zero. In the second scenario, the initial attitude is set to zero, while the desired attitude follows a sinusoidal waveform with an amplitude of 0.2 rad. In both cases, the quadrotor is initially positioned 2 m above the ground. To assess the performance of the system, the integral square error (ISE) and the integral square control input (ISCI) [14] are recorded in Table 2. These measurements provide insights into system performance and power consumption. Based on the data presented in Table 3, the ISE of the FOPID controller outperforms the PD controller by more than 50%. However, it should be noted that the FOPID controller consumes more energy according to the ISCI.

Figure 4 illustrates the state responses for attitude ϕ, θ, and ψ in scenario 1. It is evident that the FOPID controller exhibits significantly better transient response compared to the PD controller, despite both controllers having settling times of 2 s. Referring to Fig. 5, both controllers demonstrate excellent tracking of the reference trajectory, with minimal error exhibited by the PD controller.

Table 3. ISE and ISCI performance Index for Attitude simulation

Scenario	States	ISE		ISCI	
		FOPID	PD	FOPID	PD
1	ϕ	5.6500e−4	0.0030	0.0019	1.8822e−7
	θ	5.4625e−4	0.0031	0.0019	2.2996e−7
	ψ	6.6511e−4	0.0025	0.0019	2.0790e−8
2	ϕ	9.0596e−6	0.0012	1.5509e−6	7.9399e−7
	θ	1.3775e−5	0.0018	2.1306e−6	1.1678e−6
	ψ	1.0415e−7	1.2349e−5	8.9301e−8	2.7320e−8

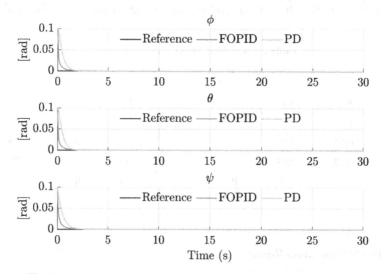

Fig. 4. Scenario 1 – attitude control for ϕ, θ, and ψ to stabilize at zero.

3.2 Altitude Simulation Results

In this simulation, two cases are conducted. In the first scenario, the initial altitude is set to zero, while the desired altitude is set to 2 m above the ground. In the second scenario,

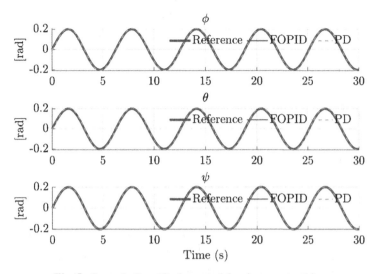

Fig. 5. Scenario 2 – attitude control for sinewave tracking

the initial attitude is set to zero, and the desired altitude follows a square wave pattern where the quadrotor hovers at 2 m and then descends to 1 m after 15 s. The attitude is set to zero for both scenarios. To evaluate the performance of the system, the integral square error (ISE) and the integral square control input (ISCI) are recorded in Table 4. These metrics provide insights into the system's performance and energy consumption.

In scenario 1, as shown in Fig. 6, the PD controller exhibits better transient response, although the FOPID controller produces a 35% overshoot. In Fig. 7, both controllers are able to track the reference signal, but the PD controller requires 5 s of settling time, while the FOPID controller takes 10 s before descending to 1 m at the 15-s mark. However, according to Table 4, the FOPID controller demonstrates better performance (ISE = 2.8485 for scenario 1, ISE = 3.5330 for scenario 2) compared to the PD controller (ISE = 3.5770 for scenario 1, ISE = 4.2095 for scenario 2). On the other hand, the FOPID controller consumes slightly more energy compared to the PD controller.

Table 4. ISE and ISCI performance Index for Altitude simulation

Scenario	States	ISE		ISCI	
		FOPID	PD	FOPID	PD
1	z	2.8485	3.5770	11.5216	11.4691
2		3.5330	4.2098	11.5490	17.1011

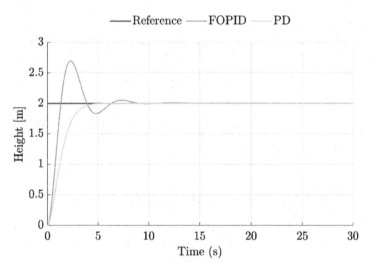

Fig. 6. Scenario 1 – altitude control to hover 2 m above ground.

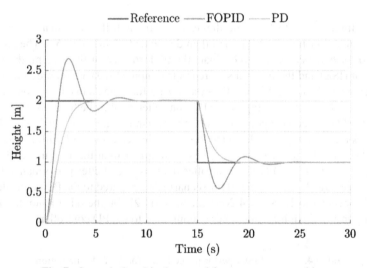

Fig. 7. Scenario 2 – altitude control for square wave tracking

4 Conclusion

A comparison study between the FOPID controller and PD controller has been presented for attitude and altitude control for quadrotor MAV. Through measurements of integral square error and integral square control input, the simulation results show that the FOPID controller's performance is very comparable to that of the PD controller. Future research can concentrate on using optimization techniques to fine-tune all parameters in fractional order control to further improve system performance. As additional potential research directions, cascade controllers and adaptive controllers in the context of fractional order PID control can be examined.

Acknowledgments. The authors would like to thank the Ministry of Higher Education (MOHE) through Fundamental Research Grant Scheme (FRGS/1/2021/TK0/UTM/02/56), Universiti Teknologi Malaysia under UTMFR (Q.J130000.3823.22H67) and Universiti Teknikal Malaysia Melaka (UTeM) for supporting this research.

References

1. Ren, J., Liu, D.-X., Li, K.: Cascade PID controller for quadrotor. In: 2016 IEEE International Conference Information Automation, pp. 120–124 (2016). https://doi.org/10.1109/ICInfA.2016.7831807
2. Noordin, A., Basri, M.A.M., Mohamed, Z., Abidin, A.F.Z.: Modelling and PSO fine-tuned PID control of quadrotor UAV. Int. J. Adv. Sci. Eng. Inform. Technol. **7**(4), 1367 (2017). https://doi.org/10.18517/ijaseit.7.4.3141
3. Fahmizal, S.A., Budiyanto, M., Arrofiq, M.: Altitude control of quadrotor using fuzzy self tuning PID controller. In: Proceedings of the 2017 5th International Conference Instrumentation, Control Automation ICA 2017, pp. 67–72 (2017). https://doi.org/10.1109/ICA.2017.8068415
4. Shamseldin, M.A., El-Samahy, A.A., Ghany, A.M.A.: Different techniques of self-tuning FOPID control for Brushless DC Motor. In: 2016 18th International Middle-East Power System Conference MEPCON 2016 – Proceedings, pp. 342–347 (2017). https://doi.org/10.1109/MEPCON.2016.7836913
5. Tajjudin, M., Johari, S.N.H., Aziz, S.A., Adnan, R.: Minimum ISE fractional-order PID (FOPID) controller for ball and beam mechanism. In: ICSGRC 2019 – 2019 IEEE 10th Control System Grad Res Colloquium, Proceeding, pp. 152–155 (2019). https://doi.org/10.1109/ICSGRC.2019.8837071
6. Baruah, A., Buragohain, M.: Design and implementation of FOPID and modified FOPID for inverted pendulum using particle swarm optimization algorithm. In: 2018 2nd International Conference Power, Energy Environ Towards Smart Technology, pp. 161–206 (2022). https://doi.org/10.1007/978-981-19-1614-4_5
7. Timis, D.D., Muresan, C.I., Dulf, E.-H.: Design and experimental results of an adaptive fractional-order controller for a quadrotor. Fractal Fract. **6**(4), 204 (2022). https://doi.org/10.3390/fractalfract6040204
8. Siresha, B.N., Swathi, N., Kiranmayi, R., Nagabhushanam, K.: Speed control of brushless DC motor using hybrid ANFIS-FOPID controller. In: 2023 Second International Conference Electronic Electronics Information Communication Technology, pp. 1–6 (2023). https://doi.org/10.1109/iceeict56924.2023.10156982
9. Zeglen-Wlodarczyk, J., Wajda, K.: Mutual influence of PID and FOPID controllers on different axes of the 3D crane. In: 2022 26th International Conference Methods Model Automation Robot MMAR 2022 – Proceedings, pp. 64–69 (2022). https://doi.org/10.1109/MMAR55195.2022.9874306
10. Haruna, Z., Mu'azu, M.B., Abubakar, Y.S., Adedokun, E.A.: Path tracking control of four wheel unmanned ground vehicle using optimized FOPID controller. In: 3rd International Conference Electronic Communication Computer Engineering ICECCE 2021, pp. 1–6 (2021). https://doi.org/10.1109/ICECCE52056.2021.9514119

11. Shao, K., Liu, J., Li, M.: Fractional order PID control of quadrotor UAV based on SA-PSO algorithm 153 (2022). https://doi.org/10.1117/12.2662180

12. Tepljakov, A., Petlenkov, E., Belikov, J.: FOMCON: a MATLAB toolbox for fractional-order system identification and control. Int. J. Microelectron. Comput. Sci. **2**, 51–62 (2011)

13. Noordin, A., Mohd Basri, M.A., Mohamed, Z.: Adaptive PID control via sliding mode for position tracking of quadrotor MAV: simulation and real-time experiment evaluation. Aerospace **10**, 512 (2023). https://doi.org/10.3390/aerospace10060512

14. Ayad, R., Nouibat, W., Zareb, M., Bestaoui Sebanne, Y.: Full control of quadrotor aerial robot using fractional-order FOPID. Iran J. Sci. Technol. – Trans. Electr. Eng. **43**, 349–360 (2019). https://doi.org/10.1007/s40998-018-0155-4

Stabilization and Altitude Tracking of Quadrotor using Backstepping Controller with a Saturation Compensator

Mohd Ariffanan Mohd Basri[1]([✉]), Aminurrashid Noordin[1,2], Mohd Saiful Azimi Mahmud[1], Fazilah Hassan[1], and Nurul Adilla Mohd Subha[1]

[1] Faculty of Electrical Engineering, Universiti Teknologi Malaysia, 81310 Johor Bahru, Johor, Malaysia
ariffanan@fke.utm.my

[2] Faculty of Electrical and Electronic Engineering Technology, Universiti Teknikal Malaysia Melaka, Hang Tuah Jaya, Durian Tunggal 76100, Melaka, Malaysia

Abstract. Researchers and control engineers have long found controlling quadrotor helicopters to be quite difficult. Due to the significant nonlinearities of this type of helicopter, numerous algorithms have been devised to control it. The dynamic model of the quadrotor has been developed in this article, and a sturdy controller has been created to handle the issue of altitude tracking and stabilization in the presence of external disturbances. Backstepping, which contains switching function as part of the controller, allows for the development of a robust control system. The switching function is utilized in the control law design to lessen the effects of external disturbances. To demonstrate the value and efficacy of the theoretical development, the proposed method is assessed in a quadrotor simulation environment. The results of the simulations demonstrate that even in the face of external perturbations, the suggested control system may produce favourable control performances for an autonomous quadrotor helicopter.

Keywords: Quadrotor Helicopter · Robust Control · Backstepping Control · Saturation Compensator

1 Introduction

Vertical take-off and landing (VTOL) vehicles are now attracting a lot of attention. They are excellent solutions for observation, inspection, and work in settings where space is at a premium and high manoeuvrability is necessary because they are able to hover and do not require runways. Four-rotor VTOL vehicles like the quadrotor have a number of advantages over conventional helicopters in terms of manoeuvrability, motion control, and price. However, the issues with stabilization control become difficult since quadrotors are intrinsically unstable, nonlinear, multi-variable, and underactuated systems.

Numerous academics have been interested in the stabilization issue with quadrotor helicopters for the past ten years. The linear quadratic regulator (LQR) control [1],

F. Hassan et al. (Eds.): AsiaSim 2023, CCIS 1912, pp. 113–125, 2024.
https://doi.org/10.1007/978-981-99-7243-2_10

proportional-integral-derivative (PID) control [2, 3], fuzzy logic (FL) control [4], feedback linearization control [5, 6], sliding mode control [7, 8], and backstepping control [9, 10] are just a few of the techniques that have been suggested to solve the quadrotor helicopter stabilization control problem. The majority of control techniques are initially built on linearized models without accounting for modelling flaws and outside disruptions. Then, in recent years, further study has been done to address the uncertainties and disturbances related to the nonlinear quadrotor dynamic model.

The stabilization issue with the quadrotor helicopter in the presence of outside disturbances is discussed in this work. Various parameters that have an impact on the dynamics of a flying structure are taken into account in the dynamical model that describes the motions of a quadrotor helicopter. The development of a new backstepping-based control method that takes into account external disturbances follows. This control strategy uses switching function as a compensatory technique. The new control strategy has advantages of robust control over the sole backstepping control scheme [9], which makes it appealing for a broad class of nonlinear systems affected by external disturbances. The synthesized control rule is illustrated by simulations that, despite the presence of external disturbances, it can produce outcomes that were largely satisfactory.

2 Quadrotor Dynamic Model

The Newton-Euler technique is used to derive the dynamic model of the quadrotor. The quadrotor's translational dynamic equations can be expressed as follows:

$$m\ddot{\xi} = -mge_z + u_T Re_z \tag{1}$$

where m denotes the quadrotor mass, g the gravity acceleration, $e_z = (0, 0, 1)^T$ the unit vector expressed in the frame E and u_T the total thrust produced by the four rotors.

$$u_T = \sum_{i=1}^{4} F_i = b \sum_{i=1}^{4} \Omega_i^2 \tag{2}$$

where F_i and Ω_i denote respectively, the thrust force and speed of the rotor i and b is the thrust factor.

The quadrotor's rotational dynamic equations can be expressed as follows:

$$I\dot{\omega} = -\omega \times I\omega - G_a + \tau \tag{3}$$

where I is the inertia matrix, $--\omega \times I\omega$ and G_a are the gyroscopic effect due to rigid body rotation and propeller orientation change respectively, while τ is the control torque obtained by varying the rotor speeds. G_a and τ are defined as:

$$Ga = i = 14Jr\omega \times ez - 1i + 1\Omega i \tag{4}$$

$$\tau = \begin{pmatrix} \tau_\phi \\ \tau_\theta \\ \tau_\psi \end{pmatrix} = \begin{pmatrix} lb(\Omega_4^2 - \Omega_2^2) \\ lb(\Omega_3^2 - \Omega_1^2) \\ d(\Omega_2^2 + \Omega_4^2 - \Omega_1^2 - \Omega_3^2) \end{pmatrix} \tag{5}$$

where J_r is the rotor inertia, l represent the distance from the rotors to the centre of mass and d is the drag factor.

Then, by recalling (1) and (3), the dynamic model of the quadrotor in terms of position (x, y, z) and rotation (ϕ, θ, ψ) is written as:

$$\begin{pmatrix} \ddot{x} \\ \ddot{y} \\ \ddot{z} \end{pmatrix} = \begin{pmatrix} 0 \\ 0 \\ -g \end{pmatrix} + \frac{1}{m} \begin{pmatrix} c_\phi s_\theta c_\psi + s_\phi s_\psi \\ c_\phi s_\theta s_\psi - s_\phi c_\psi \\ c_\phi c_\theta \end{pmatrix} u_T \tag{6}$$

$$\begin{pmatrix} \ddot{\phi} \\ \ddot{\theta} \\ \ddot{\psi} \end{pmatrix} = \begin{pmatrix} \dot{\theta}\dot{\psi}\left(\frac{I_{yy}-I_{zz}}{I_{xx}}\right) \\ \dot{\phi}\dot{\psi}\left(\frac{I_{zz}-I_{xx}}{I_{yy}}\right) \\ \dot{\theta}\dot{\phi}\left(\frac{I_{xx}-I_{yy}}{I_{zz}}\right) \end{pmatrix} - \begin{pmatrix} \frac{J_r}{I_{xx}}\dot{\theta}\Omega_d \\ -\frac{J_r}{I_{yy}}\dot{\phi}\Omega_d \\ 0 \end{pmatrix} + \begin{pmatrix} \frac{1}{I_{xx}}\tau_\phi \\ \frac{1}{I_{yy}}\tau_\theta \\ \frac{1}{I_{zz}}\tau_\psi \end{pmatrix} \tag{7}$$

Consequently, quadrotor is an underactuated system with six outputs $(x, y, z, \phi, \theta, \psi)$ and four control inputs$(u_T, \tau_\phi, \tau_\theta, \tau_\psi)$.

Finally, the quadrotor dynamic model can be written in the following form:

$$\begin{aligned} \ddot{x} &= \left(c_\phi s_\theta c_\psi + s_\phi s_\psi\right)\frac{1}{m}u_1 \\ \ddot{y} &= \left(c_\phi s_\theta s_\psi - s_\phi c_\psi\right)\frac{1}{m}u_1 \\ \ddot{z} &= -g + \left(c_\phi c_\theta\right)\frac{1}{m}u_1 \\ \ddot{\phi} &= \dot{\theta}\dot{\psi}\left(\frac{I_{yy}-I_{zz}}{I_{xx}}\right) - \frac{J_r}{I_{xx}}\dot{\theta}\Omega_d + \frac{l}{I_{xx}}u_2 \\ \ddot{\theta} &= \dot{\phi}\dot{\psi}\left(\frac{I_{zz}-I_{xx}}{I_{yy}}\right) + \frac{J_r}{I_{yy}}\dot{\phi}\Omega_d + \frac{l}{I_{yy}}u_3 \\ \ddot{\psi} &= \dot{\theta}\dot{\phi}\left(\frac{I_{xx}-I_{yy}}{I_{zz}}\right) + \frac{1}{I_{zz}}u_4 \end{aligned} \tag{8}$$

with a renaming of the control inputs as:

$$\begin{aligned} u_1 &= b\left(\Omega_1^2 + \Omega_2^2 + \Omega_3^2 + \Omega_4^2\right) \\ u_2 &= b\left(\Omega_4^2 - \Omega_2^2\right) \\ u_3 &= b\left(\Omega_3^2 - \Omega_1^2\right) \\ u_4 &= d\left(\Omega_2^2 + \Omega_4^2 - \Omega_1^2 - \Omega_3^2\right) \end{aligned} \tag{9}$$

3 Backstepping Control with Saturation Compensator

In this study, the only four controllable degrees of freedom (DOF) are the z-directional linear motion (altitude) and the three attitude angles (roll, pitch, and yaw). As a result, the dynamic model (8) can be represented by the following nonlinear dynamic equation:

$$\ddot{X} = f(X) + g(X)u + \delta \tag{10}$$

where u, X and δ are respectively the input, state and external disturbance vector given as follows:

$$u = [u_1 u_2 u_3 u_4]^T \tag{11}$$

$$X = [x_1 x_2 x_3 x_4]^T = [z\phi\theta\psi]^T \tag{12}$$

$$\delta = [\delta_1 \delta_2 \delta_3 \delta_4]^T \tag{13}$$

The bound of the external disturbance is assumed to be given, that is $|\delta| \le \beta$, where β is a given positive constant. From (8) and (12), the nonlinear dynamic function $f(X)$ and nonlinear control function $g(X)$ matrices can be written accordingly as:

$$f(X) = \begin{pmatrix} -g \\ \dot\theta\dot\psi a_1 - \dot\theta a_2 \Omega_d \\ \dot\phi\dot\psi a_3 + \dot\phi a_4 \Omega_d \\ \dot\theta\dot\phi a_5 \end{pmatrix} \quad g(X) = \begin{pmatrix} u_z \frac{1}{m} & 0 & 0 & 0 \\ 0 & b_1 & 0 & 0 \\ 0 & 0 & b_2 & 0 \\ 0 & 0 & 0 & b_3 \end{pmatrix} \tag{14}$$

with the abbreviations $a_1 = (I_{yy} - I_{zz})/I_{xx}$, $a_2 = J_r/I_{xx}$, $a_3 = (I_{zz} - I_{xx})/I_{yy}$, $a_4 = J_r/I_{yy}$, $a_5 = (I_{xx} - I_{yy})/I_{zz}$, $b_1 = l/I_{xx}$, $b_2 = l/I_{yy}$, $b_3 = 1/I_{zz}$, $u_z = (c_\phi c_\theta)$

The control objective is to design a suitable control law for the system (10) so that the state trajectory X can track a desired reference trajectory $X_d = [x_{1d} x_{2d} x_{3d} x_{4d}]^T$ despite the presence of external disturbance. For the sake of simplicity, just one of the four DOFs is taken into consideration because the description of the control system architecture of the helicopter is comparable for each DOF.

The design of ideal backstepping control (IBC) is described step-by-step as follows:
Step 1: Define the tracking error:

$$e_1 = x_{1d} - -x_1 \tag{15}$$

where x_{1d} is the desired trajectory that a reference model specifies. Following that, the derivative of tracking error can be written as:

$$\dot{e}_1 = \dot{x}_{1d} - -\dot{x}_1 \tag{16}$$

The first Lyapunov function is chosen as:

$$V_1(e_1) = \frac{1}{2} e_1^2 \tag{17}$$

The derivative of V_1 is:

$$\dot{V}_1(e_1) = e_1 \dot{e}_1 = e_1 (\dot{x}_{1d} - -\dot{x}_1) \tag{18}$$

$\dot{x}_1$ is comparable to a virtual control. Following is a definition of the intended value of virtual control, often known as a stabilizing function:

$$\alpha_1 = \dot{x}_{1d} + k_1 e_1 \tag{19}$$

where k_1 is a positive constant.
When the virtual control is changed to its intended value, Eq. (18) then becomes:

$$\dot{V}_1(e_1) = -k_1 e_1^2 \le 0 \tag{20}$$

Step 2: The virtual control's deviation from the target value is shown by:

$$e_2 = \dot{x}_1 - \alpha_1 = \dot{x}_1 - \dot{x}_{1d} - k_1 e_1 \tag{21}$$

The derivative of e_2 is expressed as:

$$\begin{aligned}
\dot{e}_2 &= \ddot{x}_1 - \dot{\alpha}_1 \\
&= f(x_1) + g(x_1)u_1 + \delta_1 - \ddot{x}_{1d} - k_1 \dot{e}_1
\end{aligned} \tag{22}$$

The second Lyapunov function is chosen as:

$$V_2(e_1, e_2) = \frac{1}{2}e_1^2 + \frac{1}{2}e_2^2 \tag{23}$$

Finding derivative of (23) yields:

$$\begin{aligned}
\dot{V}_2(e_1, e_2) &= e_1 \dot{e}_1 + e_2 \dot{e}_2 \\
&= e_1(\dot{x}_{1d} - -\dot{x}_1) + e_2(\ddot{x}_1 - \dot{\alpha}_1) \\
&= e_1(-e_2 - k_1 e_1) + e_2(f(x_1) + g(x_1)u_1 + \delta_1 - \ddot{x}_{1d} - k_1 \dot{e}_1) \\
&= -k_1 e_1^2 + e_2(-e_1 + f(x_1) + g(x_1)u_1 + \delta_1 - \ddot{x}_{1d} - \ddot{x}_d - k_1 \dot{e}_1)
\end{aligned} \tag{24}$$

Step 3: Assuming the external disturbance is well known, an IBC can be obtained as:

$$u_{IB} = \frac{1}{g(x_1)}(e_1 + k_1 \dot{e}_1 + \ddot{x}_{1d} - f(x_1) - \delta_1 - k_2 e_2) \tag{25}$$

where k_2 is a positive constant. The term $k_2 e_2$ is added to stabilize the tracking error e_1.

Substituting (25) into (24), the following equation can be obtained:

$$\dot{V}_2(e_1, e_2) = -k_1 e_1^2 - k_2 e_2^2 = -E^T K E \leq 0 \tag{26}$$

where $E = [e_1 e_2]^T$ and $K = diag(k_1, k_2)$. Since $\dot{V}_2(e_1, e_2) \leq 0, \dot{V}_2(e_1, e_2)$ is negative semi-definite.

Therefore, the IBC in (25) will asymptotically stabilize the system.

However, the IBC (25) attempt cannot guarantee the effective control performance if unforeseen external disturbance disturbances occur. As a result, supplemental control efforts should be created to reduce the impact of the unforeseen perturbations. The supplemental control effort is referred as switching control effort represented by u_{sw}.

The switching control signal is defined as:

$$u_{sw} = \lambda sign(e_2) \tag{27}$$

where λ is a constant determined by design parameter and $sign(e_2)$ is a sign function:

$$sign(e_2) = \begin{cases} 1, e_2/\rho > 0 \\ -1, e_2/\rho < 0 \end{cases} \tag{28}$$

The backstepping control effort for the nominal model ($\delta = 0$) is formulated as follows:

$$u_B = \frac{1}{g(x_1)}(e_1 + k_1\dot{e}_1 + \ddot{x}_{1d} - f(x_1) - k_2 e_2) \tag{29}$$

Totally, the robust backstepping control (RBC) law, which ensures stability and convergence for nonlinear systems in the presence of an external disturbance, can be summarized as follows:

$$\begin{aligned}u &= u_B + u_{sw} \\ &= \frac{1}{g(x_1)}(e_1 + k_1\dot{e}_1 + \ddot{x}_{1d} - f(x_1) - k_2 e_2) + \frac{\varepsilon}{g(x_1)}sign(e_2)\end{aligned} \tag{30}$$

where ε is design parameter to be determined later.

Use of the discontinuous switching function will trigger the undesirable chatter phenomenon. The most typical approach to lessen the chattering problem is to utilize the saturation function $sat(e_2/\rho)$. Thus, replacing $sign(e_2)$ by $sat(e_2/\rho)$ in (30) implies:

$$u = \frac{1}{g(x_1)}(e_1 + k_1\dot{e}_1 + \ddot{x}_{1d} - f(x_1) - k_2 e_2 + \varepsilon sat(e_2/\rho)) \tag{31}$$

The saturation function $sat(e_2/\rho)$ is defined as follows:

$$sat(e_2/\rho) = \begin{cases} sign(e_2/\rho), |e_2/\rho| > 1 \\ e_2/\rho, |e_2/\rho| \le 1 \end{cases} @ sat(e_2/\rho) = \begin{cases} 1, e_2/\rho > 1 \\ -1, e_2/\rho < -1 \\ e_2/\rho, |e_2/\rho| \le 1 \end{cases} \tag{32}$$

where ρ is a small positive constant.

4 Simulation Results

The effectiveness of the suggested technique is assessed in this section. The corresponding algorithm is implemented in the simulation environment of MATLAB/SIMULINK. The quadrotor system's model parameter values are presented in Table 1 which obtained from [11].

Two simulations have been run on the quadrotor to examine the efficacy of the suggested robust backstepping controller. The outcomes of the suggested controller in a stabilizing situation are provided in the first simulation. The effectiveness of the approach is examined in the second section with regard to the problem of altitude tracking. The nominal case (Case 1) and the external disturbance case (Case 2) are the two test conditions offered. The Dryden Wind-Gust model [12] is used to model the wind forces that cause the external disturbance.

4.1 Simulation 1: stabilizing problem

In this simulation, the goal of the control is to maintain a quadrotor at a predetermined altitude or attitude so that it can hover over a specific location. The $x_{id} = [z_d, \phi_d, \theta_d, \psi_d] =$

Table 1. Parameters of the quadrotor.

Parameter	Description	Value	Units
g	Gravity	9.81	m/s^2
m	Mass	0.5	kg
l	Distance	0.2	m
I_{xx}	Roll inertia	4.85×10^{-3}	$kg \bullet m^2$
I_{yy}	Pitch inertia	4.85×10^{-3}	$kg \bullet m^2$
I_{zz}	Yaw inertia	8.81×10^{-3}	$kg \bullet m^2$
b	Thrust factor	2.92×10^{-6}	
d	Drag factor	1.12×10^{-7}	

$[5, 0, 0, 0]^T$ indicates the desired height or attitude. The $z = 5, \phi = 0.2, \theta = 0.2$ and $\psi = 0.2$ specify the beginning states. The IBC system is initially taken into account in the simulation. In Figs. 1 and 2, respectively, the simulation results of the IBC system for stabilizing a quadrotor at Cases 1 and 2 are shown. According to the simulation results, at Case 1, IBC system can stabilize the quadrotor in hover mode. However, as seen in Fig. 2, degenerate performance response is what happens when an external disruption occurs. The RBC system is simulated using the same simulation cases. Figures 3 and 4 show, respectively, the simulation results for Case 1 and Case 2 for stabilizing a quadrotor. The suggested RBC system's strong control performance for quadrotor stabilization can be seen in the simulation results. It is evident that even when an external disturbance is applied, as shown in Fig. 4, the hovering flight is stable and the quadrotor may be kept at the required altitude or attitude. As discussed before, applying the sign switching function will result in chatter, a negative phenomenon in the control input, as shown in Fig. 5. Utilizing the saturation function is the most efficient technique to solve this issue, and the outcome is shown in Fig. 6.

4.2 Simulation 2: tracking problem

The control goal in this simulation is to keep the system tracking the target altitude trajectory while using periodic rectangular function signals. The IBC system is first simulated. Figures 7 and 8 show, respectively, the tracking responses resulting from a periodic rectangular function at the nominal condition (Case 1) and external disturbance condition (Case 2). According to the simulation results, the quadrotor may be controlled by the IBC system to follow the desired reference trajectory in Case 1 as illustrated in Fig. 7. However, as shown in Fig. 8, degenerate performance response is what happens when an external disturbance occurs. The RBC system is simulated using the same simulation cases. Figures 9 and 10 show, respectively, the simulation results for tracking a periodic rectangular function at the nominal condition and external disturbance condition. The robustness of the proposed control strategy in following the intended reference trajectory can be inferred from the simulation results, as shown in Fig. 10.

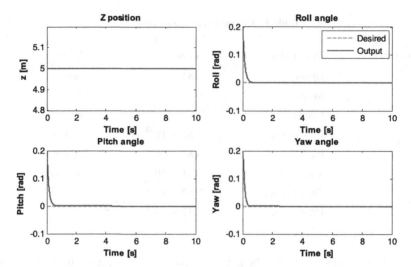

Fig. 1. Case 1: Attitude stabilization using IBC.

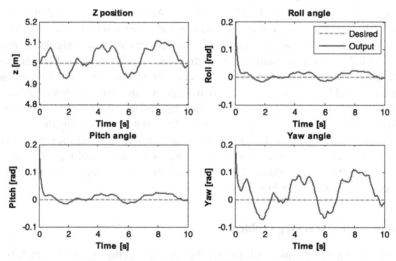

Fig. 2. Case 2: Attitude stabilization using IBC.

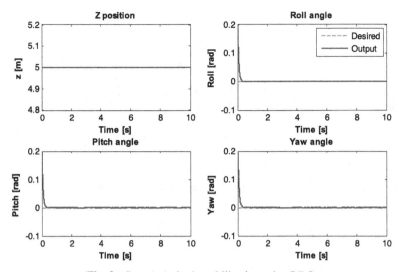

Fig. 3. Case 1: Attitude stabilization using RBC.

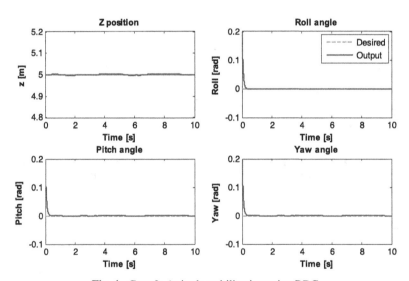

Fig. 4. Case 2: Attitude stabilization using RBC.

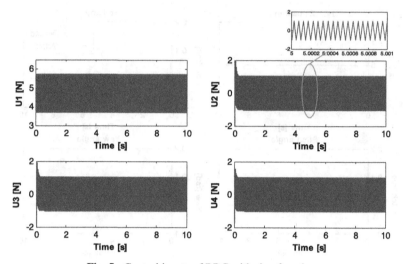

Fig. 5. Control inputs of RBC with sign function.

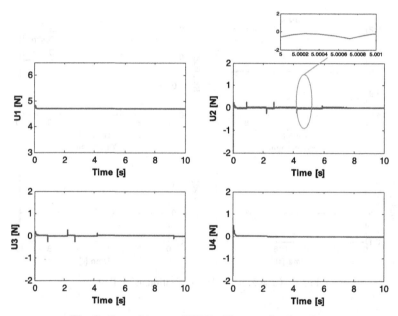

Fig. 6. Control inputs of RBC with saturation function.

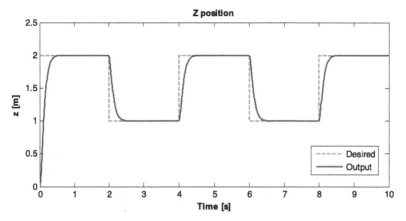

Fig. 7. Case 1: Tracking response using IBC.

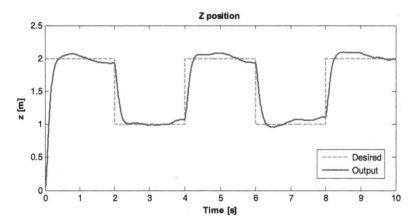

Fig. 8. Case 2: Tracking response using IBC.

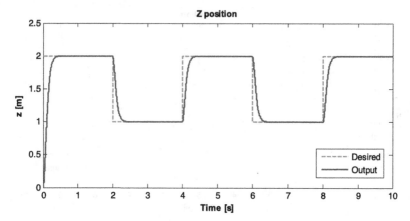

Fig. 9. Case 1: Tracking response using RBC.

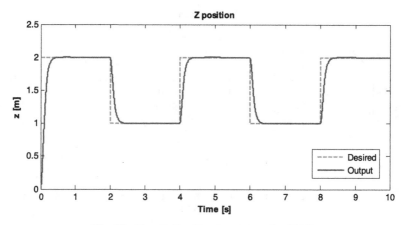

Fig. 10. Case 2: Tracking response using RBC.

5   Conclusion

The use of a robust controller for stabilization and trajectory tracking of a quadrotor helicopter affected by external disturbances is successfully demonstrated in this research. The mathematical model of the quadrotor is introduced first. The suggested robust control system includes a backstepping and a switching function. The backstepping control design is developed from the Lyapunov function to ensure system stability, while the switching function is employed to mitigate the effects of disturbances. Finally, the proposed control technique is implemented to a quadrotor helicopter. The simulation results indicate that the proposed control system can produce high-precision transient and tracking responses.

Acknowledgments. The authors would like to thank Ministry of Higher Education (MOHE) through Fundamental Research Grant Scheme (FRGS/1/2021/TK0/UTM/02/56) and Universiti Teknologi Malaysia under UTMFR (Q.J130000.3823.22H67) for supporting this research.

References

1. Martins, L., Cardeira, C., Oliveira, P.: Linear quadratic regulator for trajectory tracking of a quadrotor. IFAC-PapersOnLine. **52**(12), 176–181 (2019)
2. Sahrir, N.H., Mohd Basri, M.A.: Modelling and Manual Tuning PID Control of Quadcopter. In: Control, Instrumentation and Mechatronics: Theory and Practice. Lecture Notes in Electrical Engineering, vol 921 (2022)
3. Noordin, A., Basri, M.A., Mohamed, Z.: Simulation and experimental study on PID control of a quadrotor MAV with perturbation. Bulletin of Elec. Eng. Informatics. **9**(5), 1811–1818 (2020)
4. Ginting, A.H., Doo, S.Y., Pollo, D.E., Djahi, H.J., Mauboy, E.R.: Attitude control of a quadrotor with fuzzy logic controller on SO (3). J. Robo. Cont. (JRC). **13**;**3**(1), 01–6 (2022)
5. Ermeydan, A., Kaba, A.: Feedback linearization control of a quadrotor. In: IEEE 5th International Symposium on Multidisciplinary Studies and Innovative Technologies (ISMSIT), pp. 287–290 (2021)
6. Rahmat, M.F., Eltayeb, A., Basri, M.A.: Adaptive feedback linearization controller for stabilization of quadrotor UAV. Int. J. Integr. Eng. **30**;**12**(4), 1–7 (2020)
7. Noordin, A., Basri, M.A., Mohamed, Z.: Sliding mode control with tanh function for quadrotor UAV altitude and attitude stabilization. In: Intelligent Manufacturing and Mechatronics: Proceedings of SympoSIMM, pp. 471–491 (2021)
8. Eltayeb, A., Rahmat, M.F., Basri, M.A.: Sliding mode control design for the attitude and altitude of the quadrotor UAV. Int. J. Smart Sensing and Intell. Sys. **13**(1), 1–3 (2020)
9. Madani, T., Benallegue, A.: Backstepping control for a quadrotor helicopter. In: IEEE/RSJ International Conference on Intelligent Robots and Systems, pp. 3255–3260 (2006)
10. Huang, T., Huang, D.: Backstepping control for a quadrotor unmanned aerial vehicle. In: IEEE Chinese Automation Congress (CAC), pp. 2475–2480 (2020)
11. Voos, H.: Nonlinear control of a quadrotor micro-UAV using feedback-linearization. In: IEEE International Conference on Mechatronics (ICM 2009), pp. 1–6 (2009)
12. Waslander, S.L., Wang, C.: Wind disturbance estimation and rejection for quadrotor position control. In: AIAA Infotech@Aerospace conference and AIAA unmanned...Unlimited Conference (2009)

Noise Analysis of Miniature Planar Three-Dimensional Electrical Capacitance Tomography Sensors

Wen Pin Gooi[1] , Pei Ling Leow[1(✉)] , Jaysuman Pusppanathan[1,2] ,
Xian Feng Hor[1] , and Shahrulnizahani bt Mohammad Din[1]

[1] Faculty of Electrical Engineering, Universiti Teknologi Malaysia, 81310 UTM Skudai, Johor,
Malaysia
leowpl@utm.my

[2] Sport Innovation and Technology Centre (SiTC), Institute of Human Centered Engineering
(iHumen), Universiti Teknologi Malaysia, Block V01, 81310 Skudai Johor, Malaysia

Abstract. The planar electrical capacitance tomography (ECT) sensor has gained popularity for its ability to reconstruct two-dimensional (2D) and three-dimensional (3D) images in applications such as defect and landmines detection. Besides, miniature planar ECT has been investigated for biomedical imaging and lab-on-chip monitoring in terms of the design of electrode to improve the quality of reconstructed image. However, the study on the noise performance of the sensor is limited. Therefore, this paper compared the noise performance of peripheral, single plane distributed and dual plane distributed sensors in 3D image reconstruction using simulation approach. A detailed simulation procedure involving the modelling of sensors, generation of sensitivity map and simulating the capacitance measurement are described in this paper. The simulated capacitance data of the sensors were added with White Gaussian noise to produce noisy capacitance data that was used for 3D image reconstruction. Then, quality of the reconstructed 3D image at different noise level were compared quantitatively using the correlation coefficient. The simulation results revealed that the increase in the number of electrode pair combinations and the distributed electrode arrangement enables the miniature planar ECT sensor to reconstruct 3D image at low SNR of 20 dB.

Keywords: Noise analysis · 3D ECT · Miniature sensor

1 Introduction

The single sided assessment capability of planar ECT sensor has paved way to applications of ECT sensors such as in the landmine detection [1], subsurface or surface defect detection [2] and obstacle sensing in robotics [3]. The design of planar ECT sensor commonly consists single plane rectangular electrodes arranged in 4×3 matrix array and the size of the sensor is large at 30 cm [4]. The planar ECT sensor can be used to reconstruct two-dimensional (2D) and three-dimensional (3D) images [5–7]. However, due to the single sided assessment, the depth sensing of the sensor is limited. To address

F. Hassan et al. (Eds.): AsiaSim 2023, CCIS 1912, pp. 126–137, 2024.
https://doi.org/10.1007/978-981-99-7243-2_11

this limitation, a dual plane sensor design is introduced to and it is shown to improve the quality of reconstructed 3D image [8, 9].

The miniaturization of planar ECT sensor has expanded application of ECT in biomedical imaging and lab-on-chip monitoring. For instance [10], investigated the peripheral electrode design of miniature planar ECT sensor and successfully reconstructed the 2D image of stagnant sample. The peripheral electrode design consists of electrodes arranged around the circumference of cylindrical sensing chamber. Meanwhile [11], optimized and used the dual plane design of miniature peripheral miniature ECT sensor for 3D imaging of yeast encapsulated in agar. However, the quality of the reconstructed 3D image is inconsistent and depends on the position of samples in the sensing chamber. Apart from the peripheral electrode design [12], successfully reconstructed the 2D image of tooth surface using a miniature dual plane distributed planar ECT sensor. The distributed planar ECT sensor also showed consistent 3D image reconstruction than the peripheral planar ECT sensor [13].

The measurement noise present in capacitance data greatly impacts the quality of reconstructed 3D image [14]. The SNR limit for appropriate image reconstruction of reported planar ECT sensor ranges from 40 dB to 50 dB [4, 15]. Towards the miniaturisation of planar ECT sensor by decreasing the size of electrodes, a lower SNR capacitance measurement can be expected due to the weak signal strength [9]. Yet very few studies have examined the SNR limit for the miniature planar ECT sensor for image reconstruction.

In this paper, the noise performance of various miniature planar 3D ECT sensor designs were investigated by comparing the quality of 3D image reconstructed. The designs of miniature planar 3D ECT sensor compared in this investigation were the peripheral sensor, single plane distributed sensor and dual planar distributed sensor. The 3D image was reconstructed using simulated capacitance measurement with SNR ranging from 10 dB to 60 dB. The noisy simulated capacitance data was generated by adding White Gaussian noise to the noise free simulated capacitance data. The findings of this study will help in the following.

1. The numerical simulation of miniature planar 3D ECT sensors provides insights on SNR limit of capacitance data that is required to reconstruct 3D image. This information is beneficial as a guidance in designing hardware such as the data acquisition system.
2. This study identifies the design aspects of miniature planar 3D ECT sensor that could reduce the effect of noise on image reconstruction. The sensor design aspects considered are the electrode arrangements such as the peripheral and distributed arrangement and the number of electrodes of miniature planar 3D ECT sensor. These design aspects could produce different number of strong electrode pair combinations that is less affected by noise.

The remainder of this paper is organised as follow. In Sect. 2, we discuss the working principle of miniature planar 3D ECT sensor. In Sect. 3, we discuss the 3D modelling and numerical simulation of miniature planar 3D ECT sensors. In Sect. 4, we present the reconstructed 3D image and the analysis of noise performance of designed sensors. Finally, a conclusion is drawn based on this study in Sect. 5.

2 Working Principle of Planar 3D ECT

The planar 3D ECT sensor is a non-invasive imaging technique used to image the permittivity distribution within the sensing region of the planar ECT sensor. It operates by applying an excitation signal to the excitation electrode and measuring the resulting capacitance on the sensing electrode. Since the electrodes of planar 3D ECT sensor are flat on a surface, this forms a fringing electric field that penetrates the sensing region above the electrodes. This fringing electric field between the excitation and sensing electrodes interacts with the object presence in the sensing region and produces capacitance measurement that represents the presence of the object. The ECT involves solving the forward problem in which the electromagnetic and physical properties of the ECT sensor are modelled to obtain the relation between the measured capacitance and permittivity distribution. The forward problem of ECT involves solving the Poisson's equation

$$\nabla \cdot (\varepsilon \nabla \emptyset) = 0 \tag{1}$$

where ε is the permittivity distribution and $\emptyset$ is the potential distribution. The solving of Eq. (1) using partial differential equation computes the electric potential for a given permittivity distribution the 3D sensing region. The relationship between the capacitance and electric field distribution can be expressed as

$$C = \frac{1}{V} \iiint \varepsilon(x, y, z) E(x, y, z) dx\, dy\, dz \tag{2}$$

where C is the capacitance between the excitation and sensing electrodes, V is the potential difference between the excitation and sensing electrodes and E is the electric field distribution. Then, Eq. (2) can be linearized as

$$C = S\varepsilon \tag{3}$$

where S is the sensitivity map that relates the measured capacitance to the permittivity distribution. The formulation of sensitivity map between i^{th} and j^{th} electrode pair is given by

$$S_{ij}(x, y, z) = -\int_{\vartheta(x,y,z)} \frac{E_i(x, y, z)}{V_i} \cdot \frac{E_j(x, y, z)}{V_j} d\vartheta \tag{4}$$

where S_{ij} is the sensitivity map between i^{th} and j^{th} electrode pair, V_i and V_j are the voltage of ith and j^{th} electrodes when they are used as the excitation electrode. Meanwhile, E_i and E_j are the electric field distribution when i^{th} and j^{th} electrodes are used as the excitation electrode. In this study, the 3D sensing region is discretized into 64 × 64 × 10 voxels along the x, y and z axis. The 3D image is reconstructed using the linear back projection (LBP) algorithm. The LBP algorithm is given by

$$\varepsilon = \frac{S^T \lambda}{S^T u_\lambda} \quad u_\lambda = [1, 1, \ldots, 1] \tag{5}$$

where S^T is the transpose of S since the inverse of S does not exist and u_C is an identity vector with the same size as the number of measured capacitance. Meanwhile, λ is the normalised capacitance data that reflects the presence of sample in the sensing chamber. The formulation of λ is given by

$$\lambda = \frac{\frac{1}{C_m} - \frac{1}{C_L}}{\frac{1}{C_H} - \frac{1}{C_L}} \tag{6}$$

where C_m is the simulated capacitance when the 2 mm cube phantom is placed in the sensing chamber, C_L is the simulated capacitance when the sensing chamber is filled with low permittivity medium and C_H is the simulated capacitance when the sensing chamber is filled with high permittivity medium.

3   Simulation of Miniature Planar 3D ECT Sensor

Figure 1 illustrates the overall simulation flow of this study, starting from 3D modelling of miniature planar 3D ECT sensors to the 3D image reconstruction using noise free and noisy simulated capacitance data. Meanwhile, Fig. 2 shows the three miniature planar 3D ECT sensors being investigated which were the peripheral sensor, the single plane distributed sensor and the dual plane distributed sensor. The peripheral sensor consists of eight electrodes that surround the sensing chamber. Meanwhile, the single plane distributed sensor has nine electrodes arranged in a 3×3 matrix arrangements. The dual plane distributed sensor has the same electrode layout as the single plane distributed sensor but with an additional top electrode plane. The separation between the top and bottom electrode planes is 3 mm. These sensors have identical cylindrical sensing chambers of 17 mm diameter and 3 mm height.

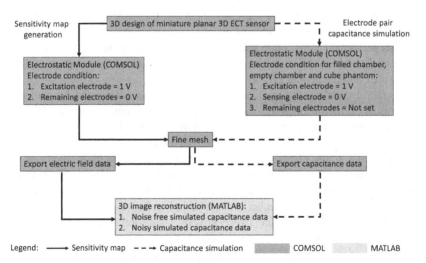

Fig. 1. Simulation flow

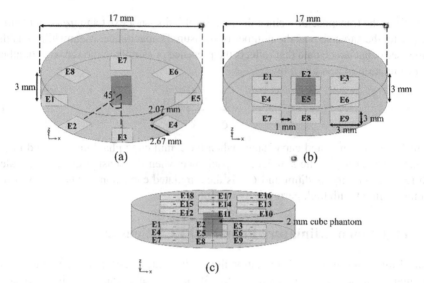

Fig. 2. 3D sensor models of miniature planar 3D ECT sensors and the position of 2 mm cube phantom for 3D image reconstruction (a) peripheral sensor (b) single plane distributed sensor (c) dual plane distributed sensor

The sensitivity map and the simulated capacitance data needed for simulation 3D image reconstruction were modelled using the Electrostatic module of COMSOL Multiphysics, utilising the Poisson's equation in Eq. (1). The sensors in Fig. 2 were modelled in 3D computational domain to simulate the fringing electric field in the z-axis direction that is required for 3D image reconstruction. The electric field distribution was simulated by setting 1 V potential to the excitation electrode while the remaining electrodes were set to 0 V potential. The setting of excitation electrode proceeded until all the electrodes acted as the excitation electrode at least once. The simulation of the electric field distribution was performed using a fine mesh. Then, the generated electric field distribution was exported to MATLAB and the sensitivity maps of the miniature planar 3D ECT sensors were computed using Eq. (4).

Next, the electrode pair capacitance measurement was simulated by setting 1 V potential to an excitation electrode and 0 V potential to a sensing electrode. The setting of excitation and sensing electrodes proceed for all possible electrode pair combination of the miniature planar 3D ECT sensors. The total electrode pair combinations for the peripheral sensor, single plane distributed sensor and dual plane distributed sensor are 28, 36 and 153 respectively. In this study, a 2 mm cube phantom and air as the background medium were used for simulation 3D image reconstruction. The position of the 2 mm cube phantom in the sensing chamber is depicted in Fig. 2. The permittivity of the 2 mm cube phantom and air were 80 and 1 respectively. Therefore, three sets of capacitance data were simulated for each miniature planar 3D ECT sensor which were the empty sensing chamber measurements, filled sensing chamber measurements and the 2 mm cube phantom in the sensing chamber measurements with air as background medium. These three sets of simulated capacitance measurement are needed for the calibration and normalised capacitance computation for 3D image reconstruction. Then, the simulated capacitance data was exported to MATLAB for 3D image reconstruction.

The 3D image of the 2 mm cube phantom was reconstructed by LBP using the simulated capacitance data and the computed sensitivity map. The simulated capacitance data was used to reconstruct the noise free 3D image of the 2 mm cube phantom. Then, the noisy capacitance data was produced by adding White Gaussian noise to the noise free simulated capacitance data using the 'awgn()' function in MATLAB [15, 16]. In this study, the 3D imaging performance of miniature planar 3D ECT sensors were investigated using noisy capacitance data of SNR between 10 dB to 60 dB with increment step of 2 dB. The quality of reconstructed 3D image was evaluated quantitatively using the correlation coefficient (CC). The formulation of CC is given by

$$CC = \frac{\sum_{i=1}^{N}(\varepsilon_i - \overline{\varepsilon})\left(\hat{\varepsilon}_i - \overline{\hat{\varepsilon}}\right)}{\sqrt{\sum_{i=1}^{N}(\varepsilon_i - \overline{\varepsilon})^2 \sum_{i=1}^{N}\left(\hat{\varepsilon}_i - \overline{\hat{\varepsilon}}\right)^2}} \tag{7}$$

where ε and $\hat{\varepsilon}$ are the reconstructed permittivity distribution and true permittivity distribution, respectively. Meanwhile, $\overline{\varepsilon}$ and $\overline{\hat{\varepsilon}}$ are the means of ε and $\hat{\varepsilon}$. The CC computes the similarity between the reconstructed 3D image and the 2 mm cube phantom.

4   Results and Discussion

4.1   3D Image Reconstruction

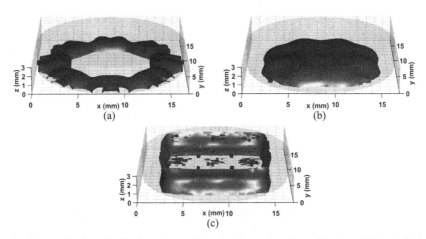

Fig. 3. 3D sensitivity distribution (a) peripheral (b) single plane distributed (c) dual plane distributed sensors

Figure 3 shows the 3D sensitivity distribution of the miniature planar 3D ECT sensors computed in MATLAB using the simulated electric field distribution. The visualized 3D sensitivity distribution is obtained the average sum of the sensitivity map of each electrode pair combination. It shows the sensitive region in the sensing chamber of the miniature planar 3D ECT sensors.

As illustrated in Fig. 3, the sensitivity distribution in the sensing chamber depends on the placement of the electrodes on the miniature planar 3D ECT sensor. For the peripheral sensor, the sensitive region is mainly distributed at the circumference of the sensing chamber where the electrodes are located and it has limited sensitivity in the middle of sensing chamber. Meanwhile, the sensitive region of the single plane distributed sensor is distributed throughout the sensing chamber. It is shown to overcome the lack of sensitivity region in the middle of sensing chamber found in the peripheral sensor. However, the height coverage of the sensitive region of single plane distributed sensor reaches half the sensing chamber only. The addition of top electrode plane in dual plane distributed sensor improved the height coverage of the sensitivity region while maintaining the horizontal sensitivity region coverage. These extended sensitive region from the bottom to the top of sensing chamber are contributed by the top and bottom electrode pair combinations.

The reconstructed 3D image using the miniature planar 3D ECT sensors are illustrated in Fig. 4. The 3D images are reconstructed using the noise free simulated capacitance data and are represented by the red isosurface region. A 2 mm reference cube representing the 2 mm cube phantom is included in the reconstructed 3D for visual comparison of the reconstructed image and the actual structure of the 2 mm cube phantom.

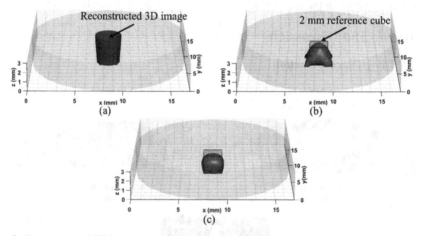

Fig. 4. Reconstructed 3D image using noise free simulated capacitance data (a) peripheral sensor (b) single plane distributed sensor (c) dual plane distributed sensor

As illustrated in Fig. 4, the miniature planar 3D ECT sensors successfully reconstructed the 3D image of the 2 mm cube phantom at the correct position. The CC of the reconstructed 3D image by peripheral sensor, single plane distributed sensor and dual plane distributed sensor are 0.7121, 0.7729 and 0.8316 respectively. However, the height of the 2 mm cube phantom is reconstructed wrongly by peripheral sensor as it extended from the bottom to the top of the sensing chamber. The error in height reconstruction of peripheral sensor can be related to the lack of sensitive region as the 2 mm cube phantom is placed in the middle of sensing chamber.

Meanwhile, the height reconstruction of the 2 mm cube phantom is improved in the single plane distributed sensor since the 2 mm cube phantom is within region sensitivity coverage. The dual plane distributed sensor further enhanced the quality of reconstructed 3D image. This is attributed to the extended sensitivity coverage between the top and bottom electrode plane contributed by the top and bottom electrode pair coupling. Therefore, it is evident that the increase in number of electrode pair combinations and thorough sensitivity coverage improves the quality of reconstructed 3D image.

4.2 Noise Analysis

Table 1 shows the reconstructed 3D image by the three sensor designs using simulated noisy capacitance data with SNR of 10 dB to 60 dB.

It appears from Table 1 that only artefacts are reconstructed by the peripheral sensor and single plane distributed sensor for SNR of capacitance data below 40 dB. Meanwhile, artefact is reconstructed by the dual plane distributed sensor for SNR of capacitance data below 20 dB. The presence of artefacts at low SNR of capacitance data implies that noise altered the capacitance data that is required for 3D image reconstruction. Figure 5 shows the CC of reconstructed 3D image at different SNR of simulated noisy capacitance data.

Figure 5 shows that the presence of noise in capacitance data significantly reduce the quality of reconstructed 3D image. The peripheral sensor is highly affected by noise as compared to the single plane distributed sensor. A capacitance data with SNR above 50 dB is needed for the peripheral sensor to reconstruct 3D image with reasonable quality as compared to above 40 dB for the single plane distributed sensor. Meanwhile, the dual plane distributed sensor can produce 3D image with high CC at a low SNR of 20 dB.

The noise analysis on the different miniature planar ECT sensor designs highlights two design aspects that reduce the effect of noise on the reconstructed image. Firstly, the effect of noise on image quality is reduced by arranging the electrodes uniformly next to one another such as the 3 × 3 matrix electrode arrangement of single plane distributed sensor. This is reflected by the better noise performance of single plane distributed sensor than peripheral sensor at low SNR of capacitance data. The 3 × 3 matrix electrode arrangement of single plane distributed sensor produced neighbouring electrode pair combinations that has high strength and are less affected by noise [5]. For instance, the electrode E5 of single plane distributed sensor has four neighbouring electrodes as compared to only two at each electrode of the peripheral sensor.

Secondly, the effect of noise on image quality can be reduced by increasing the total number of electrode pair combinations. This can be seen in the better 3D imaging performance of dual plane distributed sensor at low SNR of 20 dB. This is because it has a total of 156 electrode pair combinations. In contrast, the peripheral and single plane distributed sensor performed poorly at SNR below 50 dB since they have lower number of electrode pair combinations of 28 and 36, respectively. The 60% improvement in the performance of dual plane distributed sensor at low SNR suggests that there are sufficient necessary normalised capacitance data for 3D image reconstruction. The availability of normalised capacitance data for image reconstruction using the single and dual plane distributed sensor at different noise level is compared in Fig. 6 and Fig. 7.

Figure 6 shows that the normalised capacitance data of single plane distributed sensor at SNR of 10 dB and 30 dB is distorted and does not resemble the noise free data. Therefore, only artefacts were reconstructed by the distorted normalised capacitance data at low SNR. Similarly, the normalised capacitance data of dual plane distributed sensor

Table 1. Reconstructed 3D image using simulated noisy capacitance data

SNR (dB)	Miniature planar 3D ECT sensor		
	Peripheral	Single plane distributed	Dual plane distributed
10			
20			
30			
40			
50			
60			

at SNR of 10 dB also suffered similar distortion as illustrated in Fig. 7. Meanwhile, the normalised capacitance data of dual plane distributed sensor closely resembles the noise free normalised capacitance data, particularly between electrode pair combination 67 to 75, at SNR of 30 dB. These are the bottom and top electrode pair combinations. They

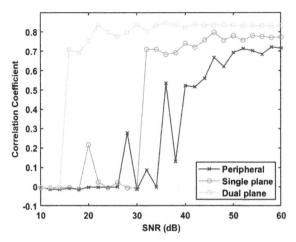

Fig. 5. Correlation coefficient of reconstructed 3D image at different SNR of simulated noisy capacitance data.

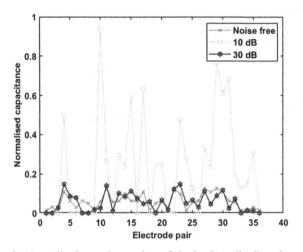

Fig. 6. Normalised capacitance data of single plane distributed sensor

are less affected by noise due to the strong signal strength between them. Therefore, it is evident that increasing the number of electrode pair combinations introduce more high strength signal that is not distorted by noise. Thus, the lower normalised capacitance data distortion resulted in high quality reconstructed image.

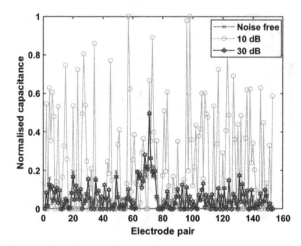

Fig. 7. Normalised capacitance data of dual plane distributed sensor

5 Conclusion

In conclusion, this paper investigated the noise performance of various design of miniature planar 3D ECT sensors. Three designs of miniature planar 3D ECT sensors were compared which include the peripheral sensor, single plane distributed sensor and dual plane distributed sensor. These sensors were modelled and simulated in COMSOL Multiphysics to obtain the sensitivity map and capacitance data required for 3D image reconstruction. To compare the 3D imaging performance of the miniature planar 3D ECT sensors at different noise levels, the White Gaussian noise was added to the noise free simulated capacitance data. The noisy simulated capacitance data has SNR ranges between 10 dB to 60 dB.

The findings from this study indicated that the influence of noise in reconstructed 3D image is affected by two sensor design aspects which are the arrangement of electrodes and the total number of electrode pair combinations. The effect of noise on image reconstruction is reduced by arranging the electrodes in 3 × 3 matrix array. This is attributed to the uniform arrangement of electrodes in 3 × 3 matrix array which produce strong neighbouring electrode pair signals that are less affected by noise. This resulted in the higher 3D image quality of single plane distributed sensor at SNR of 40 dB instead of 50 dB for the peripheral sensor. Besides, increasing the number of electrode pair combinations resulted in higher availability of normalised capacitance data that are not affected by noise for image reconstruction. This was demonstrated by the superior performance of dual plane distributed sensor at low SNR of 20 dB as it has 156 electrode pair combinations. The addition of top electrode plane produced additional high strength electrode pair between the top and bottom electrode plane that are less affected by noise. Thus, the normalised capacitance data is less distorted and can reconstruct high quality 3D image.

This study has shown that the importance of high SNR capacitance data of miniature planar 3D ECT sensor in producing high quality reconstructed. Therefore, future works involving experimental implementation of miniature planar ECT sensor should incorporate methods to increase the SNR of measured capacitance data. For instance, the coaxial cables and designing a high SNR data acquisition system should be applied for the experimental study of miniature planar ECT sensor. In terms of the fabrication of miniature ECT sensor, the dual plane sensor setup could be implemented as it provides better noise tolerance.

Acknowledgement. The authors would like to acknowledge and thank the Ministry of Higher Education Malaysia and Universiti Teknologi Malaysia for all financial support through FRGS Project code: FRGS/1/2020/TK0/UTM/02/47.

References

1. Tholin-Chittenden, C., Abascal, J.F.P.J., Soleimani, M.: Automatic parameter selection of image reconstruction algorithms for planar array capacitive imaging. IEEE Sens. J. **18**(15), 6263–6272 (2018)
2. Zhang, Y., Sun, Y., Wen, Y.: An imaging algorithm of planar array capacitance sensor for defect detection. Measurement (Lond) **168**(August 2020), 108466 (2021)
3. Ma, G., Soleimani, M.: A versatile 4D capacitive imaging array: a touchless skin and an obstacle-avoidance sensor for robotic applications. Sci. Rep. **10**(1), 1–9 (2020)
4. Suo, P., Sun, J., Tian, W., Sun, S., Xu, L.: 3-D Image Reconstruction in Planar Array ECT by Combining Depth Estimation and Sparse Representation. IEEE Trans Instrum Meas **70** (2021)
5. Wen, Y., Zhang, Z., Zhang, Y., Sun, D.: Redundancy analysis of capacitance data of a coplanar electrode array for fast and stable imaging processing. Sensors **18**(2), 31 (2017). Dec.
6. Ye, Z., Banasiak, R., Soleimani, M.: Planar array 3D electrical capacitance tomography. Insight - Non-Destructive Testing and Condition Monitoring **55**(12), 675–680 (2013). Dec.
7. Ye, Z., Wei, H.Y., Soleimani, M.: Resolution analysis using fully 3D electrical capacitive tomography. Measurement **61**, 270–279 (2015). Feb.
8. Wei, H.Y., Qiu, C.H., Soleimani, M.: Evaluation of planar 3D electrical capacitance tomography: From single-plane to dual-plane configuration. Meas. Sci. Technol. **26**(6), 65401 (2015)
9. Taylor, S.H., Garimella, S.V.: Design of electrode arrays for 3D capacitance tomography in a planar domain. Int. J. Heat Mass Transf. **106**, 1251–1260 (2017). Mar.
10. Mohd Razali, N.A.: On Chip Planar Capacitance Tomography for Two-Phase Fluid Flow Imaging (2016)
11. Hor, X.F., Leow, P.L., Ali, M.S.M., Chee, P.S., Din, S.M., Gooi, W.P.: Electrode configuration study for three-dimensional imaging of on-chip ECT. Engineering Research Express **5**(2) (Mar. 2023)
12. Ren, Z., Yang, W.Q.: Visualisation of tooth surface by electrical capacitance tomography. Biomed Phys Eng Express **3**(1), 015021 (2017)
13. Gooi, W.P., Leow, P.L., Hor, X.F., Mohammad Din, S.: Performance Comparison for On-chip 3D ECT Using Peripheral and Distributed Electrode Arrangement, vol. 921. LNEE (2022)
14. Zhang, Y., Zhao, Z., Yao, X., Wen, Y., Luo, X.: Planar array capacitive imaging method based on data optimization. Sens Actuators A Phys **347**(October), 113941 (2022)
15. Pan, Z., Wang, S., Li, P., Zhang, Y., Wen, Y.: An optimization method of planar array capacitance imaging. Sens Actuators A Phys **327**, 112724 (2021). Aug.
16. Huang, K., et al.: Effect of electrode length of an electrical capacitance tomography sensor on gas−solid fluidized bed measurements. Ind. Eng. Chem. Res. **58**(47), 21827–21841 (2019). Nov.

A Preliminary Investigation on The Correlation Between the Arrival Time of Ultrasonic Signals and The Concrete Condition

Farah Aina Jamal Mohamad[1], Anita Ahmad[1(✉)], Ruzairi Abdul Rahim[1],
Sallehuddin Ibrahim[1], Juliza Jamaludin[2], Nasarudin Ahmad[1],
Fazlul Rahman Mohd Yunus[3], Mohd Hafiz Fazalul Rahiman[4],
and Nur Arina Hazwani Samsun Zaini[1]

[1] Faculty of Electrical Engineering, Universiti Teknologi Malaysia, 81310 UTM Skudai, Johor,
Malaysia
anita@utm.my
[2] Faculty of Engineering and Built Environment, Universiti Sains Islam Malaysia, Negeri
Sembilan, 71800 Nilai, Malaysia
[3] Advance Technology Training Centre (ADTEC), Bandar Vendor Taboh Naning, 78000 Alor
Gajah, Melaka, Malaysia
[4] Tomography Imaging Research Group, Faculty of Electrical Engineering Technology,
Universiti Malaysia Perlis, Pauh Putra Campus, 02600 Arau, Perlis, Malaysia

Abstract. Concrete is a composite material that is widely used in a construction project. The evaluation of concrete structure is very important in order to determine its strength and quality. Concrete is commonly evaluated by using the ultrasonic pulse velocity (UPV) method, which adopted the concept of measuring time of a first arrival of the received signal. Hence, this paper aims to evaluate the first arrival time of the detected ultrasonic signals based on different conditions of concrete structure. A simulation study was conducted by using COMSOL Multiphysics software version 5.6. Data collected were categorized into three sections, including in concrete model with inclusion of air hole, crack, and rust. From the simulation results, concrete models with inclusion of air hole showed an increment in the arrival time as the size of air hole increase. For the concrete models with rust, the arrival time were significantly increased in 20-mm and 40-mm rust, however it turns down as the size of rust reached 60-mm. The results also indicated that transverse crack took a longer arrival time compared to other orientation of crack.

Keywords: Arrival time detection · Concrete · Ultrasonic signal · Wave propagation

1 Introduction

Concrete is one of the most used material in the construction projects, especially in large infrastructures, such as roads, bridges, and dams, due to its durability, cost, and flexibility [1]. However, concrete may degrade over time due to some reasons; environmental factors, exposure to high temperature, mechanical overloading, deicing salts, and repeated

F. Hassan et al. (Eds.): AsiaSim 2023, CCIS 1912, pp. 138–154, 2024.
https://doi.org/10.1007/978-981-99-7243-2_12

freezing-thawing cycles [2–4], which resulting in the distributed cracking in concrete, reinforcement corrosion, and eventually, a structural failure, as the strength and durability of concrete has decreases. Nowadays, the failure of concrete has becoming a major concern [3–5]. Damaged structures require a possible repair and continued services to maintain its safety. Hence, the structural inspection and evaluation is very significant to determine the quality of the concrete structures.

The most common technique to evaluate the condition of concrete structure is non-destructive methods [6]. It is the process of assessing and analyzing without inflicting any damage in the internal structure of object under test (OUT) [7]. Ultrasonic testing is one of the most popular non-destructive technique to detect any damage present in the concrete. This technique utilizes the propagation of high frequency ultrasonic waves in OUT. It is performed by using several techniques, such as ultrasonic pulse velocity (UPV), ultrasonic pulse echo (UPE), and impact echo method [8].

UPV method uses a pair of transducers at a certain distance, L, to measure the time-of-flight, t, of the ultrasonic longitudinal waves. The formula for the wave pulse velocity then, as follows:

$$V = \frac{L}{t} \tag{1}$$

UPV is introduced to evaluate the quality of concrete materials, as well as to detect any internal damages inside the concrete structure. The concept behind this technique is measuring the travel time of acoustic waves in a medium, where it reflects the internal condition of the test area. The properties of concrete can be assessed by determining the changes of time in ultrasonic signal [9]. In general, higher travel time may reflects the presence of damages in the structural components of the testing material, while shorter travel time generally correlated to fewer defects in the material [10]. Rather than UPE, UPV requires multiple points to emit short pulses. at the right angle to the face of the OUT. Hence, UPV method is more reliable and capable to produce accurate results, as the maximum pulse energy is emitted during the transmission [11].

Therefore, by using UPV method, this paper will focus on the correlation between the arrival time of ultrasonic signals with the concrete condition, based on three different cases; concrete with inclusion of air hole, rust, and crack.

2 Methodology

2.1 Flow Chart

Figure 1 shows the flow chart of this research. This project is divided into two sections, concrete modelling and analysis of wave propagation and arrival time in concrete. The concrete modelling were using COMSOL Multiphysics software version 5.6. This research is limited to the study of air hole, rust, and crack in concrete.

2.2 COMSOL Multiphysics

COMSOL Multiphysics version 5.6 was used in this study as a modelling environment since it allows direct multiphysics coupling of the description of wave propagation

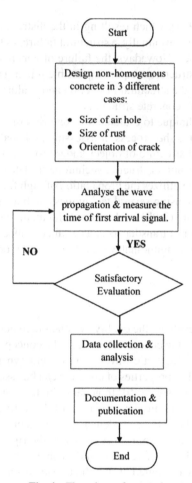

Fig. 1. Flowchart of research.

(acoustic-structure boundary) and signal detection (piezoelectric effect). For this investigation, ultrasonic through-transmission method was used as the operational mode, where the transmitter emitted the signal before being detected by the receiving transducer [12]. There are three main steps in generating a required concrete model using COMSOL Multiphysics software:

i. Pre-processing: A graphical user interface enables the desired geometry to be created based on the predefined menu in the system. The desired physics and the materials in the domain of study must be specified. To solve the generated model, different solvers can be used. In this study, time-dependent analysis was used to determine the arrival time of the ultrasonic signals.

ii. Numerical analysis: COMSOL Multiphysics software will automatically generates a solution based on the previously defined options.

iii. Post-processing: The results from the numerical solution can be studied directly from the COMSOL Multiphysics software or exported to another post-processing program.

Geometrical Modelling. A 2-D geometry has been considered in order to reduce the computational time of the simulations. The 2-D geometry was divided into three domains as illustrated in Fig. 2. The main block domain represents the concrete structure with cross-section dimensions of 300 mm × 300 mm. On the top and bottom of this domain is another block having a diameter of 20 mm × 20 mm, which acts as the transmitting and receiving transducers. The transmitting transducer was connected to an electrical circuit and was excited by a pulse of electric signal, while the second transducer which was oriented in direct transmission at the opposite side of the transmitter's position, used as a signal receiver.

A circular domain was added to represent air hole and rust in the concrete. Their diameters were used as the variable. In another cases, a rectangular domain was also included inside the concrete structure to represent the air-filled crack, where its orientation was taken as the variable.

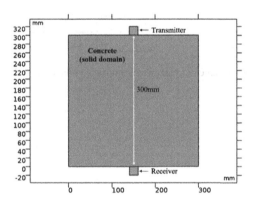

Fig. 2. The geometry of homogenous concrete (without the presence of defects).

Materials Properties. The concrete material used in this study was characterized by its own density, Poisson ratio, Young's modulus, and sound speed as shown in Table 1.

Table 1. Acoustic properties of concrete structure applied in the analysis.

Property	Value
Density (ρ)	2000 (kg/m^3)
Young's modulus (E)	3500 (GPa)
Poisson's ratio (V)	0.2
Velocity of sound	3400 (m/s)

A pair of transducers made up of lead zirconate titanate (PZT-5H) were used as the sensing element. The piezoelectric property in PZT transducers allow the generation of electrical energy from the mechanical effects including pressure and vibration energy, and vice versa.

Ultrasonic Signal. The transducers were connected to an external circuit interface using the Terminal feature to transmit ultrasonic pulse. Figure 3 shows the electrical circuit that connected to the piezoelectric device to emit short pulses into the concrete and receive the signals. A source frequency of 200 kHz was used in the simulation. Figure 4 illustrates the excitation pulse that was applied in the simulation study.

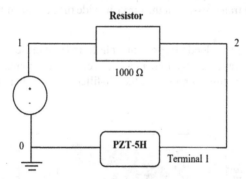

Fig. 3. External electrical circuit interface connecting to PZT-5H.

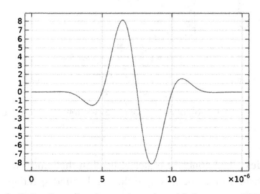

Fig. 4. Pulse force for ultrasonic probe with frequency 200 kHz.

Meshing and Solving. In order to obtain an accurate solution within 1 ms, the triangular mesh size was chosen to be in finer size for an accurate solution at all times of interest, as illustrated in Fig. 5. Once the mesh was defined, calculations were performed using a time-dependent solver.

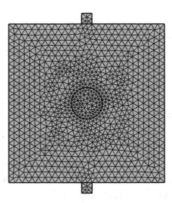

Fig. 5. Example of meshing in concrete with air hole.

2.3 Arrival Time Determination

The output data obtained from COMSOL Multiphysics software were then exported to Microsoft Excel in a text file format to determine the arrival time of ultrasonic signals in the concrete structure. Figure 6 shows an example of the determination of arrival time of the received ultrasonic signals.

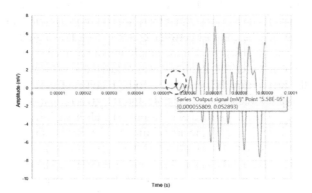

Fig. 6. Arrival time of received ultrasonic signal.

3   Results and Discussion

The simulation results included the information on the arrival time recorded, as well as the ultrasonic wave propagation for each case. In the upcoming figures illustrating the wave propagation, the waves consisted of two main colors, which is blue and red. Both colors represent the acoustic pressure values in different directions. The deeper the color is, the higher the acoustic pressure related to the color will be. The color faded during the transmission process indicating a tendency for the acoustic pressure values to drop.

The wave propagation paths were visualized in order to better understand the altered wave paths caused by the presence of changes in acoustic impedance such as cracks or other defects. The wave propagation is shown for the wave emitted from the transmitter before detected by the receiver.

The arrival time of each received ultrasonic signal was also recorded. Arrival time was indicated by the first leading edge arrival of the pulse. In next subsections, there were two figures included in the column of arrival time for each case. The upper plot is the A-scan image that demonstrated the voltage amplitude of received ultrasonic waves as a function of time. The A-scan compared two colored solid lines; blue line as the transmitted signal and green line depicted the received signal. The received signal was amplified to 300 times from its original size for better visualization. The lower plot showed the received signal with its exact amplitude.

3.1   Effect of Size of Air Hole on the Arrival Time

For the investigation of size of air hole effect, four concrete models were introduced with varying size of air hole: 20 mm, 40 mm, 60 mm and 80 mm. The results were tabulated in Table 2 and summarized in Table 3. The results show that the detected arrival times were 55.81 μs for 20-mm air hole, 56.12 μs (40 mm), 58.91 μs (60 mm), and 59.42 μs (80 mm). The visualization of wave propagation demonstrates that as the ultrasonic waves approached the air hole, they were diffracted at the edges of the hole and could only travelled around it. This behavior is consistent with what was found by Selim et. al. [12] where they assumed that there was no propagation in the air hole due to the very high impedance contrast between concrete (solid) and air. From the A-scan plot in Table 2, it can be observed that the amplitude of received signal reduced in the increment of size of air hole. The amplitude was reduced due to the wave being blocked from reaching the receiver by the presence of foreign object [13].

The relationship between size of air hole and the arrival time of ultrasonic signal can be seen in Fig. 7. The arrival time of ultrasonic signals is significantly influenced by the presence of air hole. In Fig. 7, the arrival time increased with the increasing size of air hole. Such trend was previously reported in the evaluation of additive manufactured Ti-6Al-4V part with different diameter of internal hole defects. The results showed that longitudinal wave took a longer time to arrive at the top surface of the sample when diameter of defects increased [14].

3.2   Effect of Size of Rust on the Arrival Time

To investigate the effect of size of rust on the arrival time, another four concrete models were introduced with different size of rust: 20 mm, 40 mm, 60 mm, and 80 mm. The results were tabulated in Table 4 and summarized in Table 5. The detected arrival times were 55.88 μs, 58.92 μs, 55.68 μs, and 52.76 μs, for the 20, 40, 60, and 80 mm rust, respectively. In the corroded concrete models, ultrasonic waves travelling along the path were affected by the presence of rust. Some of the waves propagated through the rust, and some were reflected. As these occurred, the amplitude of received signal is proportional to the rust size. The larger the size of rust, the smaller the amplitude of received ultrasonic signal.

Table 2. Effect of size of air hole on the wave propagation and the arrival time of ultrasonic signal

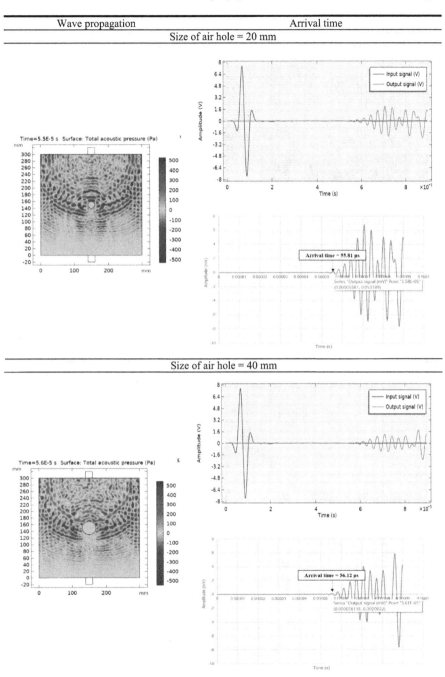

(*continued*)

Table 2. (*continued*)

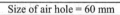

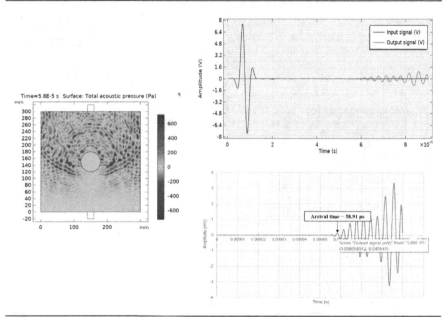

Size of air hole = 60 mm

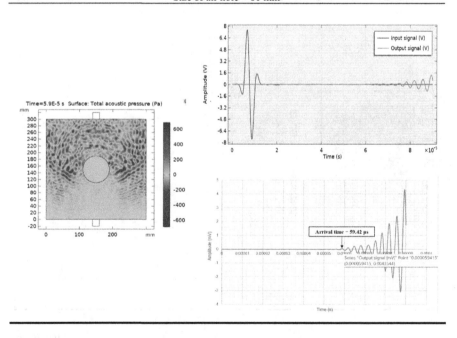

Size of air hole = 80 mm

Table 3. Arrival time of detected ultrasonic signals for different size of air hole.

Size of air hole (mm)	Arrival time (μs)
20	55.81
40	56.12
60	58.91
80	59.42

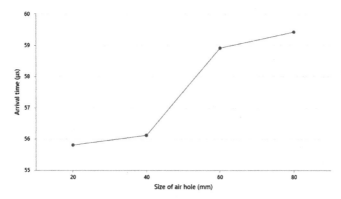

Fig. 7. Relationship between size of air hole and arrival time of ultrasonic waves.

Figure 8 shows the relationship between the size of rust and the corresponding arrival time of ultrasonic signal. As can be observed, the results returned an approximately linear relationship for 20-mm and 40-mm rust, but shifted down for 60-mm and 80-mm rust. The arrival time of ultrasonic waves in concrete with rust were delayed, due to the wave propagation velocity in material changed, as ultrasonic wave velocity is slower in rust than it is in concrete. However, the decrement showed for 60-mm and 80-mm rust probably due to the following reason. Since rust is a solid-type defect, the particles in the rust medium are closely packed. Ultrasonic waves (sound energy) conceptually, is vibration of kinetic energy passed from particle to particle. For the same medium, when size increased, the closer the particles and the tighter their bonds were, hence lesser time it takes for sound to travel [15]. Therefore, it can be concluded that, as the size of rust more than 60 mm, it becomes denser, aptly causing the ultrasonic waves velocity increased, and the arrival time were much faster.

3.3 Effect of Orientation of Crack on the Arrival time

This test was carried out to investigate the effect of crack orientations on ultrasonic detection. A 5-mm crack was positioned in the concrete structure. The results in Table 6 and Table 7 presented and analyzed the influence of transverse crack, oblique cracks of 45° and 135°, as well as vertical crack, on the wave propagation and the arrival time of ultrasonic signal. As can be seen in Table 6 and Table 7, for crack of the same

Table 4. Effect of size of rust on the wave propagation and the arrival time of ultrasonic signal.

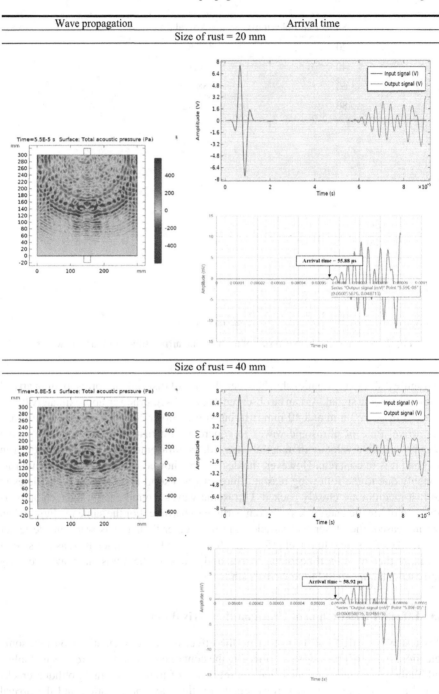

(*continued*)

Table 4. (*continued*)

Size of rust = 60 mm

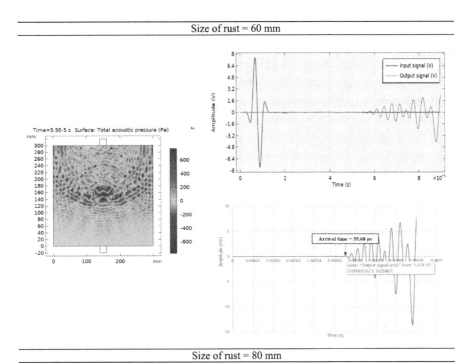

Size of rust = 80 mm

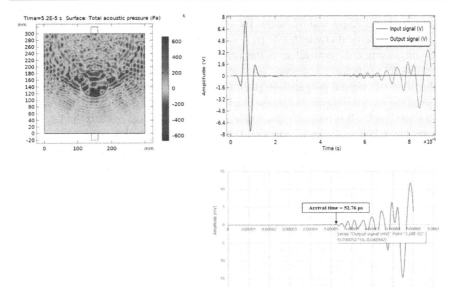

Table 5. Arrival time of detected ultrasonic signals for different size of rust.

Size of rust (mm)	Arrival time (μs)
20	55.88
40	58.92
60	55.68
80	52.76

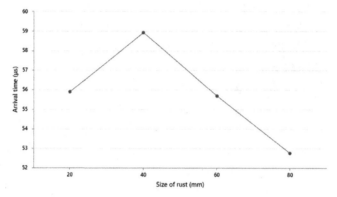

Fig. 8. Relationship between size of rust and arrival time of ultrasonic waves.

width, the transverse crack had a higher arrival time which is 75.12 μs, followed by the oblique crack (62.95 μs and 62.86 μs, for 45°- and 135°-oriented crack, respectively), and finally the vertical crack (60.29 μs). This is because the transverse crack had the largest diffracting area (equivalent as the air hole) when the ultrasonic wave propagated in the concrete structure. The same goes to the oblique cracks, which were 45° from the direction of original propagation path, took a longer time to arrive at the receiver. However, the received signal of vertical crack appeared earlier than that of the transverse and oblique cracks. It is probably due to the width of the vertical crack is not larger than the transducer, hence does not completely obstruct the propagation path.

In Fig. 9, the orientation of crack is plotted against the corresponding arrival time in order to understand the variation of arrival time with changes in orientation of crack.

Table 6. Effect of orientation of crack on the wave propagation and the arrival time of ultrasonic signal.

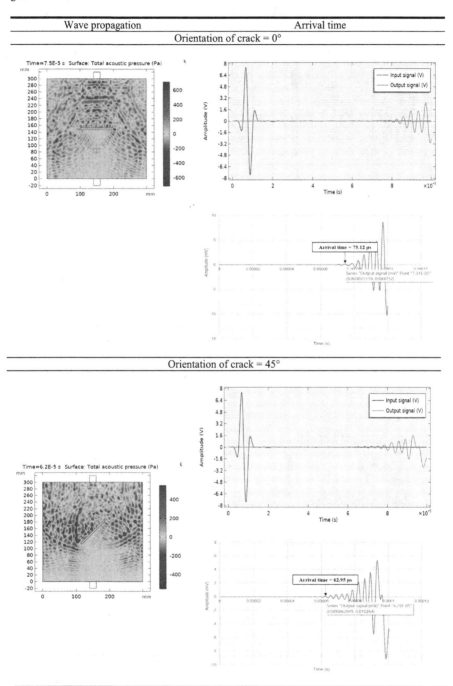

(*continued*)

Table 6. (*continued*)

Orientation of crack = 90°

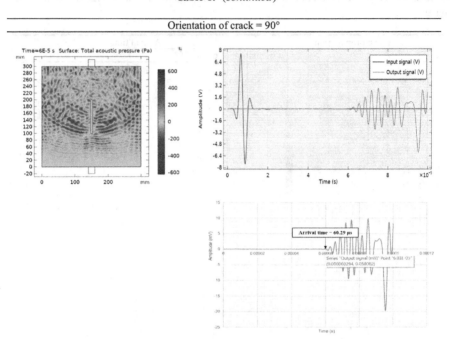

Orientation of crack = 135°

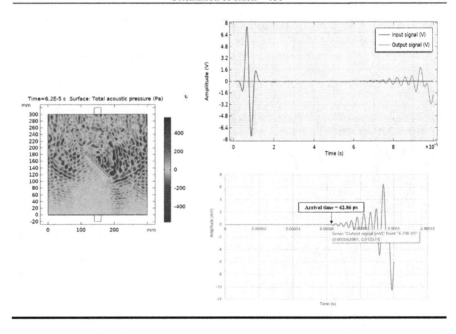

Table 7. Arrival time of detected ultrasonic signals for different orientation of crack.

Orientation of crack	Arrival time (μs)
0°	75.12
45°	62.95
90°	60.29
135°	62.86

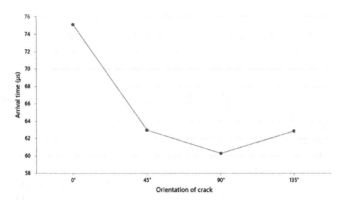

Fig. 9. Relationship between orientation of crack and arrival time of ultrasonic waves.

4 Conclusions

The first arrival time of the received ultrasonic signal is greatly important and necessary to provide an early estimation on the quality of concrete structure. In this paper, COMSOL Multiphysics software version 5.6 has been used to perform a simulation study to evaluate the concrete structure based on the arrival time of ultrasonic signals. From the results, it can be clearly seen that the increment in size of air hole affect the ultrasonic arrival time. As the size of air hole increase, the arrival time also increase. For concrete models with the presence of rust, the arrival time shows significant increment during the inclusion of rust with diameter of 20 mm and 40 mm. However, it decreased as the diameter of rust reached 60 mm. On the other cases, the orientation of crack also influenced the arrival time of the received signal. Transverse crack took a longer arrival time compared to the other orientation of crack.

For future works, data obtained can be further analyzed to determine the location and size of defects in concrete structure. Besides that, the simulation data can be integrated with in-situ real time data in order to allow more practical and reliable results obtained.

Acknowledgement. This work was supported by Universiti Teknologi Malaysia under UTM Encouragement Research Grant [Q.J130000.3851.19J63].

References

1. Harshit, J., Patankar, V.H.: Advances in Ultrasonic Instrumentation for Inspection of Concrete/RCC Structures. Elsevier, Amsterdam, The Netherlands (2019)
2. Davis, A.G., et al.: Nondestructive test methods for evaluation of concrete in structures. American Concrete Institute **8** (1998). https://doi.org/10.1002/9781118745977
3. Azari, H., Nazarian, S., Yuan, D.: Assessing sensitivity of impact echo and ultrasonic surface waves methods for nondestructive evaluation of concrete structures. Computers and Chemical Engineering **71** (2014). https://doi.org/10.1016/j.conbuildmat.2014.08.056
4. Dwivedi, S.K., Vishwakarma, M., Soni, P.A.: Advances and researches on non destructive testing: a review. Materials Today: Proceedings **5**(2) (2018). https://doi.org/10.1016/j.matpr.2017.11.620
5. Hertlein, B., Davis, A.: Nondestructive Testing of Deep Foundations. John Wiley & Sons (2007)
6. Ongpeng, J.: Non-destructive testing using ultrasonic waves in reinforced. Concrete (2015). https://doi.org/10.13140/RG.2.1.1381.1922
7. Gholizadeh, S.: A review of non-destructive testing methods of composite materials. Procedia Structural Integrity **1** (2016). https://doi.org/10.1016/j.prostr.2016.02.008
8. Kalyan, T.S., Kishen, J.M.C.: Experimental Evaluation of Cracks in Concrete by Ultrasonic Pulse Velocity. Proceedings of the APCNDT (2013)
9. Xu, J., Wei, H.: Ultrasonic testing analysis of concrete structure based on S transform. Shock. Vib. (2019). https://doi.org/10.1155/2019/2693141
10. Mohamad, F.A.J., et al.: A Study on the Arrival Time of Ultrasonic Waves in Concrete Material. Control, Instrumentation and Mechatronics: Theory and Practice, pp. 83–92. Springer (2022)
11. Rahiman, M.H.F., Rahim, R.A., Ayob, N.M.N.: The front-end hardware design issue in ultrasonic tomography. IEEE Sensors Journal **10**(7) (2010). https://doi.org/10.1109/JSEN.2009.2037602
12. Selim, H., Picó, R., Trull, J., Prieto, M.D., Cojocaru, C.: Directional ultrasound source for solid materials inspection: Diffraction management in a metallic phononic crystal. Sensors **20**(21) (2020). https://doi.org/10.3390/s20216148
13. Khairi, M.T.M, Ibrahim, S., Yunus, M.A.M., Faramarzi, M.: Two-dimensional and pseudo three-dimensional visualisation of foreign objects in milk carton using ultrasonic tomography. Measurement: Food **6** (2022)
14. Yu, J., et al.: Detection of internal holes in additive manufactured Ti-6Al-4V part using laser ultrasonic testing. Applied Sciences **10**(1) (2020). https://doi.org/10.3390/app10010365
15. Connacher, W., et al.: Micro/nano acoustofluidics: Materials, phenomena, design, devices, and applications. Lab on a Chip **18**(14) (2018). https://doi.org/10.1039/c8lc00112j

A Fast Algorithm for Satellite Coverage Window Based on Segmented Dichotomy

Fusheng Li[1], Wei He[2], Tao Chao[1(✉)], Weibo Sun[1], and Shenming Quan[2]

[1] Harbin Institute of Technology, Harbin 150000, China
chaotao2000@163.om
[2] Electro-Mechanical Engineering Institute, Shanghai 201109, China

Abstract. In order to meet the requirement of fast calculation of the ground coverage time window of observation satellites, a method of fast calculation of the ground coverage time window of satellites based on the segmented dichotomy is proposed. Firstly, the method uses a 2D map with the geographic longitude of ascending node and the argument of latitude as the horizontal and vertical axes, and constructs a mathematical model for the satellite trajectory on the 2D map based on the SGP4. Subsequently, an applicable segmented dichotomy method is proposed according to the characteristics of window calculation, which can solve all intersections of satellite trajectories with the coverage area on the 2D map quickly and accurately. Finally, the intersection correction formula and the window calculation formula are proposed, and through the above two formulas, the satellite to ground coverage window is deduced. The accuracy of the algorithm is evaluated by taking one satellite to multiple points of coverage as an example, and the results show that the method is equal to the calculation results of commercial software within the error tolerance; in addition, 100 observation satellites are selected for algorithm efficiency evaluation, and Harbin city is selected as the coverage area for coverage window calculation and timing, and the results show that the calculation time of the method in this paper is reduced by more than 62% on average compared with the fixed-step method.

Keywords: SGP4 · 2D map · Satellite coverage window

1 Introduction

The satellite ground coverage time window is the period during which a satellite can complete its observations of a specified area on the ground while in orbit. It is closely related to the coverage performance of the observation constellation and determines the performance indicators such as the coverage area, observation frequency and observation duration of the observation constellation. Therefore, the calculation of the satellite's time window for ground coverage is essential.

The conventional method for calculating satellite-to-ground coverage windows involves recursively determining the coverage time of each satellite in the constellation at every moment, integrating the coverage situation of each satellite to obtain the

F. Hassan et al. (Eds.): AsiaSim 2023, CCIS 1912, pp. 155–170, 2024.
https://doi.org/10.1007/978-981-99-7243-2_13

coverage window of the constellation. However, this method's accuracy heavily depends on the time step size, which presents a challenge for optimizing the design of observation constellations. Short time steps increase the computation time of coverage window calculations, slowing down optimization speed, while long time steps introduce large errors that affect the final optimization results. Therefore, there is a need for a method to calculate satellite-to-ground coverage windows quickly and accurately.

Alfano proposed a method using a hybrid quadratic polynomial fit, which achieved high accuracy with a time step of 200s [1]. A more complex hybrid quartic polynomial fit was also proposed, but showed no significant advantage over the quadratic method [2]. Radzik and Maral assumed a stationary ground target within one orbit of a low Earth orbit satellite to compute coverage windows, but this method suffered from computational instability [3]. Ali approximated the satellite ground track as a great circle arc and computed coverage windows using the intersection of the arc and the target's visible area, but with limited accuracy [4]. Tang R proposed a compensation strategy based on the mean anomaly to compute coverage windows, but this method is only applicable to two-body problems [5].

Ulybyshev pioneered the use of two-dimensional spatial mapping in constellation coverage analysis, using the argument of latitude and the right ascension of ascending node to analyze constellation coverage without considering perturbations [6]. He developed a 2D map with right ascension of ascending node and time to calculate revisit times for constellations with short and discontinuous coverage of target regions, defining the coverage band theory and analyzing global coverage of the Walker constellation as well as Russia [7]. Ulybyshev proposed a new method for revisit time calculation based on the 2D map, enabling constellation coverage analysis without considering constellation motion [8, 9]. However, his analysis was limited to circular orbits and did not consider orbital perturbations. Researchers have proposed new approaches based on 2D map theory, including using transformed reference frames to calculate coverage performance of constellations on circular orbits [10], considering orbital perturbations to calculate satellite visibility windows [11], and using relative and constellation field mapping theory to determine maximum revisit time for constellations [12]. In addition, Han chao et al. proposed a metamodel-based framework for fast prediction of satellite visibility to satellites and ground points [13].

The current methods for calculating satellite-to-ground coverage windows are inaccurate, as they only consider circular orbits with simple perturbations. To address these limitations, this article proposes a new method for fast and accurate calculation of coverage windows, which has the following features:

1. The method constructs a mathematical model for satellite trajectories on near-circular orbits, considering the effects of orbital perturbations, using 2D map theory and the SGP4 orbit model.
2. It proposes a method of segmented dichotomy to determine all intersection points between satellite trajectories and coverage areas quickly and accurately.
3. It introduces an intersection point correction formula to avoid errors caused by the task duration not being an integer multiple of the orbit period. The window is then calculated using the corrected intersection points and the coverage window formula.

The paper is structured as follows: Sect. 1 reviews previous research. Section 2 describes the ground coverage model and the 2D map theory used in this paper. Section 3 describes the modelling of satellite trajectories in 2D maps and the process of the segmental dichotomy method, while giving the specific steps of the window calculation. Numerical calculations are performed in Sect. 4, which include a comparison of the accuracy of the calculations as well as the speed of the algorithm. Section 5 concludes the paper.

2 Basic Theory

2.1 Satellite to Ground Coverage Model

For the convenience of the analysis, the effect of the shape of the earth is not considered in this paper and the earth is considered as a sphere with the mean radius of the earth $R_e = 6371.30$ km.

Firstly, the satellite ground coverage model is introduced. In this paper, the simplest conical sensor is selected and the ground coverage model is shown in Fig. 1. According to the geometric relationship in the figure, the relationship equation between the orbital inclination i, the orbital height h, the half-tensor angle of the satellite ground coverage beams θ and the ground coverage area angle d can be derived.

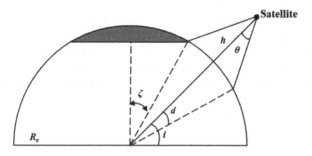

Fig. 1. Satellite-to-ground coverage model

$$d = \arcsin\left(\frac{\sin\theta(R_e + h)}{R_e}\right) - \theta \qquad (1)$$

To analyze satellite coverage of an area, the ground coverage area angle from a coverage model is used. To check if a ground point is covered, calculate α using Eq. (2) with the ground point's latitude and longitude (φ_W, λ_W) and the satellite subsatellite point's latitude and longitude (φ_S, λ_S). The ground point is covered if $\alpha \le d$.

$$\alpha = \arccos\{\sin(\varphi_W)\sin(\varphi_S) + \cos(\varphi_W)\cos(\varphi_S)\cos(\lambda_S - \lambda_W)\} \qquad (2)$$

2.2 Two-Dimensional Map Theory

In order to perform the satellite window calculation, it is first necessary to define the coverage area in the 2D map [13], and Eq. (3) is the curve equation for the boundary of the coverage area in the 2D map.

$$\Omega(u) = \lambda_W - \arctan 2\left(\frac{\cos i \sin u}{\cos u}\right) \pm \arccos \frac{\cos d - \sin \varphi_W \sin i \sin u}{\cos \varphi_W \sqrt{1 - (\sin i \sin u)^2}} \qquad (3)$$

where Ω is the right ascension of ascending node, u is the argument of latitude, i is the orbital inclination and $\arctan 2(\cdot)$ is a custom inverse trigonometric function that can be calculated from Eq. (4).

$$\arctan 2(y/x) = \begin{cases} \arctan(y/x) & x \geq 0 \\ \arctan(y/x) + \pi & x < 0 \end{cases} \qquad (4)$$

Equation (5) yields the range of u, further according to the inverse trigonometric function value domain, it is known that Eq. (3) corresponds to the interval of existence of the curve satisfy $u \in [-\pi/2, 3\pi/2]$.

$$\begin{cases} u \in [u_{\min}, u_{\max}] \cup [\pi - u_{\max}, \pi - u_{\min}] \\ u_{\substack{\max \\ \min}} = \text{sgn}(\varphi_W \pm d) \cdot \arcsin \frac{\sin[\min(i, |\varphi_W \pm d|)]}{\sin i} \end{cases} \qquad (5)$$

The function defining a 2D map is composed of four curves, which form different regions depending on the parameters. These regions are illustrated in Fig. 2 and can be classified as: two discrete regions when meeting the conditions of Eq. (a) in Eq. (6), a concave region when meeting the conditions of Eq. (b), a convex region when meeting the conditions of Eq. (c), and a ribbon region when meeting the conditions of Eq. (d).

$$\begin{cases} (a) & |\varphi_W| \leq i - d & (b) & i - d < |\varphi_W| \leq i \\ (c) & i < |\varphi_W| \leq i + d & (d) & \pi - i - d < |\varphi_W| \leq \pi/2 \end{cases} \qquad (6)$$

The λ_W in Eq. (3) is the equatorial longitude, which is defined in the Earth-Centered Inertial Frame (ECI). Equation (3) changes continuously due to the Earth's rotation, making it difficult to solve for a moving area. To simplify this, Eq. (7) is derived by sub-tracting the Greenwich time angle from both sides of Eq. (3). And Ω_G is the geographic longitude of the ascending node and λ_W is the geographical longitude. The 2D maps referred to in the following analysis use the geographic longitude of ascending node and the argument of latitude as horizontal and vertical coordinates. (See in Fig. 3)

$$\Omega_G(u) = \lambda_W - \arctan 2\left(\frac{\cos i \sin u}{\cos u}\right) \pm \arccos \frac{\cos d - \sin \varphi_W \sin i \sin u}{\cos \varphi_W \sqrt{1 - (\sin i \sin u)^2}} \qquad (7)$$

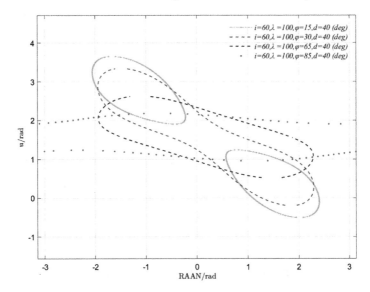

Fig. 2. Four Types of 2D Mapping of Ground Points

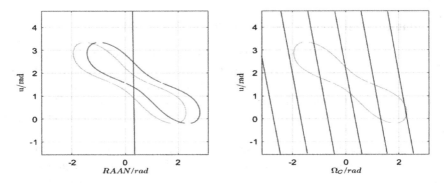

Fig. 3. Coverage area translation and horizontal coordinate transformation

And the intersection of the curve formed by the satellite's orbital parameters and Eq. (7) marks the start and end of ground coverage as the satellite moves and can see the ground point. (See in Fig. 4).

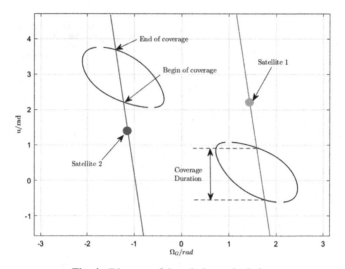

Fig. 4. Diagram of the window calculation

3   Quick Calculation of the Satellite Window

3.1   Mathematical Model of the Satellite Trajectory in 2D Map

The SGP4 model (Simplified General Perturbations 4), developed by Ken Cranford in 1970, takes into account the effects of perturbations such as solar and lunar gravity, solar radiation pressure and atmospheric drag, as opposed to simple two-body motion models, and can be applied to near-earth objects with orbital periods of less than 225 min. The model predicts the position and velocity of a satellite at any given moment from its Two-Line Element (TLE).

The geographic longitude of the satellite's ascending node varies uniformly and the argument of latitude varies in the range $[0, 2\pi]$ without taking any regression into account. The trajectory of a satellite in a two-dimensional map is a series of evenly spaced inclined straight lines with a spacing of $\Delta\lambda = \omega_e \cdot T_S$, $\omega_e \cdot T_S$ being the angular velocity of the Earth's rotation and the period of the satellite's motion, respectively.

When the SGP4 orbit extrapolation model is used, the satellite trajectories in the two-dimensional map will change. Two satellites with different eccentricities were selected from the TLE public data [14] and 86640 s were calculated from the given time of the TLE data to obtain the trajectories in the 2D map as shown in Fig. 5, where the green dotted line indicates the starting orbit of the satellite at the beginning of the simulation. As can be seen from the figure, when the eccentricity is small, the satellite's trajectory in the two-dimensional map is approximated as a series of equidistant and inclined straight lines; when the eccentricity increases to a certain value, the straight lines become curves.

Based on the above analysis, when the eccentricity is small, a linear mathematical model can be used, which means that the satellite trajectory obtained using the SGP4 orbit extrapolation can be replaced by an inclined straight line in a two-dimensional map, and the calculation process is given below:

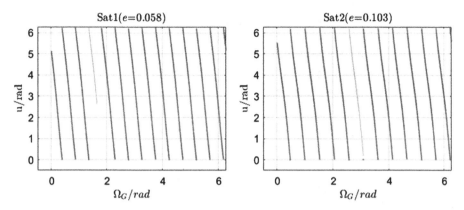

Fig. 5. Satellite trajectories in 2D maps

Step1. The set of satellite parameter coordinates $\{(\lambda_i, u_i)\}_l$ (the geographic longitude of ascending node, the argument of latitude) is obtained by extrapolation of the SGP4 orbit over one cycle, and l represents the track number corresponding to the set.

Step2. A linear fit is performed on all coordinates in the set, and the linear fit formula is shown in Eq. (8). The slope $k_{(l)}$ corresponding to each period and the correlation coefficient $R^2_{(l)}$ are obtained after the fit.

$$k_{(l)} = \frac{\sum u_i \cdot \lambda_i - n\bar{u}\bar{\lambda}}{\sum \lambda_i^2 - n\bar{\lambda}^2}, \bar{u} = \frac{\sum u_i}{n}, \bar{\lambda} = \frac{\sum \lambda_i}{n}$$

$$R^2_{(l)} = \left[\frac{\sum \lambda_i u_i - n \sum \frac{\lambda_i}{n} \sum \frac{u_i}{n}}{\sqrt{\left[\sum \lambda_i^2 - n\left(\sum \frac{\lambda_i}{n}\right)^2\right]\left[\sum u_i^2 - n\left(\sum \frac{u_i}{n}\right)^2\right]}} \right]^2 \tag{8}$$

Step3. Calculate the slope of the satellite trajectory curve k and the correlation coefficient R^2 according to Eq. (9).

$$k = \sum k_{(l)}/m, R^2 = \sum R^2_{(l)}/m \quad (m \text{ is the number of loops}) \tag{9}$$

Step4. Based on the initial argument of latitude $u_{initial}$, the initial geographic longitude of ascending node $\lambda_{initial}$ and the slope k of the satellite, the intercept b_l of each line is obtained by Eq. (10).

$$\begin{cases} b_{(l)} = u_{initial} - k\lambda_{initial} & (l = 1) \\ b_{(l)} = b_{(l-1)} - 2\pi & (l > 1) \end{cases} \tag{10}$$

The slope and correlation coefficients are calculated for different eccentricities of the orbits. The correlation coefficient varies with eccentricity as shown in Table 1 from which the correlation coefficient decreases with increasing eccentricity and is approximately 1 when $e < 0.1$. Therefore, when the orbit is a sub-circular orbit, the trajectory of a satellite with an SGP4 orbital extrapolation can be approximated as a series of parallel straight lines on a two-dimensional map. That is, the satellite's trajectory in a two-dimensional map is modelled as $u = k\lambda + b_{(l)}$.

Table 1. The relationship between e and R^2

Orbital eccentricity (e)	R^2
0.001451	0.999997
0.003586	0.999996
0.006943	0.999985
0.058007	0.998529
0.090953	0.997638
0.103332	0.988134
0.311847	0.968832

3.2 Segmented dichotomy

According to the two-dimensional map theory, since the equations of the ground point coverage area and the satellite mapping curve are known, the problem of solving the coverage window is transformed into the problem of solving the intersection of the two curves. Further the problem of solving the intersection point is converted into the problem of the zero point of the function $f(u)$.

$$f(u) = \lambda_W - \arctan 2\left(\frac{\cos i \sin u}{\cos u}\right) \pm \arccos \frac{\cos d - \sin \varphi_W \sin i \sin u}{\cos \varphi_W \sqrt{1 - (\sin i \sin u)^2}} - \frac{1}{k}(u - b)$$

The set of zeros of $f(u)$ is shown in Eq. (11), where $f_1(u), f_2(u), f_3(u), f_4(u)$ are the four functions represented by $f(u)$ respectively. Equation (12) gives different cases of the number of zeros of $f(u)$, In the Eq. (12), $card(\cdot)$ denotes the number of elements in the set, which represents the number of zeros of the curve corresponding to the set. Therefore, the number of zeros in this problem is unknown, and the existing zeros solution algorithm is usually given an initial value near the zero point for iterative calculation, which eventually converges to only one zero point and is not applicable to this problem. Therefore, a new solution method is needed to solve this problem.

$$\begin{aligned} I_1 &= \{(\lambda, u)|f_1(u) = 0\}, I_2 = \{((\lambda, u)|f_2(u) = 0\} \\ I_3 &= \{(\lambda, u)|f_3(u) = 0\}, I_4 = \{(\lambda, u)|f_4(u) = 0\} \end{aligned} \tag{11}$$

$$\begin{cases} (a)(i_1 = 1\|i_2 = 1\|i_3 = 1\|i_4 = 1)\&\& \sum_{l=1}^{4} i_l = 2 \\ (b)i_1 = 0\&\&i_2 = 0\&\&i_3 = 0\&\&i_4 = 0 \\ (c)(i_1 = 2\|i_2 = 2\|i_3 = 2\|i_4 = 2)\&\& \sum_{l=1}^{4} i_l = 2 \\ (d)(i_1 = 1\|i_2 = 1\|i_3 = 1\|i_4 = 1)\&\& \sum_{l=1}^{4} i_l = 1 \end{cases} \tag{12}$$

The traditional dichotomy requires the existence of an interval of zeros corresponding to the value of the function for a different sign, and ultimately can only find a single zero point in the interval. For this problem, a segmented dichotomy is proposed. The basic idea of the algorithm is to divide the possible interval of zero points into a few segments, use the dichotomy for each segment, and then integrate the calculation results to obtain all the zero points in the interval. The algorithm flows as follows. The algorithm finds all the zeros in the interval quickly and accurately without needing to be given an initial value.

Algorithm: Segmented Dichotomy
Input: A function $f(u)$, the interval $[u_{\min}, u_{\max}]$, the number of segments t, the error err
Output: The roots array arr. If there is no root, return None.

1:    $gap \leftarrow (u_{\max} - u_{\min})/t$; $ide \leftarrow 1$
2: **FOR** $i \leftarrow 1$ **TO** t
3:        $u_1 = u_{\min} + (i-1) \cdot gap$; $u_2 = u_{\min} + i \cdot gap$
4:        **IF** $f(u_1) \cdot f(u_2) < 0$ **THEN**
5:            **WHILE** $u_2 - u_1 > err$ **DO**
6:                $mid \leftarrow (u_1 + u_2)/2$
7:                **IF** $f(mid) == 0$ **THEN**
8:                    $arr[ide] \leftarrow mid$; $ide \leftarrow ide + 1$
9:                **ELSE IF** $f(u_1) \cdot f(mid) < 0$ **THEN**
10:                  $u_2 \leftarrow mid$
11:                **ELSE IF** $f(u_2) \cdot f(mid) < 0$ **THEN**
12:                  $u_1 \leftarrow mid$
13:             **END IF**
14:        **END WHILE**
15:        $arr[ide] \leftarrow (u_1 + u_2)/2$; $ide \leftarrow ide + 1$
16:        **ELSE IF** $f(u_1) == 0$ **THEN**
17:            $arr[ide] \leftarrow u_1$; $ide \leftarrow ide + 1$
18:        **ELSE IF** $f(u_2) == 0$ **THEN**
19:            $arr[ide] \leftarrow u_2$; $ide \leftarrow ide + 1$
20:        **END IF**
21: **NEXT** i
22: **IF** arr is empty **THEN**
23:        **RETURN** None
24: **ELSE**
25:        **RETURN** arr
26: **END IF**

3.3 Coverage Window Calculation

From the intersection points the coverage window of the satellite to the ground point can be calculated. Equation (14) gives the formula for the window calculation, where $u_1^{(*)}$ and $u_2^{(*)}$ are the calculated intersection points ($u_1^{(*)} < u_2^{(*)}$) and $(*)$ represents the serial number of a line. All the coverage windows are combined to obtain the coverage window of the satellite to the ground point during the whole mission cycle.

$$\begin{cases} t_{start}^{(l)} = \left(u_1^{(l)} - u_{initial}\right) \cdot T/(2\pi) + (i-1) \cdot T \\ t_{end}^{(l)} = t_{satrt}^{(l)} + \left(u_2^{(l)} - u_1^{(l)}\right) \cdot T/(2\pi) \\ t = \underset{i}{\cup}\left\{\left[t_{start}^{(l)}, t_{end}^{(l)}\right]\right\} \end{cases} \tag{13}$$

When the satellite operates in its first cycle, it does not necessarily satisfy $u_1^{(1)} \geq u_{initial}$; similarly, and when the satellite operates in its last cycle, it does not necessarily satisfy $u_2^{(last)} \leq u_{last}$. u_{last} is the argument of latitude of the satellite at the end of the mission; when the satellite has less than one cycle of mission time, there are four cases shown in Fig. 6, and the two previous cases are also included, which lead to deviations in the window calculation, So the variables in Eq. (13) are corrected according to Eq. (14).

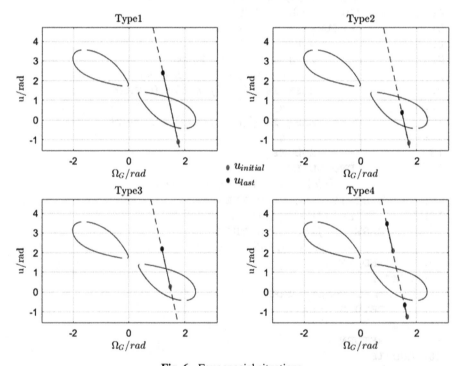

Fig. 6. Four special situations

$$\left(u_1^{(*)}, u_2^{(*)}\right) = \begin{cases} u_1^{(*)} = u_1^{(*)}, u_2^{(*)} = u_2^{(*)} & \left(u_1^{(*)} \geq u_{initial} \& u_2^{(*)} \leq u_{last}\right) \\ u_2^{(*)} = u_{last} & \left(u_1^{(*)} \geq u_{inital} \& u_2^{(*)} > u_{last}\right) \\ u_1^{(*)} = u_{initial} & \left(u_1^{(*)} < u_{initial} \& u_2^{(*)} \leq u_{last}\right) \\ \emptyset & (\text{else}) \end{cases} \tag{14}$$

The complete process of the window calculation is given below and the flow chart is shown in Fig. 7:

Step 1. Mesh the proposed analysis coverage area.

Step 2. Calculate the slope k and intercept of the satellite's straight-line trajectory in 2D map according to Sect. 3.1.

Step 3. Determine the intersection existence interval according to Eq. (5) and calculate the intersection using the segmented dichotomy method.

Step 4. Perform the intersection correction by Eq. (14) and obtain the single star to single point window according to Eq. (13).

Step 5. Iterate over the ground feature points and repeat **Step 3–Step 4** to obtain multiple windows.

Step 6. Integrate the windows to obtain a satellite coverage window for the whole area.

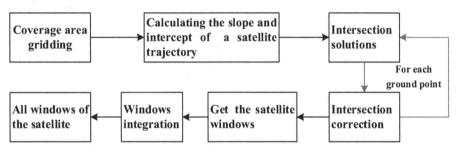

Fig. 7. General flow of window calculations

4   Simulation Results

This section performs simulation validation and evaluates the accuracy and computational efficiency of the computational results of this paper's methodology based on the results. All simulation results were calculated using Intel Core i7-9750H (CPU) on a DELL Inspiron 7590 notebook.

4.1   Assessment of the accuracy of the calculation results

Table 2 gives the TLE of a satellite chosen to first verify the validity of the segmented dichotomy. Given $d = 45°$, the satellite runs for 86640 s and the latitudes and longitudes of ground points are $\lambda = 100°$, $\varphi = 5°15°60°85°$. The results are shown in Fig. 8, The colored lines in the figure represent the trajectory of the satellite in the 2D map, the black curves represent the coverage area in the 2D map, and * is the intersection point solution result. As can be seen from the figure, all the intersection locations are at the junction of the two curves, proving the accuracy of the segmented dichotomy results.

Further, given the satellite sensor field of view $\theta = 140°$, Four ground points of the same longitude and different latitudes were selected from low to high latitudes,

Table 2. The TLE data

Row	The Two-Line Elements
01	1 44717U 19074E 22213.78591446 .00002323 00000-0 17476-3 0 9996
02	2 44717 53.0569 353.9322 0001515 76.0318 284.0840 15.06403855150130

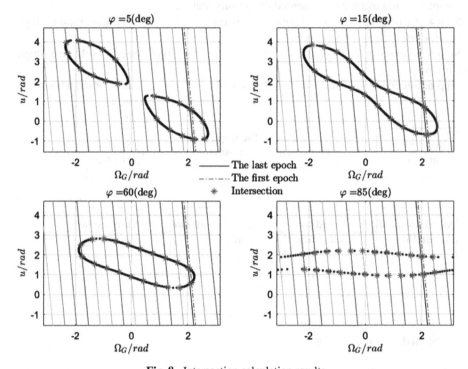

Fig. 8. Intersection calculation results

and numerical simulations were carried out using commercial software and this paper's method respectively. The commercial software also used the SGP4 orbit extrapolation model to calculate the satellite positions.

The results of the window calculation between the method in this paper and the commercial software are shown in Fig. 9 (In Sect. 4.2). As can be seen from the figure, the number of windows obtained by the method proposed in this paper and the commercial software are the same, and the length of windows are approximately equal; comparing the specific values, the window errors obtained by the two calculations are within 1.8 min, and most of the window errors are within 1 min. In such a short period of time, it is not possible to complete the re-planning of the observation mission and the adjustment of the satellite status, so the error is within the acceptance range. Therefore, the results of this method are consistent with the results of commercial software within the error tolerance. The above two simulation results prove that the window calculation method proposed in this paper has a high accuracy.

4.2 Computational Efficiency Evaluation

100 low-orbiting and near-circular observation satellites were selected for computational efficiency assessment: given the satellite sensor field of view $\theta = 140°$, the observation area was chosen as Harbin city, the area range was $[44°N, 46°N]$, $[125°E, 130°E]$, the step size of the fixed-step method was chosen as $30\,s$, $60\,s$, $90\,s$, and the window calculation was carried out using the fixed-step method and the method of this paper respectively, and the time used for the statistical calculation method.

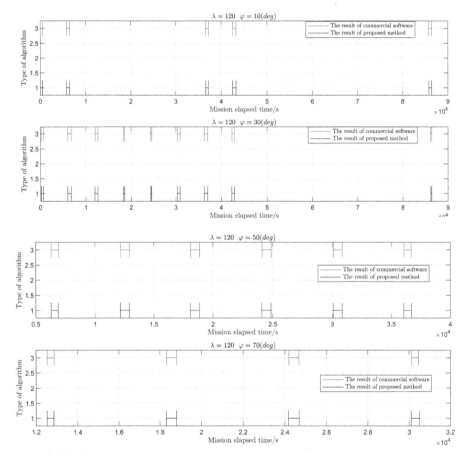

Fig. 9. Numerical simulation results of commercial software and proposed method

The calculation times are shown in Fig. 10. From the figure, it can be seen that the time consumed by the method in this paper is less than that of the fixed-step method. As the simulation time increases, the time taken by the fixed-step method increases exponentially; at the same time, as the step size decreases, the simulation time tends to increase significantly. Although the time consumed by this method increases as the simulation time increases, the increase is much smaller than that of the fixed-step method. The accuracy of the method is independent of the time step, so the time taken by the fixed-step method is much greater than that of the method in this paper, provided that the accuracy is the same; As can be seen from the data in Table 3, the time used by the window calculation method proposed in this paper is reduced by more than 62% on average compared to the fixed-step method, with the speed advantage of the method in this paper gradually increasing as the task time is extended.

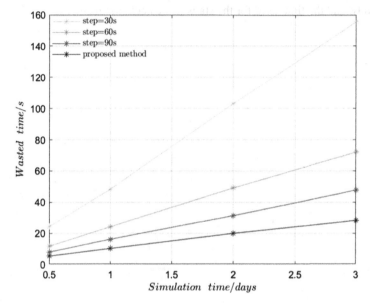

Fig. 10. Comparison of calculation times

Table 3. The results of fixed step method and proposed method

Region	Simulation Time	Fixed Step Method			Proposed Method	Average Reduction
	units: s	30	60	90	/	/
Harbin	43200	24.25	11.72	8.00	5.44	62.88%
	86400	48.30	24.11	16.14	10.24	65.31%
	172800	103.09	49.12	31.30	19.92	67.44%
	259200	155.58	72.15	47.79	28.31	69.17%

5 Conclusion

Based on two-dimensional map theory, this paper proposes a quick window calculation method for satellites based on the SGP4 orbit model. This method considers the satellite orbit perturbation compared to existing methods and is suitable for fast and accurate calculation of all satellites in near-circular orbits with a period of less than 225 min. We have performed numerical calculations with a single satellite covering multiple ground points, and the results show that the ground coverage time window error of this method is within 1.8 min, which is highly accurate compared to commercial software. In addition, we also used 100 observation satellites to observe Harbin city and counted the time taken to calculate the ground coverage time window, the results showed that this method takes much less time than the fixed-step method. As the mission time increases, the time difference between the two methods gradually increases. After quantitative calculations, the average time reduction of this method compared to the fixed-step method is more than 62%.

There are shortcomings in this paper, that is, the satellite trajectory in the 2D map is modelled using a linear model, which deviates from the actual satellite trajectory model in the 2D map, resulting in some deviation in the calculation of the satellite's time window for ground coverage. This can be improved in the future to improve the accuracy of the calculation. In addition, we also plan to extend the method to the calculation of ground coverage time windows for satellites in regression orbit to further extend the application of the method.

References

1. Alfano, S.: Rapid-determination of satellite visibility periods. In: 1st Annual Spaceflight Mechanics Meeting, pp. 1289–1304 (1991)
2. Alfano, S.: Rapid generation of site/satellite timetables. J. Spacecr. Rocket. **30**, 760–764 (1993)
3. Radzik, J.: A methodology for rapidly evaluating the performance of some low-earth-orbit satellite systems. IEEE J. Sel. Areas Commun. **13**, 301–309 (1995)
4. Ali, I.: Predicting the visibility of LEO sate. IEEE Trans. Aerosp. Electron. Syst. **35**, 1183–1190 (1999)
5. Tang, R.: TAIC algorithm for the visibility of the elliptical orbits' satellites. In: IEEE International Geoscience and Remote Sensing Symposium (IGARSS), pp. 781–785 (2007)
6. Ulybyshev, Y.: Satellite constellation design for complex coverage. J. Spacecr. Rocket. **45**, 843–849 (2008)
7. Ulybyshev, Y.: Geometric analysis and design method for discontinuous coverage satellite constellations. In: AIAA/AAS Astrodynamics Specialist Conference (2012)
8. Ulybyshev, Y.: Geometric analysis and design method for discontinuous coverage satellite constellations. J. Guid. Control. Dyn. **37**, 549–557 (2014)
9. Ulybyshev, Y.: A general analysis method for discontinuous coverage satellite constellations. In: AIAA/AAS Astrodynamics Specialist Conference (2014)
10. He, X.: Analytical solutions for earth discontinuous coverage of satellite constellation with repeating ground tracks. Chin. J. Aeronaut. **35**, 275–291 (2022)
11. Han, C.: Geometric analysis of ground-target coverage from a satellite by field-mapping method. J. Guid. Control. Dyn. **44**, 1469–1480 (2021)

12. Zhang, Y.: Geometric analysis of a constellation with a ground target. Acta Astronaut. **191**, 510–521 (2022)
13. Wang, X.: Onboard satellite visibility prediction using metamodeling-based framework. Aeros. Sci. Technol. **94**, 105377 (2019)
14. The TLE data. http://celestrak.org/NORAD/elements

A Simulation of Single and Segmented Excitation in Electrical Capacitance Tomography System

Nur Arina Hazwani Samsun Zaini[1], Herman Wahid[1]([✉]), Ruzairi Abdul Rahim[1],
Juliza Jamaludin[2], Mimi Faisyalini Ramli[3], Nasarudin Ahmad[1],
Ahmad Azahari Hamzah[4], and Farah Aina Jamal Mohamad[1]

[1] Faculty of Electrical Engineering, Universiti Teknologi Malaysia, 81310 UTM, Skudai, Johor,
Malaysia
herman@utm.my
[2] Faculty of Engineering and Built Environment, Universiti Sains Islam Malaysia, 71800 Nilai,
Negeri Sembilan, Malaysia
[3] Faculty of Technology Engineering, Universiti Tun Hussein Onn, Campus Pagoh, Panchor,
Johor, Malaysia
[4] Section of Chemical Engineering, TechnologyUniKL Micet, 78000Alor Gajah, Melaka,
Malaysia

Abstract. Soft field sensor is common with their forward problem which is the distribution of the sensitivity field throughout the target medium is not uniform. This paper concise the effect of single and segmented excitation in Electrical Capacitance Tomography for Gas-solid flow. The electrodes excitation for Protocol 1, Protocol 2, Protocol 3, and Protocol 4 were done to observe the potential value reading at the center of the pipe flow. The analysis on the potential value at the center of ECT sensor was done in COMSOL Multiphysics to determine the best excitation method. Results indicate that protocol 4 has 299.83% difference when compared to the single excitation in both homogenous and non-homogenous ECT system.

Keywords: Electrical Capacitance Tomography · Segmented excitation · Potential value

1 Introduction

The word "tomography" is used often to refer to all measuring and visualization techniques that use measurements taken at the item's edge to provide information about the spatial distribution of certain attributes within the object being examined [1]. Tomography is a potent imaging method that has revolutionized a number of industries, including non-destructive testing, geophysics, medicine, and geophysics [2]. Tomography uses a variety of imaging techniques, including electrical capacitance, electrical impedance, electrical resistance, and ultrasound tomography, and more [3]. By reassembling cross-sectional pictures from several projections, it makes it possible to see and characterize

F. Hassan et al. (Eds.): AsiaSim 2023, CCIS 1912, pp. 171–179, 2024.
https://doi.org/10.1007/978-981-99-7243-2_14

an object's interior structures and features. Tomographic imaging methods must be used to alter remote sensor data in process tomography in order to get precise quantitative data from limited areas [4].

Electrical Capacitance Tomography (ECT) is a soft-field sensor. The distribution of the sensitivity field throughout the target medium is not uniform. The region nearer the excitation has a higher sensitivity and a pair of detecting electrodes. On the other hand, the center area has significantly less sensitivity. Consequently, this soft-field effect's disadvantage will have an impact on the measurements made in the center region [5]. Soft-field sensor ECT offers low quality pictures, especially in the pipe's center [6, 7]. This paper compares the effect of single and segmented excitation in homogenous and non-homogenous systems. A 16-electrode ECT is used for the analysis in both homogenous and non-homogenous configuration. For single excitation, only one electrode was excited at one time. It is also known as Protocol 1. For segmented excitation, the simulation involved Protocol 2, Protocol 3, and protocol 4.

2 Principle of ECT

Due to its non-invasive nature, cheap cost, and instantaneous flow visualization, the ECT technique has been applied in several industrial applications where fast-moving multi-phase flows are frequently found [7]. Typically, the ECT system will include 8 to 12 electrodes [8]. One electrode at a time is stimulated to a predetermined voltage while the other electrodes are grounded in ECT. The electrical capacitances between the excited electrode and the grounded electrodes are then measured. This is also the measurement method applied in this study. Several electrodes should be placed around the area's edges in order to see the medium present in the pipes or vessel. The capacitance between all of the electrode combination pairs may then be calculated. Any acceptable reconstruction procedure can be used to create a picture from the data acquired [7, 9–11].

Figure 1 below shows the diagram of ECT system. The data acquisition system (DAS) and computer system are further components of the ECT system, as shown in Fig. 1. Capacitance variations are monitored by the DAS, which then transforms the measured capacitance into a digital signal and sends the information to a computer system. The computer system is used to both create the needed interpretation information, such as estimates of void fraction, and to regulate the capacitance measurement processes.

The independent measurement (M) for single electrode excitation is given by Eq. (1) [13]:

$$M = \frac{N(N-1)}{2} \tag{1}$$

where N denotes the total number of electrodes and M denotes the complete collection of capacitance measurements. N value for Protocol 1 is 16. The independent measurement (M) for Protocol 2, 3 and 4 can be calculated using Eq. (2) [14]:

$$M = \frac{N(N-(2P-1))}{2} \tag{2}$$

where P denotes the protocol number, which can be presented as the number of electrodes grouped together to be excited at the same time, field rotations for segmented excitation

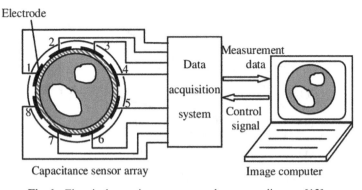

Fig. 1. Electrical capacitance tomography system diagram [12].

become finer than the single excitation approach with more independent measurement, which raises the resolution of the pictures [15].The N value for Protocol 2, 3, and 4 are 32, 48 and 64.

3 Methodology

To simulate the 16-channel ECT model, in COMSOL Multiphysics software, a two-dimensional (2-D) configuration for homogenous and non-homogenous ECT model has been constructed. In this simulation, the Finite Element Method (FEM) and the Electrostatics (es) module were used.

3.1 FEM Modelling and Setup

One homogeneous and one non-homogeneous ECT model were both used in the simulation. The parameters for this simulation were set. Figures 2 and 3 below show the diagram of the geometry model for both homogeneous and non-homogenous ECT. The initial values for the Electric Potential, $V = 0\ V$.The non-homogeneous model is made of air plus a ball of polymethylmethacrylate (PMMA), whereas the homogenous model is made up entirely of air. The PMMA material was used as the solid in the non-homogenous setting. A sinusoidal waveform function with the following parameters was utilized for this simulation: amplitude $= 1$, phase $= 0$, and angular frequency $= 2*pi*10e6$.

3.2 Excitation Method

Single Excitation. Single excitation also known as protocol 1, where only one electrode was excited at one time until one complete cycle. In this simulation, Electrode 1 (E1) was excited. For the single excitation, the configuration in both homogenous and non-homogenous ECT system for the excitation electrode are shown in Fig. 4. The independent measurement for single excitation is 120.

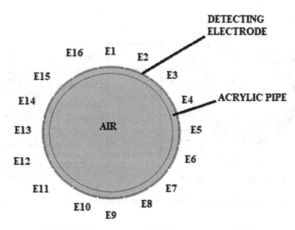

Fig. 2. Homogenous ECT geometry model.

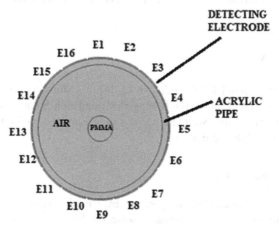

Fig. 3. Non-homogenous ECT geometry model.

Segmented Excitation. Segmented excitation is the excitation of more than one electrode at one time. In this simulation Protocol 2, Protocol 3, and Protocol 4 were used as the segmented excitation. For Protocol 2, E1 and E2 were excited and for Protocol 3, E1, E2, and E3 were excited. Lastly for Protocol 4, E1, E2, E3, and E4 were excited. The independent measurement for segmented excitation is different from single excitation. The value of M for Protocol 2, 3, and 4 are 464, 1032, and 1824; refer Eq. (1). Figures 5, 6, and 7 show the excitation electrodes configuration for Protocol 2, Protocol 3, and Protocol 4 in a non-homogenous ECT system.

E1

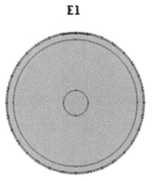

Fig. 4. Protocol 1 excitation electrode configuration.

E1 **E2**

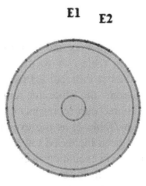

Fig. 5. Protocol 2 excitation electrode configuration

E1 **E2** **E3**

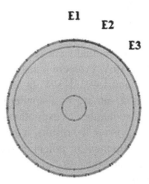

Fig. 6. Protocol 3 excitation electrode configuration.

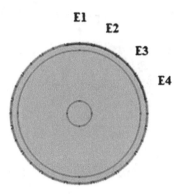

Fig. 7. Protocol 4 excitation electrode configuration.

4 Result and Discussion

The simulated result for both homogenous and non-homogenous ECT system were shown in Table 1 and Table 2 below. Table 1 shows the electric potential distribution for different excitation electrodes, which is Protocol 1, Protocol 2, Protocol 3, and Protocol 4. From the electric potential diagram, the white line represents the electric field inside the piping area, whereas the coloured part represents the electric potential distributed throughout the piping area. From Table 1, as the excitation electrode increased, the electric potential distribution has been increased towards the center area. The electric field line also increases as the number of excitation electrodes excites at one time increase. From the diagram, Protocol 1(E1) has the lowest electric potential distribution compared to others. While Protocol 4 (E1, E2, E3, E4) showed the intensified electric potential distribution towards the center of the piping region.

The collected data for potential value (V) were shown in Table 2. From the tabulated data in both homogenous and non-homogenous ECT system, the higher the number of electrode excited at one time, the higher the potential value located at the center of the pipeline. Protocol 1 has the lowest potential value reading in both system configurations with 3.5224E-10 V and 3.5163E-10 V. Protocol 4 has the highest potential value at the center region with 1.4086E-9 V and 1.4058E-9 V. In addition to an increasing number of electrodes, segmented excitation introduces a smaller rotational step of electric field compared to single excitation. The number of excitation electrodes excited at one time affects the potential value reading. From the results, it can be concluded that segmented excitation can intensify the measurement of potential value at the center region of an ECT system. Hence, it can overcome the soft-field sensor drawbacks and improve the image quality of an ECT system.

To quantify the effectiveness of segmented excitation compared to single excitation method, the percentage differences between protocol 2, 3, and 4 with respect to the Protocol 1 potential value reading in homogenous ECT system were recorded in Table 3. Protocol 4 with 299.83% gave the highest percentage difference compared to Protocol 2 and 3. Protocol 2 has the lowest percentage difference 99.89% compared to other protocols. The percentage differences between Protocol 2, 3, and 4 with respect to the Protocol 1 potential value reading in non-homogenous ECT system were recorded in

Table 1. Electric potential distribution for different excitation electrode.

Excitation electrode	Homogenous	Non-homogenous
E1		
E1, E2		
E1, E2, E3		
E1, E2, E3, E4		

Table 4. Protocol 4 with 299.83% gave the highest percentage difference compared to Protocol 2 and 3. In the non-homogenous ECT system, Protocol 2 also has the lowest percentage difference compared to others. This shows that when more electrode is excited

Table 2. Potential value reading for different excitation electrode.

Excitation electrode	Homogenous	Non-homogenous
E1	3.5224E-10 V	3.5163E-10 V
E1, E2	7.0409E-10 V	7.0292E-10 V
E1, E2, E3	1.0563E-9 V	1.0543E-9 V
E1, E2, E3, E4	1.4086E-9 V	1.4058E-9 V

Table 3. Percentage difference for homogenous ECT system.

Excitation electrode	Potential value (V)	Percentage difference (%)
Protocol 2	7.0409E-10	99.89
Protocol 3	1.0563E-9	199.83
Protocol 4	1.4086E-9	299.83

Table 4. Percentage difference for non-homogenous system.

Excitation electrode	Potential value (V)	Percentage difference (%)
Protocol 2	7.0292E-10	99.90
Protocol 3	1.0543E-9	199.86
Protocol 4	1.4058E-9	299.83

at one time, it can enhance the electrical potential value reading at the center region of piping system.

5 Conclusion

The homogenous and non-homogenous ECT system were simulated for single and segmented excitation method. ECT is a soft- field sensor and the electrical field weakens as it moves towards the pipe's center due to the uneven distribution of electrical fields there. Segmented excitation functions as an approach for the issue that modifies current system as minimally as possible. From the simulated results, it shows that Protocol 4 in both homogenous and non-homogenous system have the highest potential value 1.4086E-9 V and 1.4058E-9 V. On the other hand, Protocol 4 has the highest percentage difference compared to Protocol 2 and 3. It has 299.83% difference when compared to the single excitation in both homogenous and non-homogenous ECT system. Segmented excitation enhanced the potential value in both homogenous and non-homogenous systems. This proves that segmented excitation can improve the reading at the center area. Hence, it also can improve the image quality of the ECT system.

Acknowledgement. The authors of this paper would like to express their gratitude to Universiti Teknologi Malaysia for funding the work through the UTMFR Research Fund with vote number 22H59.

References

1. Tapp, H.S., Peyton, A.J., Kemsley, E.K., Wilson, R.H.: Chemical engineering applications of electrical process tomography. Sens. Actuators, B Chem. **92**, 17–24 (2003)
2. Beck, M., Williams, R.: Process tomography: a European innovation and its applications. Meas. Sci. Technol. **7**, 215 (1999)
3. Rymarczyk, T., Sikora, J.: Applying industrial tomography to control and optimization flow systems. Open Phys. **16**, 332–345 (2018)
4. Williams, R.A., Beck, M.S.: Process Tomography. Elsevier (1995)
5. Ameran, H.L.M., Hamzah, A.A., Rahim, R.A., et al.: Sensitivity mapping for electrical tomography using finite element method. Int. J. Integr. Eng. **10**, 64–67 (2018)
6. Wang, A., Marashdeh, Q., Teixeira, F., Fan, L.-S.: Electrical capacitance volume tomography: a comparison between 12- and 24-channels sensor systems. Prog. Electromagn. Res. M. **41**, 73–84 (2015)
7. Cui, Z., Xia, Z., Wang, H.: Electrical capacitance tomography sensor using internal electrodes. IEEE Sens. J. **20**, 3207–3216 (2020)
8. Abd Rahman, N.A., Abdul Rahim, R., Mohd Nawi, A., et al.: A review on electrical capacitance tomography sensor development. J. Teknol. **73** (2015)
9. Jamaludin, J., Zawahir, Z., Rahim, R.A., et al.: A review of tomography system. Jurnal Teknologi **64**, 2180–3722 (2013)
10. Rahim, R.A.: Electrical Capacitance Tomography Principal, Techniques and Applications. Penerbit UTM Press, Johor Bahru (2011)
11. Zainal-Mokhtar, K., Mohamad-Saleh, J.: An oil fraction neural sensor developed using electrical capacitance tomography sensor data. Sensors (Switzerland) **13**, 11385–11406 (2013)
12. Wu, X., Huang, G., Wang, J., Xu, C.: Image reconstruction method of electrical capacitance tomography based on compressed sensing principle. Meas. Sci. Technol. **24**, 075401 (2013)
13. Deabes, W., Sheta, A., Bouazza, K.E., Abdelrahman, M.: Application of electrical capacitance tomography for imaging conductive materials in industrial processes. J. Sensors **2019**, 1–22 (2019)
14. Din, S.M., Ling, L.P., Rahim, R.A., et al.: Electrical potential and electrical field distribution of square electrical capacitance tomography. J. Teknol. **77**, 159–163 (2015)
15. Din, S.M., Nam, R., Azmi, A., et al.: Comparison of single and segmented excitation of electrical capacitance tomography. In: 2015 10th Asian Control Conf Emerg Control Tech a Sustain World, ASCC (2015)

Robust Input Shaping for Swing Control of an Overhead Crane

Amalin Aisya Mohd Awi[1,2], Siti Sarah Nabila Nor Zawawi[1,2], Liyana Ramli[1,2(✉)], and Izzuddin M. Lazim[1,2]

[1] Universiti Sains Islam Malaysia, 71800 Nilai, Malaysia
lyanaramli@usim.edu.my

[2] Department of Electrical and Electronic Engineering, Faculty of Engineering and Built Environment, 71800 Nilai, Negeri Sembilan, Malaysia

Abstract. Underactuated robotic and mechatronic systems have been employed in many practical applications for a long time. It is crucial to increase crane efficiency for practical applications; yet the primary factor limiting crane efficiency is the payload swing driven on by inertia or outside disturbances. The swing of the crane's payload mass, which moves like a pendulum, has created numerous challenges since it can collide with the operator and result in accidents. This paper presents the simulation implementation of an open-loop input-shaper controller to control the swing angle of an overhead crane. A mathematical model of the two-dimensional overhead crane and input shaper controller was constructed. The model of the overhead crane and the input shaper was created in MATLAB/Simulink and the simulation was executed. This paper evaluated the performance and robustness of input shaping techniques with constant cable length using the zero vibration (ZV), zero vibration derivative (ZVD), zero vibration derivative-derivative (ZVDD), and zero vibration derivative-derivative-derivative (ZVDDD). The payload mass varied in two cases which are 1 kg and 0.3 kg. Based on the simulation results, ZVDDD controller showed the highest reductions in the overall and residual payload swing with 91% for both cases. It is envisaged that the proposed method can be used for improving the robustness of input shapers for payload swing suppression of an overhead crane.

Keywords: ZVDDD · ZV · Overhead Crane · Input Shaper · Payload Swing

1 Introduction

One type of crane that is frequently used in industrial settings is the overhead crane. The two parallel runways of an overhead crane are joined by a moving bridge. The hoist, or lifting part, of a crane moves across the bridge. Overhead crane was used widely at the industry and construction site. Conventional crane was operated by the human operator and is still used up until now[1, 2]. Due to its flexible nature, it is susceptible to excessive swing if the crane's human operator is inexperienced in operating it or if unanticipated external disturbances occur. Besides, payload hoisting which is a fundamental crane operation, might cause higher payload swing during the operation [3]. As a result, it's

© The Author(s), under exclusive license to Springer Nature Singapore Pte Ltd. 2024
F. Hassan et al. (Eds.): AsiaSim 2023, CCIS 1912, pp. 180–187, 2024.
https://doi.org/10.1007/978-981-99-7243-2_15

essential to develop robust controllers that can remove the unwanted payload swing caused by these variables to accomplish efficient operation. There are two ways to reduce the swing angle by utilizing open-loop and closed loop system. For closed-loop system, variety of controller had been used such as PID [4, 5], sliding mode control [6–9], heuristic optimization [10, 11], nonlinear feedback control [12, 13] and observer-based control [14]. However, those controllers required complex system dynamic model.

Input shaping is one of the simple methods that may be used to reduce the payload swing in a flexible system [15]. The input shaping method's fundamental concept is to combine the baseline command with a series of Dirac impulses. The impulses should be delivered at predetermined times and with predetermined amplitudes [16]. A wide range of control strategies had been used to dampen the payload swing by using the input shaper controller. Distributed zero vibration were proposed to reduce the payload swing [17–19]. However, the payload swing was not significantly suppressed by the controller despite the design and testing for crane system in open-loop and closed-loop input shaping schemes. Besides, the ZV scheme's robustness characteristics and computational efficiency are carried over into the adjustable ZV shaper, which improves the shaper's design flexibility and operational domain [20]. However, this system's configuration makes it difficult to control because it requires significant input variation. Next, the input shaper also had been tune by optimization method. It can reduce the payload swing but it cannot increase the robustness of the system [10, 21, 22]. The vibration generated on by both operator commands and external disturbances can be lessened using a sensor-less combination of input shaping and radial spring-damper [22]. However, the difference is low between the hybrid method and non-hybrid method. Time optimal control also been used with input shaper for gantry crane, but the maneuver time increase with the robustness of the system [23]. Same goes with input shaper and H∞ region in [24].

The study was carried out in order to achieve an efficient crane control system. The efficient crane control system consists of minimum or without swings to ensure the safety of the workers at the working site and to have a safe crane operation without any workers injured or getting hurt. Moreover, the efficient crane control system can increase production volume as the faster the loaded material process, the more materials can be loaded in a period of time and can help minimize time operation. The limitation of the ZVDDD input shaper is that it takes a few seconds longer compared to ZV, ZVD and ZVDD to reach a steady state. ZVDDD have a longer response time compared to another type of input shaper. In this paper, an input shaper is used to regulate the swing of an overhead crane.

2 Model Description

2.1 Dynamic Model for Overhead Crane

The overhead crane modelling was constructed and then simulated it in SIMULINK.

A non-linear system, the two-dimensional overhead crane. The equations used in the controllers are often linearized and simplified (Table 1).

Although the real crane's swing reaction will have some residual swing, an input-shaping method may completely eliminate residual swing in a simple second-order crane model. In this work, a non-linear model is built and utilized for simulation.

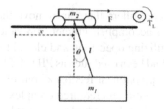

Fig. 1. Schematic diagram of overhead crane model [25].

Table 1. System parameters for the two-dimensional overhead crane.

Variables	Values
Mass of payload, m_1	1 kg
Mass of trolley, m_2	5 kg
Cable length, l	0.5 m
Gravitational constant, g	9.8 m/s
Damping coefficient, B	12.32 Ns/m

The mathematical equation for modeling the system is derived from the Lagrange equation [25, 26]. There are two separate generalized coordinates of the gantry crane system such as the displacement of the trolley given as x and payload swing angle as θ. The nonlinear model of the gantry crane system is modelled based on the schematic diagram shown in Fig. 1. The equation of motion of the overhead crane model is:

$$F = (m_1 + m_2)\ddot{x} + m_1 l\ddot{\theta}\cos\theta - m_1 l\dot{\theta}^2\sin\theta + B\dot{x} \tag{1}$$

$$l\ddot{\theta} + \ddot{x}\cos\theta + g\sin\theta = 0 \tag{2}$$

2.2   Input Shaping Controller

Input shaping is a feed-forward control approach that involves using an input shaper to filter the required command [27]. A general input shaper's mathematical description is as follows:

$$IS(t) = \sum_{j=1}^{m} A_j\delta(t - t_j) \tag{3}$$

where $\delta(t)$ represents the Dirac delta function, t_j is time of a jth pulse and a non-negative value and A_j is the amplitude of jth pulse and a non-zero value. The system will be driven by the shaped input that arises from the convolution, and the shaped command will lessen the oscillatory system's negative consequences. A superposition of second-order systems, each with a transfer function, can be used to simulate an oscillating system [19].

Zero Vibration

Below shows the transfer function

$$G(s) = \frac{\omega_n^2}{s^2 + 2\xi\omega_n s + \omega_n^2} \tag{4}$$

where ω_n is the natural frequency and ξ is the damping ratio of the system. Thus, the impulse response of a single mode system at time t is:

$$y(t) = \frac{A\omega_n}{\sqrt{1-\xi^2}} e^{-\xi\omega_n(t-t_0)} \sin(\omega_n\sqrt{1-\xi^2}(t-t_0)) \tag{5}$$

where A and t_0 are the amplitude and time of the impulse, respectively. The superposition concept may also be used to determine the response to a series of impulses. For M impulses, with $\omega_d = \omega_n\sqrt{1-\xi^2}$, the impulse response can be expressed as

$$y(t) = M \sin(\omega_d t + \alpha 1) \tag{6}$$

where,

$$M = \sqrt{(\sum_{j=1}^{m} B_j \cos \emptyset_j)^2 + (\sum_{j=1}^{m} B_j \sin \emptyset_j)^2}$$

$$B_j = \frac{A\omega_n}{\sqrt{1-\xi^2}} e^{-\xi\omega_n(t-t_j)}$$

$$\emptyset_j = \omega_d t_j$$

$$\alpha 1 = \tan^{-1}(\sum_{j=1}^{m} \frac{B_j \cos \emptyset_j}{B_j \sin \emptyset_j})$$

A_j and t_j are the magnitudes and times at when the impulses occur. At the moment of the final impulse, t_m, the residual single mode vibration amplitude of the impulse response is determined. It is determined as,

$$V = \sqrt{V_1^2 + V_2^2} \tag{7}$$

where,

$$V_1 = \sum_{j=1}^{m} \frac{A\omega_n}{\sqrt{1-\xi^2}} e^{-\xi\omega_n(t-t_j)} \cos(\omega_d t_j)$$

$$V_2 = \sum_{j=1}^{m} \frac{A\omega_n}{\sqrt{1-\xi^2}} e^{-\xi\omega_n(t-t_j)} \sin(\omega_d t_j)$$

In order to attain ZV following the final impulse, both V_1 and V_2 must be independently zero. Furthermore, the total of the amplitudes of the impulses must be unity to ensure that the shaped command input causes the same rigid body motion as the unshaped command. To avoid response delay, the first impulse is selected at time $t_1 = 0$. Hence, by setting V_1 and V_2 to zero, $\sum_{j=1}^{m} A_j = 1$ and produce a two-impulse sequence with parameters as (Fig. 2)

$$t_1 = 0, t_2 = \frac{\pi}{\omega_d}, A_1 = \frac{1}{1+K}, A_2 = \frac{K}{1+K} \tag{8}$$

where,

$$K = e^{-\frac{\xi\pi}{\sqrt{1-\xi^2}}}$$

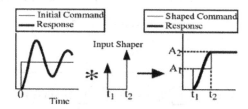

Fig. 2. ZV input shaping process.

The other robust input shapers of ZVD, ZVDD and ZVDDD were designed based on [28, 29].

3   Simulation Result

The simulations were conducted in two cases which are payload mass with 0.3 kg and 1 kg. For the graph, ZV and ZVDDD were analyzed to see the efficiency and reduction of the swing for both type of the input shapers.

3.1   Payload Swing Response

Figure 3 shows the graph of the swing of the payload mass of 1 kg and 0.3 kg using the ZV shaper. Both Figures show that the oscillation is better than the graph of the swing of the payload mass without input shaper as its highest peak of the oscillation graph is about 0.55° for both payload masses. Figure 4 shows the graph of the swing of the payload masses of 1 kg and 0.3 kg using the ZVDDD shaper. The ZVDDD shaper was shown to have a better swing as compared to the ZVDD as the maximum swing angle using ZVDDD recorded 0.29° for case 1. For case 2, the maximum swing angle recorded roughly 0.27°. It also can be seen that the graph in Fig. 4 (a) is smooth and has no oscillation at about 6.5 s which means that the payload has completely stopped swinging. The graph obtained has no oscillation as the crane system is implemented with the third order of zero vibration. Additionally, the smoothness and lack of oscillation of the line in Fig. 4 (b) at roughly 7.6 s indicate that the payload mass is not swinging. Since the crane system is constructed with the third order of zero vibration, the obtained graph is oscillation-free. The percentages of the swing reduction for both payload masses utilizing ZV, ZVD, ZVDD and ZVDDD are shown in Table 2.

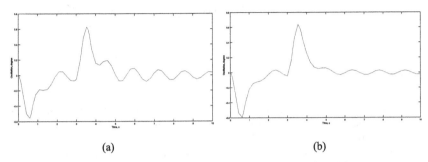

(a)   (b)

Fig. 3. Payload swing response with ZV (a) 1 kg (b) 0.3 kg

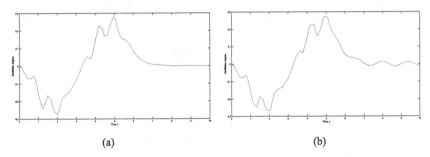

(a)   (b)

Fig. 4. Payload swing response with ZVDDD (a) 1 kg (b) 0.3 kg

Table 2. The percentage of the reduction payload swing.

Percentage of the reduction, %	without shaper	ZV	ZVD	ZVDD	ZVDDD
1 kg	0	79.13	86.78	91.05	91.1
0.3 kg	0	84.14	90.73	89.22	91.06

4 Conclusion

In conclusion, to minimize the payload swing of the overhead crane, the robust input shapers were implemented. The input shapers were designed in four types which are zero vibration (ZV), zero vibration derivative (ZVD), zero vibration derivative-derivative (ZVDD), and zero vibration derivative-derivative-derivative (ZVDDD). From the result obtained, it is shown that the input shaper controller is capable in controlling the overhead crane to achieve a minimum payload swing or small oscillation. Based on the simulation results, the zero-vibration derivative-derivative-derivative (ZVDDD) input shaper shows that it achieved the highest reductions in the overall and residual payload swing responses.

Acknowledgement. The authors gratefully acknowledge the Malaysian Education Ministry for the financial support it provided via the Fundamental Grant Research Scheme (FRGS/1/2020/TK0/USIM/02/2).

References

1. Ramli, L., Mohamed, Z., Abdullahi, A.M., Jaafar, H.I., Lazim, I.M.: Control strategies for crane systems: A comprehensive review. Mech. Syst. Signal Process. **95**, 1–23 (2017). https://doi.org/10.1016/j.ymssp.2017.03.015

2. Mojallizadeh, M.R., Brogliato, B., Prieur, C.: Modeling and control of overhead cranes: A tutorial overview and perspectives. Annu. Rev. Control March, 100877 (2023). https://doi.org/10.1016/j.arcontrol.2023.03.002

3. Ramli, L., Mohamed, Z., Efe, M., Lazim, I.M., Jaafar, H.I.: Efficient swing control of an overhead crane with simultaneous payload hoisting and external disturbances. Mech. Syst. Signal Process. **135**(1), 106326 (2020). https://doi.org/10.1016/j.ymssp.2019.106326

4. Zheng, F., Yang, C.H., Hao, G., Wang, K.C., Hong, H.L.: Vision-based fuzzy proportional–integral–derivative tracking control scheme for gantry crane system. Sensors Mater. **33**(9), 3333–3344 (2021). https://doi.org/10.18494/SAM.2021.3403

5. Azmi, N.I.M., Yahya, N.M., Fu, H.J., Yusoff, W.A.W.: Optimization of the PID-PD parameters of the overhead crane control system by using PSO algorithm. MATEC Web Conf. **255**, 04001 (2019). https://doi.org/10.1051/matecconf/201925504001

6. Wang, T., Tan, N., Zhou, C., Zhang, C., Zhi, Y.: A novel anti-swing positioning controller for two dimensional bridge crane via dynamic sliding mode variable structure. Procedia Comput. Sci. **131**, 626–632 (2018). https://doi.org/10.1016/j.procs.2018.04.305

7. Wu, X., Xu, K., Lei, M., He, X.: Disturbance-compensation-based continuous sliding mode control for overhead cranes with disturbances. IEEE Trans. Autom. Sci. Eng. **17**(4), 2182–2189 (2020). https://doi.org/10.1109/TASE.2020.3015870

8. Wu, Y., Sun, N., Chen, H., Fang, Y.: Adaptive output feedback control for 5-DOF varying-cable-length tower cranes with cargo mass estimation. IEEE Trans. Ind. Informatics **17**(4), 2453–2464 (2021). https://doi.org/10.1109/TII.2020.3006179

9. Wu, Q., Wang, X., Hua, L., Xia, M.: Modeling and nonlinear sliding mode controls of double pendulum cranes considering distributed mass beams, varying roped length and external disturbances. Mech. Syst. Signal Process. **158**, 107756 (2021). https://doi.org/10.1016/j.ymssp.2021.107756

10. Mohammed, A., Alghanim, K., Andani, M.T.: An optimized non-linear input shaper for payload oscillation suppression of crane point-to-point maneuvers. Int. J. Dyn. Control **7**(2), 567–576 (2019). https://doi.org/10.1007/s40435-019-00536-7

11. Liu, H., Cheng, W.: Using the bezier curve and particle swarm optimization in trajectory planning for overhead cranes to suppress the payloads' residual swing. Math. Probl. Eng. **2018** (2018). https://doi.org/10.1155/2018/3129067

12. Wu, X., Xu, K., He, X.: Disturbance-observer-based nonlinear control for overhead cranes subject to uncertain disturbances. Mech. Syst. Signal Process. **139**, 106631 (2020). https://doi.org/10.1016/j.ymssp.2020.106631

13. Mary, A.H., Miry, A.H., Kara, T., Miry, M.H.: Nonlinear state feedback controller combined with RBF for nonlinear underactuated overhead crane system. J. Eng. Res. **9**(3), 197–208 (2021). https://doi.org/10.36909/jer.v9i3A.9159

14. Miranda-Colorado, R.: Robust observer-based anti-swing control of 2D-crane systems with load hoisting-lowering. Nonlinear Dyn. **104**(4), 3581–3596 (2021). https://doi.org/10.1007/s11071-021-06443-x

15. Conker, C., Yavuz, H., Bilgic, H.H.: 2097 A review of command shaping techniques for elimination of residual vibrations in flexible-joint manipulators, pp. 2947–2958 (2016)

16. Gniadek, M., Brock, S.: Basic algorithms of input shaping autotuning. MM Sci. J. **2015**(1), 627–630 (2015 October). https://doi.org/10.17973/MMSJ.2015_10_201509

17. Maghsoudi, M.J., Mohamed, Z., Sudin, S., Buyamin, S., Jaafar, H.I., Ahmad, S.M.: An improved input shaping design for an efficient sway control of a nonlinear 3D overhead crane with friction. Mech. Syst. Signal Process. **92**, 364–378 (2017). https://doi.org/10.1016/j.ymssp.2017.01.036

18. Maghsoudi, M.J., Nacer, H., Tokhi, M.O., Mohamed, Z.: A Novel Approach in S-Shaped Input Design for Higher Vibration Reduction (2018)

19. Iplikci, S., Maghsoudi, M.J., Mohamed, Z., Tokhi, M.O., Husain, A.R., Abidin, M.S.Z.: Control of a gantry crane using input-shaping schemes with distributed delay. Trans. Inst. Meas. Control. **39**(3), 361–370 (2017). https://doi.org/10.1177/0142331215607615

20. Mohammed, A., Alghanim, K., Taheri Andani, M.: An adjustable zero vibration input shaping control scheme for overhead crane systems. Shock Vib. **2020** (2020). https://doi.org/10.1155/2020/7879839

21. Jaafar, M.S.H.I., Mohame, Z., Ahmad, M.A., Wahab, N.A., Ramli, L.: Control of an underactuated double-pendulum overhead crane using model reference command shaping: Design. Simulation and Experiment **151**, 1–18 (2020). https://doi.org/10.1016/j.ymssp.2020.107358

22. La, V.D., Nguyen, K.T.: Combination of input shaping and radial spring-damper to reduce tridirectional vibration of crane payload. Mech. Syst. Signal Process. **116**, 310–321 (2019). https://doi.org/10.1016/j.ymssp.2018.06.056

23. Stein, A., Singh, T.: Minimum time control of a gantry crane system with rate constraints. Mech. Syst. Signal Process **190**(August 2022), 110120 (2023). https://doi.org/10.1016/j.ymssp.2023.110120

24. Goubej, M., Schlegel, M., Vyhlídal, T.: Robust controller design for feedback architectures with signal shapers. IFAC-PapersOnLine **53**(2), 8650–8655 (2020). https://doi.org/10.1016/j.ifacol.2020.12.461

25. Solihin, M.I., Wahyudi, Kamal, M.A.S., Legowo, A.: Optimal PID controller tuning of automatic gantry crane using PSO algorithm. In: Proceeding 5th Int. Symp. Mechatronics its Appl. ISMA 2008, pp. 25–29 (2008). https://doi.org/10.1109/ISMA.2008.4648804

26. Ramli, L., Mohamed, Z., Jaafar, H.I.: A neural network-based input shaping for swing suppression of an overhead crane under payload hoisting and mass variations. Mech. Syst. Signal Process. **107**(March), 484–501 (2018). https://doi.org/10.1016/j.ymssp.2018.01.029

27. Piedrafita, R., Comín, D., Beltrán, J.R.: Simulink® implementation and industrial test of Input Shaping techniques. Control. Eng. Pract. **79**(June), 1–21 (2018). https://doi.org/10.1016/j.conengprac.2018.06.021

28. Alhassan, A., Mohamed, Z., Abdullahi, A.M., Bature, A.A., Haruna, A., Tahir, N.M.: Input shaping techniques for sway control of a rotary crane system. J. Teknol. **80**(1), 61–69 (2018). https://doi.org/10.11113/jt.v80.10297

29. Richiedei, D.: Adaptive shaper-based filters for fast dynamic filtering of load cell measurements. Mech. Syst. Signal Process. **167**(PA), 108541 (2022). https://doi.org/10.1016/j.ymssp.2021.108541

Sway Control of a Tower Crane with Varying Cable Length Using Input Shaping

Haszuraidah Ishak[1,2](✉), Zaharuddin Mohamed[1], S. M. Fasih[3], M. M. Bello[1], Reza Ezuan[1], and B. W. Adebayo[1]

[1] Faculty of Electrical Engineering, Universiti Teknologi Malaysia, 81310 Johor Bahru, Johor, Malaysia

haszuraidah@ump.edu.my

[2] Faculty of Electrical and Electronics Engineering Technology, Universiti Malaysia Pahang Al-Sultan Abdullah, 26600 Pekan, Pahang, Malaysia

[3] Department of Electronic Engineering, Faculty of Engineering, The Islamia University of Bahawalpur, Bahawalpur, Punjab, Pakistan

Abstract. This paper presents the development of input shaping techniques for effective sway control of a tower crane, considering varying cable lengths. The nonlinear dynamic model of the tower crane is derived using the Lagrange equation, and simulations confirm its close agreement with experimental results. To design and assess the input shaping control schemes, an unshaped square-pulse input is initially employed to determine the system's characteristic parameters. Zero Vibration (ZV) and Zero Vibration Derivative (ZVD) shapers are specifically devised based on the tower crane's inherent properties, and their performance is examined under various operating conditions, encompassing payload hoisting scenarios. The effectiveness of these shapers is evaluated through criteria such as maximum oscillation and reduction of mean squared error (MSE) for overall oscillations. The outcomes of this study demonstrate that implementing input shapers can offer practical and advantageous solutions for enhancing tower crane performance and ensuring smoother operation.

Keywords: Tower Crane · Input Shaping · Sway Control

1 Introduction

Cranes are typically utilized to transport loads from one point to another as quickly as possible, without causing any swaying motion in the load during the operation. However, during the operation, the load may experience a pendulum-like motion and the sway may exceed a safe limit. Furthermore, under-actuated systems like cranes possess a lower number of actuators than their degree of freedom, making it more challenging to control their complex dynamics and nonlinear behaviors [1].

In contrast to gantry and overhead cranes, a tower crane operation involves trolley translational and jib rotational motions. The rotational motion results in increased difficulty in control of the crane. in addition, the tower crane exhibits intricate behavior when

F. Hassan et al. (Eds.): AsiaSim 2023, CCIS 1912, pp. 188–200, 2024.
https://doi.org/10.1007/978-981-99-7243-2_16

performing the hoisting up and down operations, where variations in cable length during payload hoisting result in changes in the natural frequency and damping ratio of the payload sway. These factors, along with other issues such as employing different payload masses, can lead to unforeseen payload movement including excessive swaying, which can adversely impact the accuracy of the payload's placement and the stability of the crane [2]. Therefore, the main objective of tower crane system control is to accomplish precise trolley positioning while minimizing payload oscillation angles, to enable the load to be transported to its destination in the shortest time possible with low sway [3].

Several methods for managing crane systems, utilizing both feed-forward and feedback control methods, have been suggested. In feedback control technique, the Proportional-Integral-Derivative (PID) is widely used to control crane systems. Its purpose is to manage the payload swing angle and compensate for external disturbances. In the work of [4], they designed separate PID controllers for the trolley and jib of a tower crane to achieve smooth movement and minimize payload oscillation. The parameters of the PID controllers were optimized using Genetic Algorithm (GA). On the other hand, [5] demonstrated that full state feedback Linear Quadratic Regulator (LQR) effectively and quickly positioned the jib and trolley while minimizing payload swing caused by external wind disturbance. In [6], the same authors improved the system by adding a compensator actuator at the joint between the trolley and payload cable. To minimize disturbances caused by unknown dynamics, they utilized LQR-Disturbance Rejection Observer (DRO) control with the Luenberger-based Extended State Observer. However, linear controllers have limitations since they require a linearized model of the crane system, which does not provide an accurate assessment of the system dynamics, leading to the degraded performance of linear controllers.

Numerous studies have also explored the implementation of fuzzy logic controller (FLC) on crane systems. One of the advantages of FLC is that it does not require an accurate model of the system, which is especially useful for systems with complex architectures and nonlinearities like tower cranes. Several variations of FLC design for tower crane systems have been proposed [7–9]. Nevertheless, it should be noted that this approach necessitates a large set of rules, leading to substantial computational requirements.

A different method for managing the crane system involves the utilization of feed-forward control techniques. These techniques are primarily designed to minimize swaying and involve creating control inputs based on an understanding of the system's physical properties and tendency to sway. By considering these factors, the aim is to decrease the swaying of the system at its response modes. Input shaping is a type of feed-forward control technique that is suitable for real-time application. This technique is a commonly utilized approach to reduce vibration or oscillation of under-actuated systems, such as crane systems. The ZV and ZVD techniques are commonly used input shaping techniques in crane systems. Input shaping has also been utilized for a tower crane, but due to its nonlinear behavior and coupling between different motions, most research only considered a maximum of two simultaneous motions as demonstrated in works by [10, 11].

The command smoothing technique, similar to input shaping, is another feed-forward method that considers the natural frequency and damping ratio of the system and has been applied to tower crane system [12–15]. In a study conducted by [12], the effectiveness of the S-curve command smoothing technique was compared with the ZV and Extra Sensitive (EI) shapers in suppressing oscillations in a portable tower crane. The experimental results indicated that the input shaping technique not only suppressed oscillations more effectively but was also more robust against modeling errors.

This paper presents the development of input shaping control schemes for sway control of a tower crane. A nonlinear tower crane system is considered, and the dynamic model of the system is derived using the Lagrange approach. To design and evaluate input shaping control techniques for sway control in a tower crane system, an unshaped square pulse input is initially employed to determine the system's characteristic parameters. Subsequently, input shapers are designed based on the system's properties. The results depict the response of the tower crane system to the shaped inputs. The performance of the input shapers is evaluated in the experimental environment in terms of reducing swing angle, and its robustness. Finally, a comparative assessment of the control techniques is presented.

2 Dynamic Modeling of a Tower Crane

Different modeling approaches have been employed in various studies. However, the Lagrangian method has been predominantly utilized for crane modeling and controller design. This method involves determining the system's kinematic and potential energies and solving the Lagrange equation to obtain the mathematical equations that represent the system [3]. Numerous mathematical models for different cranes have been developed by researchers using the Lagrangian method including tower cranes [16–18]. This paper presents a concise formulation for modeling a tower crane using the Lagrangian method.

The tower crane is composed of a tall vertical mast and a long jib that is attached to it. Additionally, there is a trolley that moves along the jib and connects to a payload via an inextensible cable. The jib can perform a slew motion along the mast, while the trolley can traverse along the length of the jib. By combining the radial movement of the jib with the tangential movement of the trolley, the crane operator can position a payload in three-dimensional space within the crane's parameter. Figure 1 illustrates the 3-D model of the tower crane. The crane's motion causes significant oscillations of the payload in a circular motion. These pendulations are characterized by two angles: the in-plane angle, ϕ, and the out-of-plane angle, θ. The angle ϕ is measured along the z-axis in the xz-plane, while θ represents the angle between the cable and the xz-plane.

The crane consists of four independent generalized coordinates namely the trolley position, x, the jib angular position, γ, ϕ, and θ. M, m and L are the mass of the trolley, the mass of the payload and the massless cable length between the trolley and payload, respectively. From Fig. 1, the payload, $\vec{r_L}$ and the trolley, $\vec{r_T}$ position vectors can be written as:

$$\vec{r_L} = [x - L\cos\theta\sin\varnothing]\hat{i} + [L\sin\theta]\hat{j} - [L\cos\theta\cos\varnothing]\hat{k} \tag{1}$$

$$\vec{r_T} = [x]\hat{i} \tag{2}$$

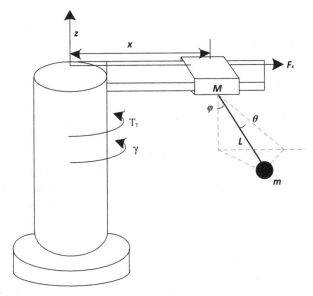

Fig. 1. Schematic diagram of a laboratory tower crane.

The velocities of the trolley and the load can be calculated using

$$\dot{r} = \frac{\partial \vec{r}}{\partial t} + \vec{\omega} \times \vec{r} \tag{3}$$

where

$$\vec{\omega} = [\dot{\gamma}]\hat{k} \tag{4}$$

and $\dot{\gamma}$ is the angular velocity of the jib. The kinetic energy, K_E and potential energy, P_E are given by:

$$K_E = \frac{1}{2}M\vec{\dot{r_T}} \cdot \vec{\dot{r_T}} + \frac{1}{2}m\vec{\dot{r_L}} \cdot \vec{\dot{r_L}} \tag{5}$$

$$P_E = -mgL\cos\theta\cos\varnothing \tag{6}$$

where g is the gravitational force acting on the payload. The generalized forces corresponding to the generalized displacement vector are described by:

$$\vec{F} = \{F_x, 0, T_\gamma, 0\} \tag{7}$$

$$\vec{q} = \{x, \varnothing, \gamma, \theta\} \tag{8}$$

Solving the Lagrange equations yields:

$$\begin{aligned}
&M\ddot{x} + m(\ddot{x} - \ddot{L}\cos\theta\sin\varnothing + \dot{L}\dot{\theta}\sin\theta\sin\varnothing - \dot{L}\dot{\varnothing}\cos\theta\cos\varnothing + \dot{L}\dot{\theta}\sin\theta\sin\varnothing \\
&+ L\ddot{\theta}\sin\theta\sin\varnothing + L\dot{\theta}^2\cos\theta\sin\varnothing + L\dot{\theta}\dot{\varnothing}\sin\theta\cos\varnothing - \dot{L}\dot{\varnothing}\cos\theta\cos\varnothing \\
&- L\ddot{\varnothing}\cos\theta\cos\varnothing + L\dot{\theta}\dot{\varnothing}\sin\theta\cos\varnothing + L\dot{\varnothing}^2\cos\theta\sin\varnothing - \dot{L}\dot{\gamma}\sin\theta - L\ddot{\gamma}\sin\theta \\
&- L\dot{\gamma}\dot{\theta}\cos\theta) - Mx\dot{\gamma}^2 - m\dot{\gamma}(\dot{L}\sin\theta + L\dot{\theta}\cos\theta + x\dot{\gamma} - L\dot{\gamma}\cos\theta\sin\varnothing) = F_x
\end{aligned} \tag{9}$$

$$
\begin{aligned}
&- L\ddot{\varnothing}\cos^2\theta + L\dot{\theta}\dot{\varnothing}\sin2\theta - \tfrac{1}{2}L\ddot{y}\sin2\theta\cos\varnothing - 2L\dot{y}\dot{\theta}\cos^2\theta\cos\varnothing \\
&- x\dot{y}^2\cos\theta\cos\varnothing + \tfrac{1}{2}L\dot{y}^2\cos^2\theta\sin2\varnothing + \ddot{x}\cos\theta\sin\varnothing - 2\dot{L}\ddot{\varnothing}\cos^2\theta \\
&- \dot{L}\dot{y}\sin2\theta\cos\varnothing - g\cos\theta\sin\varnothing = 0
\end{aligned} \tag{10}
$$

$$
\begin{aligned}
&M\dot{x}^2\dot{y} + Mx^2\ddot{y} + m(x - L\cos\theta\sin\varnothing)(\ddot{L}\sin\theta + \dot{L}\dot{\theta}\cos\theta + \dot{L}\dot{\theta}\cos\theta + L\ddot{\theta}\cos\theta \\
&- L\dot{\theta}^2\sin\theta + \dot{x}\dot{y} + x\ddot{y} - \dot{L}\dot{y}\cos\theta\sin\varnothing - L\ddot{y}\cos\theta\sin\varnothing + L\dot{\theta}\dot{y}\sin\theta\sin\varnothing \\
&- L\dot{\varnothing}\dot{y}\cos\theta\cos\varnothing) + m(\dot{L}\sin\theta + L\dot{\theta}\cos\theta + x\dot{y} - L\dot{y}\cos\theta\sin\varnothing)\dot{x} \\
&- \dot{L}\cos\theta\sin\varnothing + L\dot{\theta}\sin\theta\sin\varnothing - L\dot{\varnothing}\cos\theta\cos\varnothing) - mL\sin\theta(\ddot{x} - \ddot{L}\cos\theta\sin\varnothing \\
&+ \dot{L}\dot{\theta}\sin\theta\sin\varnothing - \dot{L}\dot{\varnothing}\cos\theta\cos\varnothing + \dot{L}\dot{\theta}\sin\theta\sin\varnothing + L\ddot{\theta}\sin\theta\sin\varnothing \\
&+ L\dot{\theta}^2\cos\theta\sin\varnothing + L\dot{\theta}\dot{\varnothing}\sin\theta\cos\varnothing - \dot{L}\dot{\varnothing}\cos\theta\cos\varnothing - L\ddot{\varnothing}\cos\theta\cos\varnothing \\
&+ L\dot{\theta}\dot{\varnothing}\sin\theta\cos\varnothing + L\dot{\varnothing}^2\cos\theta\sin\varnothing - \dot{L}\dot{y}\sin\theta - L\ddot{y}\sin\theta - L\dot{y}\dot{\theta}\cos\theta) - m(\dot{L}\sin\theta \\
&+ L\dot{\theta}\cos\theta)(\dot{x} - \dot{L}\cos\theta\sin\varnothing + L\dot{\theta}\sin\theta\sin\varnothing - L\dot{\varnothing}\cos\theta\cos\varnothing - L\dot{y}\sin\theta) = T_y
\end{aligned} \tag{11}
$$

$$
\begin{aligned}
&L\ddot{\theta} + 2\dot{L}\dot{\theta} - 2\dot{L}\dot{\varnothing}\dot{y}\cos^2\theta\cos\varnothing + \tfrac{1}{2}L\dot{\varnothing}^2\sin2\theta - \tfrac{1}{2}L\dot{y}^2\sin2\theta\cos^2\varnothing \\
&+ 2\dot{x}\dot{y}\cos\theta - x\dot{y}^2\sin\theta\sin\varnothing + \ddot{x}\sin\theta\sin\varnothing - 2\dot{L}\dot{y}\sin\varnothing + x\ddot{y}\cos\theta - L\ddot{y}\sin\varnothing \\
&+ g\sin\theta\cos\varnothing = 0
\end{aligned} \tag{12}
$$

Equations (9) – (12) demonstrate that the dynamics of tower crane are highly nonlinear, and they are coupled. Therefore, payload swing control of both angles is challenging.

3 Input Shaping Control Schemes

This paper discusses two types of input shapers that aim to achieve the design goals of determining the amplitudes and time positions of impulses. The objective is to minimize the negative impacts of system flexibility by obtaining these parameters from the system's natural frequencies and damping ratios.

3.1 Zero Vibration

According to [19], the technique of zero vibration can be applied to reduce or eliminate vibration in any system that can be modeled as a second order system. The response of such system can be represented as:

$$
y_0 = \left[\frac{A_0\omega}{\sqrt{1 - \zeta^2}} e^{-\zeta\omega(t - t_0)} \right] \sin(\omega_n\sqrt{1 - \zeta^2}(t - t_0)) \tag{13}
$$

where A_0 is the amplitude of the applied impulse at time t_0, ζ denotes the system's damping ratio, and ω_n represents the system's natural frequency. If a sequence of impulses is given as input to the system, its response can be expressed as:

$$
y_\Sigma = \sum_{i=1}^{n} \left[\frac{A_i\omega_n}{\sqrt{1 - \zeta^2}} e^{-\zeta\omega_n(t - t_i)} \right] \sin(\omega_d(t - t_i)) \tag{14}
$$

where

$$
\omega_d = \omega_n\sqrt{1 - \zeta^2} \tag{15}
$$

The amplitude and time location of each impulse are represented by A_i and t_i, respectively. To ensure that the final position of the system is the same as that of the unshaped input, the sum of the amplitudes of all the applied impulses must be equal to one.

$$\sum_{i=1}^{n} A_i = 1 \tag{16}$$

A shaper that utilizes two impulses is referred as ZV shaper. The amplitude and time location of these impulses can be determined as:

$$\begin{bmatrix} A_i \\ t_i \end{bmatrix} = \begin{bmatrix} \frac{1}{1+K} & \frac{K}{1+K} \\ 0 & \Delta T \end{bmatrix} \tag{17}$$

where

$$\Delta T = \frac{\pi}{\omega_n \sqrt{1 - \zeta^2}} \tag{18}$$

$$K = e^{\left(\frac{-\zeta \pi}{\sqrt{1-\zeta^2}} \right)} \tag{19}$$

To avoid response delay, the first impulse in the ZV shaper is applied at time zero, while the second pulse is delayed by half of the system's damped vibration period.

3.2 Zero Vibration Derivative

The effectiveness of the ZV shaper in eliminating vibration is heavily reliant on the precise estimation of the system parameters, and even a minor error can lead to increased vibration levels. To enhance the robustness of the ZV shaper, a constraint can be added by setting the derivative of the residual vibration with respect to frequency equal to zero. This constraint can be expressed as:

$$\frac{\partial}{\partial \omega_n} V(\omega_n, \zeta) = 0 \tag{20}$$

By incorporating the constraint, a shaper can be designed using three impulses, which is referred as ZVD shaper. Figure 2 provides a visual depiction of the ZVD shaper with three impulses. The first impulse is always applied at time $t = 0$ to avoid any response delay. The amplitude and time locations of the ZVD shaper can be determined using Eq. (21).

$$\begin{bmatrix} A_i \\ t_i \end{bmatrix} = \begin{bmatrix} \frac{1}{1+2K+K^2} & \frac{2K}{1+2K+K^2} & \frac{K^2}{1+2K+K^2} \\ 0 & \Delta T & 2\Delta T \end{bmatrix} \tag{21}$$

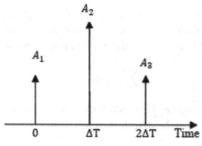

Fig. 2. ZVD input shaper

4 Implementation and Results

The experimental setup for this study comprises a laboratory tower crane, an interface module, a laptop, and desktop computer. Figure 3 illustrates the entire experimental setup and the dimensions of the crane are 1.2 m × 1.2 m × 1.5 m for its length, height, and width respectively. Three DC motors are used for actuating the motions of the jib, trolley, and payload. The crane is equipped with five encoders, two of which measure the trolley and jib positions in the polar plane, two measure the payload oscillation angles, and one measure the length of the cable attached to the payload. These encoders have a high resolution of 4096 pulses per rotation, providing an accuracy of 0.0015 rad for measuring the payload sway. The information gathered from all encoders is transmitted in real-time to a computer using a Peripheral Component Interconnect (PCI) card, while MATLAB and Simulink are employed for real-time control implementation.

To validate the nonlinear dynamic model of the tower crane described in Sect. 2, simulation and experimental results of payload swing responses are obtained and compared. Figure 4 depicts the payload swing responses when the crane was subjected to pulse inputs that have magnitudes of 1 N and 0.45 N for tangential and radial motions, respectively, with a pulse period of 2 s. A tower crane with a payload mass of 300 g and a cable length of 0.4 m is considered. The results in the x and y-angles demonstrate a satisfactory agreement between the simulation and experiment, especially on the sway frequencies of the payload swing.

The oscillation frequencies of the payload through simulation were calculated to be 4.21 rad/s for both x and y-angles, whereas a frequency of 4.22 rad/s was obtained in the experiments. Thus, the accuracy of the nonlinear model is confirmed, allowing for confident utilization in controller design. However, a slight disparity in the x and y-axes could be atributed to various factors, such as air drag and friction, which were not adequately accounted in the dynamic model.

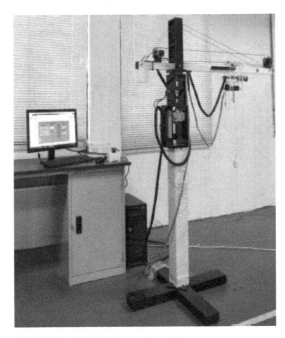

Fig. 3. Experimental setup for a laboratory tower crane.

4.1 Experimental Results Using Input Shaper Without Hoisting

In this case, the tower crane performs two simultaneous motions of trolley and jib. Figure 5 depicts the payload pendulations along the x and y-axes respectively, with a cable length of 0.1 m. The oscillation responses in both directions reveal that the ZV and ZVD shaper produced significant reduction in oscillation as compared to the unshaped response. As expected, the ZVD shaper outperforms the ZV shaper, where the ZVD shaper significantly reduced the maximum oscillation compared to the unshaped response by 63.2% and 63.5% for x-angle and y-angle respectively. Table 1 summarizes the overshoot (OS) and mean squared error (MSE) for both angles.

4.2 Experimental Results using Input Shaper with Payload Hoisting

In this case, the tower crane performs two simultaneous motions of trolley and jib with payload hoisting. Figure 6 depicts the payload pendulations along the x and y-axes respectively, with payload hoisting from 0 m to 0.4 m. On the other hand, Fig. 7 illustrates the spherical pendulation of a payload during maximum hoisting operation, which is from 0 m to 0.7 m. Comparing the unshaped responses with the ZV and ZVD shapers, it

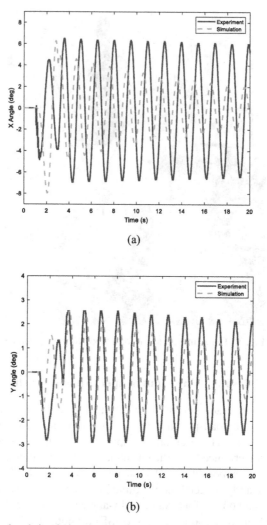

Fig. 4. Experimental and simulation results of payload swing response (a) *x*-angle (b) *y*-angle.

becomes evident from the oscillation responses in both directions that significant reductions in oscillations were achieved with the application of these shapers. As expected, the ZVD shaper provided better performance as compared to the ZV shaper, where the ZVD shaper significantly reduced the maximum oscillation compared to the unshaped response by 52% and 55.5% for *x*-angle and *y*-angle respectively. Table 2 summarizes the overshoot (OS) and mean squared error (MSE) for both angles.

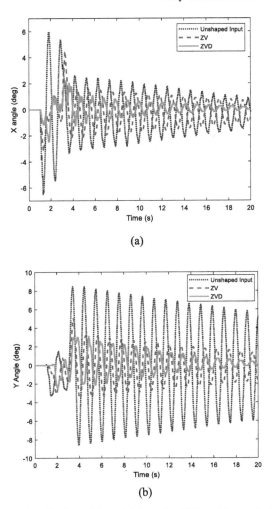

Fig. 5. Payload oscillation with a cable length of 0.1 m (a) x-angle (b) y-angle.

Table 1. Performance comparison for the tower crane system between the input shapers and the unshaped input

	x-angle		y-angle	
	θ_{max}	MSE	θ_{max}	MSE
Unshaped Input	5.98	2.29	8.44	23.69
ZV Shaper	4.40	1.21	5.01	3.77
ZVD Shaper	2.20	0.25	3.08	1.30

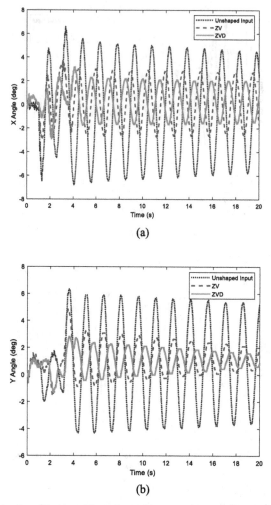

Fig. 6. Payload oscillation with payload hoisting to 0.4 m (a) *x*-angle (b) *y*-angle.

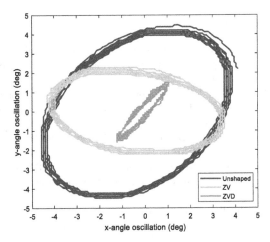

Fig. 7. Spherical pendulation of a payload during maximum hoisting operation

Table 2. Performance comparison for the tower crane system between the input shapers and the unshaped input

	x-angle		y-angle	
	θ_{max}	MSE	θ_{max}	MSE
Unshaped Input	6.59	13.68	6.33	14.26
ZV Shaper	3.25	3.91	4.40	1.59
ZVD Shaper	3.16	2.12	2.81	1.07

5   Conclusion

The development of ZV and ZVD shapers for sway control of a tower crane that involves trolley translational and jib rotation as well as under varying cable length has been presented. The performance of these control schemes is evaluated based on the reduction in maximum oscillation and overall oscillation for both angles. Both control strategies demonstrate satisfactory sway control capability during tower crane operations. A comparison of the results reveals that the ZVD shapers outperform the ZV shaper and provided the lowest overall oscillations for various hoisting operations.

References

1. Chen, H., Fang, Y., Sun, N.: Optimal trajectory planning and tracking control method for overhead cranes. IET Control Theory Appl. **10**(6), 692–699 (2016)
2. Jiang, W., Ding, L., Zhou, C.: Digital twin: stability analysis for tower crane hoisting safety with a scale model. Autom. Constr. **138**, 104257 (2022)
3. Ramli, L., et al.: Control strategies for crane systems: a comprehensive review. Mech. Syst. Signal Process. **95**, 1–23 (2017)

4. Nguyen, M.T., et al.: pid control to decrease fluctuation of load for tower crane. Robotica Manage. **26**(2) (2021)
5. Win, T., Hesketh, T., Eaton, R.: Construction tower crane simmechanics visualized modelling, tower vibration impact on payload swing analysis and LQR swing control. Int. Rev. Model. Simul. **7**(6), 979–993 (2014)
6. Win, T., Hesketh, T., Eaton, R.: Robotic tower crane modeling and control (RTCMC) with LQR-DRO and LQR-LEIC for linear and nonlinear payload swing minimization. Int. Rev. Autom. Control **9**(2), 72–87 (2016)
7. Ramli, L., et al.: Modelling and fuzzy logic control of an underactuated tower crane system. Appl. Model. Simul. **4**, 1–11 (2020)
8. Yang, T., Sun, N., Fang, Y.: Adaptive fuzzy control for a class of MIMO underactuated systems with plant uncertainties and actuator deadzones: design and experiments. IEEE Trans. Cybern. **52**(8), 8213–8226 (2021)
9. Li, M., Chen, H., Zhang, R.: An input dead zones considered adaptive fuzzy control approach for double pendulum cranes with variable rope lengths. IEEE/ASME Trans. Mechatron. **27**(5), 3385–3396 (2022)
10. Blackburn, D., et al.: Radial motion assisted command shapers for nonlinear tower crane rotational slewing. Control Eng. Pract. **18**(5), 523–531 (2010)
11. Lawrence, J., Singhose, W.: Command shaping slewing motions for tower cranes. J. Vib. Acoust. **132**(1), (2010)
12. Singhose, W., Eloundou, R., Lawrence, J.: Command generation for flexible systems by input shaping and command smoothing. J. Guidance Control Dyn. **33**(6), 1697–1707 (2010)
13. Caporali, R.P.: Iterative method for controlling the sway of a payload on tower (slewing) cranes using a command profile approach. Int. J. Innovative Comput. Inf. Control **14**(4), 1169–1187 (2018)
14. Peng, J., Huang, J., Singhose, W.: Payload twisting dynamics and oscillation suppression of tower cranes during slewing motions. Nonlinear Dyn. **98**, 1041–1048 (2019)
15. Al-Fadhli, A., Khorshid, E.: Payload oscillation control of tower crane using smooth command input. J. Vib. Control **29**(3–4), 902–915 (2023)
16. Al-Mousa, A.A., Nayfeh, A.H., Kachroo, P.: Control of rotary cranes using fuzzy logic. Shock. Vib. **10**(2), 81–95 (2003)
17. Omar, H.M.: Control of gantry and tower cranes. Diss. Virginia Polytechnic Institute and State University (2003)
18. Tinkir, M., et al.: Modeling and control of scaled a tower crane system. In: 2011 3rd International Conference on Computer Research and Development, vol. 4, pp. 93–98. IEEE (2011)
19. Singhose, W.: Command shaping for flexible systems: a review of the first 50 years. Int. J. Precis. Eng. Manuf. **10**(4), 153–168 (2009). https://doi.org/10.1007/s12541-009-0084-2

Numerical Study of Sensitivity Enhancement in Optical Waveguide Sensor Due to Core Diameter

Thienesh Marippan[1] , Nur Najahatul Huda Saris[1(✉)] ,
Ahmad Izzat Mohd Hanafi[1] , and Nazirah Mohd Razali[2]

[1] Faculty of Electrical Engineering, Universiti Teknologi Malaysia, 81310 Skudai, Johor,
Malaysia
nurnajahatulhuda@utm.my
[2] Malaysia-Japan International Institute of Technology, Universiti Teknologi Malaysia, 54100
Kuala Lumpur, Malaysia

Abstract. The need for sensing technology has arisen as a result of the growing environmental concern over microplastic pollution in water systems. In this work, a potential microplastic sensor based on an optical planar doped polymer waveguide was theoretically investigated by using Wave Optics Module-COMSOL Multiphysics® software over a range of 2 μm to 16 μm waveguide core diameters. The microplastic refractive index of the analytes employed in this research ranged from 1.48 RIU to 1.50 RIU reflecting the microplastic refractive index. The results revealed that a 2 μm core diameter of the waveguide sensor achieved the best sensitivity value of 5.12×10^{-5} at fixed cladding depth and wavelength of 0 μm and 617 nm, respectively. The simulation work offers useful information for designing a potential waveguide for real world applications to detect microplastics in water. This study is believed to contribute towards the development of cutting-edge sensing technologies through the optimized optical waveguide sensor design, which shows promise in addressing the serious problem of microplastic pollution in water systems.

Keywords: Microplastics · Water · Optical waveguide sensor · Sensitivity · COMSOL Multiphysics

1 Introduction

Microplastics are basically plastics that are very small in size which are invisible to the naked eye [1]. It is defined as plastics that can form in terms of surface area of not more than 5 mm in size [2, 3], often broken down from either plastic products or from any commercial product development factories. The density of microplastics can vary depending on the type of plastic it is made of. There are five major commodities of plastics, which are common sources of microplastics in water bodies including polypropylene (PP), polyvinyl chloride (PVC), polyethylene (PE), polystyrene (PS) and polyethylene terephthalates (PET) [4, 5]. Recently, microplastic pollution has become a serious environmental issue with detrimental effects on human health and aquatic ecosystems if it is not properly disposed of [6–8]. This has become critical in the current era as many lives of living creatures, both land and marine are being threatened [9–11].

To date, several prevalent techniques have been explored related to identifying microplastics in water including electron microscopy, infrared and Raman spectroscopies. Although the techniques aforementioned are literally well established, they are nevertheless expensive, complicated and require complex in-laboratory equipment, making them less efficient in detecting microplastics. Recently, optical detection using optical waveguide sensors for microplastics detection are attracting considerable interest and pose as an alternative due to its excellent interaction of the analyte medium via evanescent waves, on top of its exceptional properties from corrosion resistance, portability, to flexible designs. In 2019, Asamoah et al. [12] had successfully developed a prototype of a portable optical sensor for the detection of transparent and translucent microplastics in freshwater, which has the capability of measuring the specular laser light reflection and transmission from microplastic particles, simultaneously. Iri et al. [13] in 2021 had also developed a portable optical system to sense microplastics with concentrations of less than 0.015% w/v. However, these findings have not treated design development in much detail to detect microplastics in water.

Until recently, too little attention has been paid to numerical analysis for theoretical investigation of the potential optical waveguide sensor for microplastics detection in water [14, 15]. Hence, this paper seeks to further enhance the simulation on a potential optical waveguide sensor, namely europium aluminium doped polymer [15] with the variation of core diameter from 2 μm to 16 μm via Wave Optics Module-COMSOL Multiphysics® software. Apart from that, the simulation on the evanescence wave at different refractive index of the analyte and electric field intensity for all waveguide cores are demonstrated. Note that similar ranges of analyte are being implemented reflecting the refractive index of the microplastics, which is from 1.480 RIU up to 1.500 RIU with an incremental of 0.002 RIU. The operating wavelength in this simulation is fixed at 617 nm due to the fact that it is within the low attenuation region for the waveguide material [16]. Meanwhile, the cladding thickness is fixed at 0 μm to allow high exposure of evanescent wave towards the analyte medium.

The findings from this research will contribute to the technological development of advanced optical devices for sensing with optimized sensitivity and linearity performance in monitoring and combating microplastic pollution in water systems to produce clean water and promote good health, which aligns with Sustainable Development Goals (SDGs) of good health and wellbeing, clean water and sanitation, and life below water. Accordingly, this paper has been divided into four parts. The first part deals with the overview of the current trend of research. The second part is concerned with the methodology and simulation settings used in this research via the Wave Optics Module-COMSOL Multiphysics® simulation software. It will then proceed to the findings of the research. Finally, areas for further research are identified.

2 Methodology

This research is mainly simulated by using the Wave Optics Module-COMSOL Multiphysics® software for the waveguide cross section analysis. The program provides a range of finite element analysis, solver, and simulation tools for numerous physics and engineering applications, particularly those that involve coupled phenomena and multiphysics. A typical physics-based user interface is provided by the program, which allows users to deal with linked systems of partial differential equations. Figure 1 shows the graphical user interface (GUI) of Wave Optics Module-COMSOL Multiphysics® software utilized in this research.

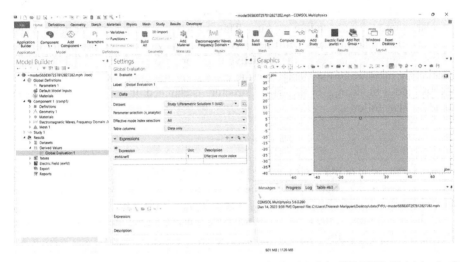

Fig. 1. The Graphical User Interface (GUI) of Wave Optics Module-COMSOL Multiphysics® software

The structure of the waveguide sensor is constructed [15, 16] as illustrated in Fig. 2. The waveguide is designed with planar waveguide type with circular cross-section shapes of the core diameter, where the surface of the waveguide comes into contact with the analyte.

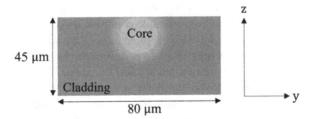

Fig. 2. The structure of europium aluminum doped polymer waveguide sensor

Fundamentally, the interaction between light and the analyte occurs at the sensing region of the waveguide. When the analyte comes into contact with the sensing region, it causes a change in the refractive index, absorption, or scattering properties, which modulates the behavior of the guided light. The change in the optical properties of the waveguide due to the presence of the analyte can be detected using various techniques. One common method is to measure the change in the transmitted or reflected optical power. When the refractive index or absorption properties of the sensing region are altered by the analyte, it affects the propagation characteristics of the guided light, resulting in a change in the transmitted or reflected intensity.

In this work, the theory applied for designing the proposed optical waveguide sensor is based on the principle of evanescent wave sensing. When light propagates through the waveguide, a small fraction of the optical field extends beyond the waveguide's boundaries into the surrounding medium. This is known as the evanescent field. When the evanescent field interacts with the microplastic analyte in close proximity to the waveguide surface, changes in the refractive index or absorption of the analyte can be detected by monitoring the changes in the evanescent field intensity. As a result, the effective index mode, n_{eff} of the sensor structure is changed. It can be expressed as in Eq. (1) [17] as follows:

$$n_{eff} = \frac{c}{v_g} \tag{1}$$

where c is the speed of light and v_g is the group velocity of light that propagates through the planar waveguide. Subsequently, the sensitivity of the waveguide sensor can be determined by plotting the graph of n_{eff} versus n_{sample} [18]. The variation of n_{sample} affect the penetration of the evanescent wave, d_p thus, changing the n_{eff} as value as in described in Eq. (2) as follow:

$$Penetration\ Depth = \frac{\lambda}{2\pi \sqrt{n_{eff}^2 - n_{sample}^2}} \tag{2}$$

The optimization of the geometry design of the proposed waveguide using simulation settings is necessary to enhance the sensitivity of the waveguide sensor prior to the fabrication process. Thus, this work could be dedicated as a preliminary work to contribute towards theoretical understanding on evanescent field theory of the waveguide sensor. In the simulation setting, the template of the proposed waveguide sensor is created in the body builder to initiate the geometry in the software. The geometry is divided into several components namely Core (c1), Square 1 (sq1), Rectangle 1 (r1) and Rectangle 2 (r2). These components indicate the waveguide core, side length of the waveguide, the size of the planar structure, and the cladding dimension of the waveguide, respectively. Once the waveguide's cross-section template is designed, the value of core diameter variation from 2 μm to 16 μm are inserted at "size and shape" component. After that, the "Result" component in model builder is clicked to set the operating wavelength to be 617 nm in mode analysis prior to initiate the program. The program is ready to run simultaneously, and the sensitivity can be calculated accordingly based on the effective mode refractive index result in mode analysis section. A summary of parameter used in this research can be found in Table 1.

Table 1. Summary of parameters used

Parameter	Details
Core Material	Europium Aluminum doped polymer
Core Refractive Index	1.51
Cladding Refractive Index	1.501
Analyte Refractive Index	1.48–1.50
Wavelength	617 nm
Cladding Depth	0 μm
Core Diameter	2 μm–16 μm
Waveguide Cladding Dimension	80 μm × 80 μm
Waveguide Core Shape	Circular

3   Results and Discussion

Figure 3 demonstrates the geometry cross-section of the waveguide sensor built in the simulation program with variation of core diameter from 2 μm to 16 μm. It consists of the analyte as a sensing region on top of the waveguide, circular core and surrounded by the cladding.

Accordingly, the output of the simulation for the light propagation through the different core diameters of waveguide sensor are depicted in Fig. 4. At the right end of the intensity of electric field image, the bar's color variation indicates the relative intensity of light. The red color in the core shows that the electric field is concentrated at the core. This indication shows that at the center of the core, the intensity of the light emission is the highest and gradually diminishes throughout the analyte and cladding. The electric field gradually decreases due to the attenuation of light energy upon reaching the core/cladding or core/analyte boundaries. On top of that, it can be deduced that strong evanescent field is found in the smaller core diameter with strong light–analyte interactions [18].

Correspondingly, the effective mode index against analyte medium obtained in the mode analysis of the simulation for all waveguide core diameters are depicted in Fig. 5. It is shown that by increasing analyte refractive index of the microplastic from 1.48 RIU to 1.50 RIU results in the increasing of n_{eff} exponentially. The non-linear response occurs due to non-linear evanescent wave interaction between the analyte and the core. It is based on the fact that the more robust the nonlinearity of the evanescent wave energy is, the closer the environmental RI is to the RI of the NCF. As the evanescent wave penetration depth increases, the effective change in RI of the core mode increases, resulting in a higher sensitivity [19]. Here, the linear trendline of the effective mode for each analyte medium indicated in Fig. 5 is tabulated in Table 2. The trendline is used to calculate the sensitivity exhibited by the core diameter. Clearly, the 2 μm core diameter of the waveguide resulted in the highest gradient in the trendline equation which is 0.0256 in comparison to the other waveguide core diameters.

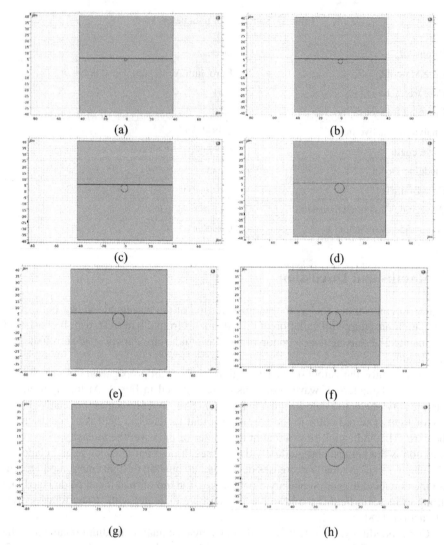

Fig. 3. The geometry of the proposed waveguide sensor with variation of core diameter from 2 μm to 16 μm at fixed cladding depth of 0 μm

Figure 6 shows the sensitivity of the proposed waveguide sensor for the influence of core diameter on the waveguide structure and geometry. As expected from the trendline equation, the highest sensitivity is observed at 2 μm core diameter followed by 4 μm and 6 μm with value of 5.12×10^{-5}, 0.9×10^{-5} and 0.28×10^{-5}, respectively. The lowest sensitivity is recorded at 16 μm with 0.012×10^{-5}. It is apparent that the sensitivity decreases as the core diameter of the proposed waveguide increased. This could be attributed to a larger evanescent field resulting from smaller core radius.

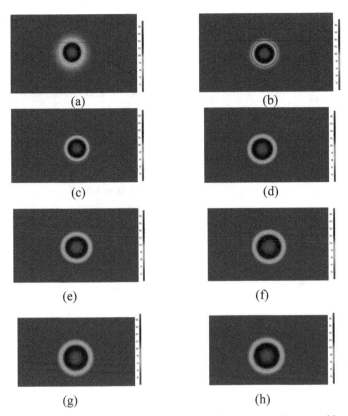

Fig. 4. The electric field intensity around the sensor for different core diameter of 2 μm to 16 μm at fixed cladding depth of 0 μm

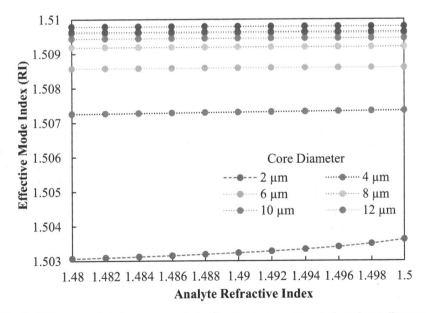

Fig. 5. Effective mode index versus analyte refractive index with variation of core diameters

Table 2. The trendline of the effective mode index for variation of core diameters

Core Diameter (μm)	Trendline Equation
2	$y = 0.0256x + 1.4651$
4	$y = 0.0045x + 1.5006$
6	$y = 0.0014x + 1.5065$
8	$y = 0.0006x + 1.5083$
10	$y = 0.0003x + 1.5090$
12	$y = 0.0002x + 1.5093$
14	$y = 0.00009x + 1.5095$
16	$y = 0.00006x + 1.5097$

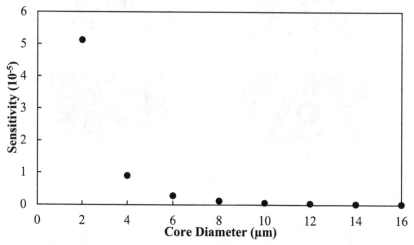

Fig. 6. The sensitivity of the proposed waveguide sensor with variation of core diameter from 2 μm to 16 μm at 617 nm wavelength and 0 μm cladding thickness

4 Conclusion

In this work, the simulation of the proposed waveguide sensor has been successfully simulated with variations of core diameter from 2 μm to 16 μm by using Wave Optics Module-COMSOL Multiphysics® software. The core diameter of 2 μm has shown the best sensitivity, 5.12×10^{-5} in dimensionless unit, as compared to the other core diameters at fixed cladding depth and wavelength of 0 μm and 617 nm, respectively. As a result of this research, advanced sensing technologies could be developed and microplastic pollution in water systems could be addressed.

Acknowledgement. This research is funded by Universiti Teknologi Malaysia Fundamental Research Grant (UTMFR) under Vote No: Q.J130000.3851.21H76 as well as Leave a Nest research grant. The authors would like to thank Lightwave Communication Research Group (LCRG) as well as Optical Device System Laboratory (ODESY) members for their help and encouragement.

References

1. Thushari, G.G.N., Senevirathna, J.D.M.: Plastic pollution in the marine environment. Heliyon **6**(8) (2020)
2. Asamoah, B.O., et al.: Towards the development of portable and in situ optical devices for detection of micro-and nanoplastics in water: a review on the current status. Polymers **13**(5), 730 (2021)
3. Kye, H., et al.: Microplastics in water systems: a review of their impacts on the environment and their potential hazards. Heliyon (2023)
4. Munier, B., Bendell, L.: Macro and micro plastics sorb and desorb metals and act as a point source of trace metals to coastal ecosystems. PLoS ONE **13**(2), e0191759 (2018)
5. Kankanige, D., Babel, S.: Smaller-sized micro-plastics (MPs) contamination in single-use PET-bottled water in Thailand. Sci. Total. Environ. **717**, 137232 (2020)
6. Karbalaei, S., et al.: Occurrence, sources, human health impacts and mitigation of microplastic pollution. Environ. Sci. Pollut. Res. **25**, 36046–36063 (2018)
7. Yuan, Z., Nag, R., Cummins, E.: Human health concerns regarding microplastics in the aquatic environment-from marine to food systems. Sci. Total. Environ. **823**, 153730 (2022)
8. Choudhury, A., et al.: Microplastic pollution: an emerging environmental issue. J. Entomol. Zool. Stud. **6**(6), 340–344 (2018)
9. Silva, A.B., et al.: Microplastics in the environment: challenges in analytical chemistry-a review. Anal. Chim. Acta **1017**, 1–19 (2018)
10. Padervand, M., et al.: Removal of microplastics from the environment. Rev. Environ. Chem. Lett. **18**, 807–828 (2020)
11. Zhang, S., et al.: Microplastics in the environment: a review of analytical methods, distribution, and biological effects. TrAC Trends Anal. Chem. **111**, 62–72 (2019)
12. Asamoah, B.O., et al.: A prototype of a portable optical sensor for the detection of transparent and translucent microplastics in freshwater. Chemosphere **231**, 161–167 (2019)
13. Iri, A.H., et al.: Optical detection of microplastics in water. Environ. Sci. Pollut. Res., pp. 1–7 (2021)
14. Battula, S., et al.: In-situ microplastic detection sensor based on cascaded microring resonators. In: OCEANS 2021, San Diego–Porto. IEEE (2021)
15. Razali, N.M., Saris, N.N.H.: Simulation of optical planar waveguide sensor for microplastics detection in water. In: 2022 International Conference on Numerical Simulation of Optoelectronic Devices (NUSOD). IEEE (2022)
16. Saris, N.N.H., et al.: Waveguide length and pump power effects on the amplification of europium aluminum doped polymer. Optik **239**, 166670 (2021)
17. Syuhada, A., et al.: Single-mode modified tapered fiber structure functionalized with GO-PVA composite layer for relative humidity sensing. Photonic Sens. **11**(3), 314–324 (2021)
18. Wang, W., et al.: Sensitive evanescence-field waveguide interferometer for aqueous nitro-explosive sensing. Chemosensors **11**(4), 246 (2023)
19. Xiao, G., et al.: Graphene oxide sensitized no-core fiber step-index distribution sucrose sensor. In: Photonics. MDPI (2020)

Stress Simulation
of Polydimethylsiloxane-Coated Fiber Bragg
Grating Bend Sensor

Nazirah Mohd Razali[1,2] ⓘ, Nur Najahatul Huda Saris[3](✉) ⓘ,
Shazmil Azrai Sopian[1] ⓘ, Noor Amalina Ramli[2] ⓘ,
and Wan Imaan Izhan Wan Iskandar[2] ⓘ

[1] Malaysia-Japan International Institute of Technology, Universiti Teknologi Malaysia, 54100
Kuala Lumpur, Malaysia
[2] CAD-IT Consultants (M) Sdn Bhd, 47301 Petaling Jaya, Selangor, Malaysia
[3] Faculty of Electrical Engineering, Universiti Teknologi Malaysia, 81310 Skudai,
Johor, Malaysia
nurnajahatulhuda@utm.my

Abstract. Fiber Bragg Grating (FBG) has garnered significant interest in the field
of bend sensors. However, one major drawback of the FBG is its inherent structure fragility due to structural deformation. Hence, this paper seeks to.remedy this
problem by presenting a stress simulation of a polydimethylsiloxane (PDMS)-
coated FBG by utilizing Ansys Workbench 2023 R1 software. The FBG's design
optimization was performed by varying the PDMS thickness from 0 mm to 2 mm.
To evaluate the stress performance of the FBG, the coated and uncoated FBGs
were compared with those subjected to an additional force of 0.010 N. The simulation results revealed that increasing the PDMS thickness slightly reduces structural deformation but significantly improves the maximum combined stress. The
uncoated FBG exhibited the largest deformation of 0.28926 mm and the highest maximum combined stress of 73.309 MPa, whereas the smallest deformation
of 0.27419 mm and the lowest maximum combined stress of 61.885 MPa were
achieved with a PDMS thickness of 2 mm. Therefore, the minimal changes in
deformation, up to 5.21% of the PDMS-coated FBG compared with the uncoated
FBG, make it a suitable candidate to sustain the curvature structure of the FBG.
Meanwhile, the maximum combined stress of the PDMS-coated FBG was 15.58%
higher than that of the uncoated FBG, proving that the presence of PDMS
improves the mechanical performance of the FBG by providing enhanced physical
robustness, thereby showcasing its potential as a new bend sensor.

Keywords: Stress · Simulation · FBG · PDMS · Ansys Workbench

1 Introduction

Fiber Bragg Grating (FBG) has been extensively utilized to measure physical parameters
[1]. It offers bend (curvature/deformation) measurements for quality control and real-
time monitoring purposes. Typically, stress is applied to the FBG, resulting in the bending

© The Author(s), under exclusive license to Springer Nature Singapore Pte Ltd. 2024
F. Hassan et al. (Eds.): AsiaSim 2023, CCIS 1912, pp. 210–220, 2024.
https://doi.org/10.1007/978-981-99-7243-2_18

of its structure. The physical properties are then measured by observing changes in the strained grating-induced Bragg wavelength [2]. FBG possesses numerous advantages, including strong signal reflection, resistance to optical signal intensity modulation due to their spectral coding, excellent multiplexing, and self-referencing capabilities, allowing addressing of multiple sensors with a single optical cable [3, 4]. Thus far, the bent FBG sensor has found its way into varieties of applications in structural monitoring [5–9].

To function as a bend sensor, the FBG is fixed between two points, with axial or transverse force being applied to create a curved structure [10, 11]. However, it is widely recognized that FBGs made of silica are structurally fragile and highly susceptible to damage [12]. Excessively bending the FBG sensor with a smaller bend radius will cause the fiber to break. Hence, for practical, real-world applications, it becomes crucial to ensure the protection of the FBG. It requires adequate safeguarding to withstand the rigors of harsh-field usage.

To address these limitations, a solution has been proposed for the FBG by applying a coating of polydimethylsiloxane (PDMS), which is a silicone-based elastomer [13]. PDMS is chosen due to its numerous advantageous material properties, such as flexibility, biocompatibility, ease of use, and excellent thermal and chemical stability, which is suitable for harsh environment applications [14]. Moreover, PDMS offers convenient patterning techniques of replica molding [15, 16], which can replace the sophisticated coating technique. Coating FBGs with PDMS also provides improved protection, resulting in increased elasticity of the FBG during physical stretching, compression, and twisting, as well as enhanced strain sensitivity [11].

In recent years, research on PDMS-coated FBGs for bending (including stress/strain/deformation) has increased. Morais et al. [15] presented a liquid-level sensor system based on a pair of FBGs embedded in a circular PDMS diaphragm. The hydrostatic pressure caused by the liquid level variation leads to pressure variations in the diaphragm, thereby changing the strain value of the FBG. Meanwhile, Li et al. [17] reported using FBGs embedded in PDMS tubes for joint movement monitoring. The structure was sewn on the finger guard band and tested on the index finger, demonstrating its ability to detect bending motions of the finger knee joint with good repeatability. In their latest work, Lee et al. [11] introduced a novel force-sensing smart textile by inlaying four silicone-embedded FBGs into a knitted undergarment. The results demonstrated that the PDMS-coated FBGs achieved a more stable and linear sensing response compared to uncoated FBGs when force was applied.

Despite the limited available literature on simulation for this structure design, its fabrication process, and advantages, as mentioned before duly prove the feasibility of a PDMS coating layer in enhancing the mechanical properties of the FBG as a bend sensor. In this paper, the FBG is simulated via Ansys Workbench 2023 R1, a simulation integration platform software. This study will examine the effect of the force applied to the FBG and PDMS-coated FBG. Furthermore, the structural deformation and stress analysis of the structure are studied, and the underlying physics are discussed. Finally, both structures will be compared to analyze their mechanical performance.

2 Methodology

2.1 Theory

In optics, FBG is sensitive to strain changes due to its nanograting period. When stress is applied to the FBG, the structure will bend, forming a curvature that results in a change in the FBG length, as illustrated in Fig. 1. The bending effect will either stretch or compress the grating period, depending on how the bend is formed, leading to a change in the Bragg wavelength [18]. Stretched gratings will have a longer grating period, causing the Bragg wavelength to shift to longer wavelengths (red shift), while compressed gratings

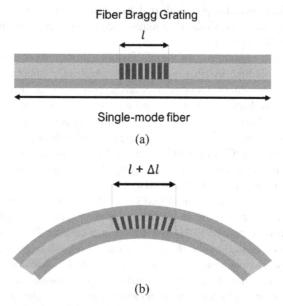

Fig. 1. Bending mechanism of the FBG during (a) stress-free and (b) stress applied and bend

In mechanics, the strain, ε of the material can be described as in Eq. 1 [19].

$$\varepsilon = \frac{\Delta l}{l} \tag{1}$$

Here, l is the FBG length, and Δl is the changes in FBG length. According to optical theory, the relationship between the Bragg wavelength, λ_B or change in the Bragg wavelength, $\Delta\lambda_B$ and ε can be measured using Eq. 2 [20].

$$\frac{\Delta\lambda_B}{\lambda_B} = (1 - P_e)\varepsilon \tag{2}$$

Here, P_e is the photoelasticity of the FBG. Strain represents the amount of deformation the FBG undergoes due to stress. Stress is defined as the force F, applied to an area, A, as described in Eq. 3.

$$E = \frac{F}{A} \tag{3}$$

The ratio of stress and strain represents the Young's modulus, which is described by Eq. 4. This parameter allows us to study the property of the material and determines how easily it can deform.

$$E = \frac{\sigma}{\varepsilon} \tag{4}$$

However, this study will focus solely on the mechanics rather than the optics of the PDMS-coated FBG. Its aim is to investigate how the presence of PDMS can impact the deformation of the FBG.

2.2 Simulation Parameters

The simulation workflow relies on Ansys Workbench 2023 R1, which efficiently integrates data from engineering simulations to create models. To initiate the process, the static structural toolbox is selected. In the second step, the material data is defined by inputting the relevant parameter values. For FBG, the values for density, Young's modulus, and Poisson's ratio are set to 2500 kg/m³, 7.2×10^{10} Pa, and 0.18, respectively [21]. On the other hand, for PDMS, the values for density, Young's modulus, and Poisson's ratio are set to 970 kg/m³, 7.5×10^{5} Pa, and 0.49, respectively [22]. A summary of all the material data used can be found in Table 1.

Table 1. Material specifications

Parameter	FBG [21]	PDMS [22]
Material	Silica	Elastomer
Density (kg/m³)	2500	970
Young's Modulus (Pa)	7.2×10^{10}	7.5×10^{5}
Poisson's Ratio	0.18	0.49

Figure 2 illustrates the structure under investigation, which consists of an FBG and a PDMS-coated FBG, both with a length of 3 cm. The FBG has a diameter of 125 μm, while the PDMS diameter ranges from 0 mm to 2 mm for optimization purposes. It is important to note that certain assumptions have been made due to constraints in the design of the FBG owing to its microscale structure which are listed as follows:

1. The FBG is assumed to have a diameter of 125 μm and a length of 5 mm. Typically, the FBG is inscribed in a single-mode fiber (SMF), so to simulate the entire structure, the total length considered is approximately 3 cm, which includes the SMF with the FBG centered along the structure.
2. Despite the FBG being composed of a core with nano pitch grating size and a cladding with slightly doped material, it is assumed that the core has similar mechanical properties to the cladding. This assumption is based on the extremely small diameter size of the core, which is 12 times smaller than the cladding.
3. Therefore, the FBG (with the SMF) is designed as a solid silica fiber.

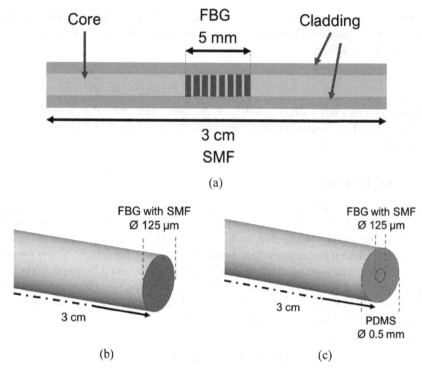

Fig. 2. The FBG structure is depicted in (a) actual design, (b) design assumption, and (c) when coated with PDMS. The structure is designed using Ansys SpaceClaim

To analyze the stress, a fixed support is applied to both vertices of the structure. Additionally, a force of 0.010 N in the form of a point load is applied along the negative y-axis (−y). The fixed support location and force direction are illustrated in Fig. 3.

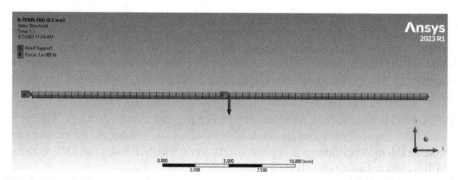

Fig. 3. The fixed support and force direction position. The model is set up using Ansys Mechanical.

3   Results and Discussion

Figure 4 illustrates the total deformation and maximum combined stress of the FBG. Deformation refers to the change in size or shape of an object when an external load is applied to an engineering assembly. It is useful in determining the amount of distortion a structure may experience under various loading conditions. The maximum combined stress represents a combination of direct stress and maximum bending stress. In the case of a pure bending problem, the maximum combined stress is considered as the maximum bending stress. As the force is applied to the FBG, the structure bends, resulting in a maximum deformation of 0.28926 mm and a maximum combined stress of 73.399 MPa when a force of 0.010 N is applied.

On the other hand, Fig. 5. And Fig. 6 show the total deformation and maximum combined stress of the PDMS-coated FBG at different PDMS thickness ranging from 0.5 mm to 2 mm, respectively. As the PDMS becomes thicker, the total deformation reduces from 0.28967 mm to 0.27419 mm with total changes of 0.01548 mm while the maximum combined stress is reduced from 72.406 MPa to 61.885 MPa with total changes of 10.521 MPa.

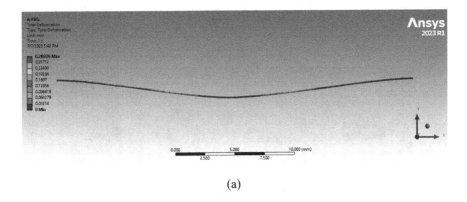

(a)

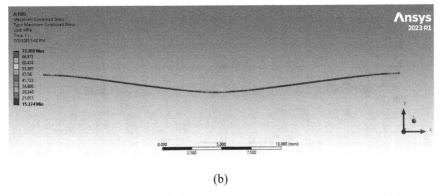

(b)

Fig. 4. Simulation results of FBG in (a) total deformation and (b) maximum combined stress

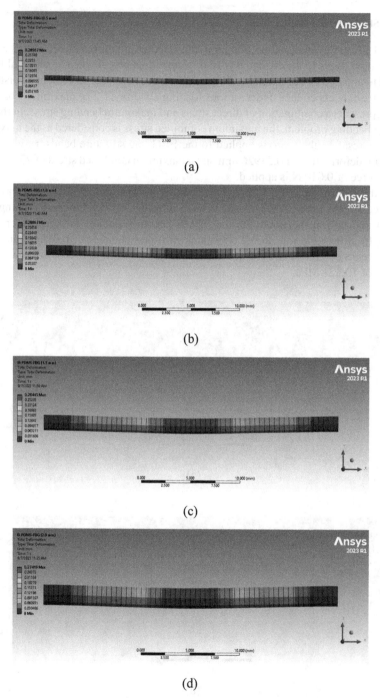

Fig. 5. Total deformation of PDMS-coated FBG with PDMS thickness of (a) 0.5 mm (b) 1.0 mm (c) 1.5 mm and (d) 2.0 mm

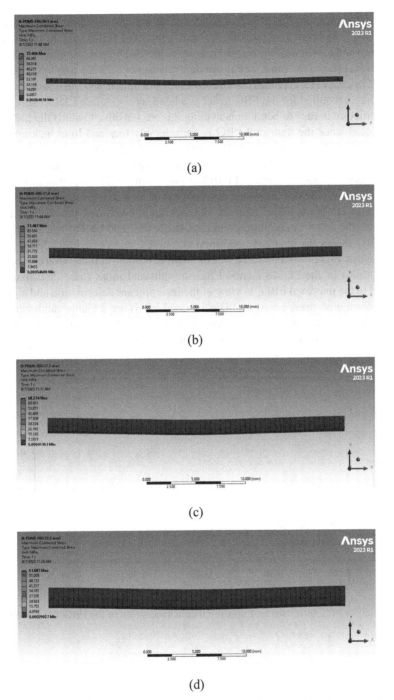

(a)

(b)

(c)

(d)

Fig. 6. Maximum combined stress of PDMS-coated FBG with PDMS thickness of (a) 0.5 mm (b) 1.0 mm (c) 1.5 mm and (d) 2.0 mm

Figure 7 shows the overall performance for all simulated designs for comparison. As the PDMS thickness increases from 0 mm (uncoated FBG) to 2.00 mm, the total deformation and maximum combined stress are reduced. Through the graph analysis, the uncoated FBG shows the most significant deformation of 0.28926 mm when compared to the PDMS-coated FBG (2 mm thickness) of 0.27419 mm. It is worth noting that a large deformation is required for the FBG to act as a bend sensor. A large deformation means that the structure is easy to bend, thus making it more sensitive towards the measurand. It is expected since the uncoated FBG is considerably tiny and light making it more flexible to bend. The presence of PDMS only gives a slight effect on the deformation as the structure becomes thicker, making it harder to bend. The deformation only shows small changes with total changes of 0.01507 mm or only 5.21% deterioration after the FBG is coated with PDMS; therefore, the changes are considered insignificant. However, the PDMS layer significantly improves the maximum combined stress. The uncoated FBG shows the largest maximum combined stress of 73.309 MPa, while the PDMS-coated FBG shows the smallest maximum combined stress of 61.885 MPa, giving a total change of 11.424 MPa or 15.58%. The small maximum combined stress represents the capabilities of the PDMS-coated FBG to withstand more force (and load) when compared to the uncoated FBG. In view of the insignificant deformation and significant maximal combined stress changes, PDMS-coated FBGs are a viable candidate for use as a bend sensor when compared to uncoated FBGs. Table 2 summarizes the mechanical performance of both uncoated and PDMS-coated FBGs.

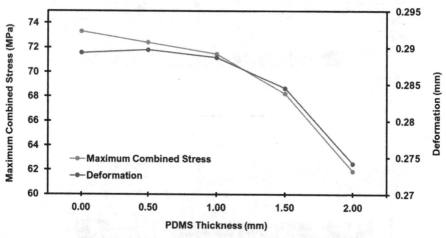

Fig. 7. Analysis of deformation and maximum combined stress of the FBG and PDMS-coated FBG

Table 2. Mechanical performance of the FBG and PDMS-coated FBG

Parameter	Total Deformation (mm)	Maximum Combined Stress (MPa)
FBG	0.28926	73.309
PDMS-coated FBG (2mm Thickness)	0.27419	61.885
Results	Deterioration = 5.21%	Improvement = 15.58%

4 Conclusion

This work demonstrated the simulation of PDMS-coated FBG for different PDMS thicknesses ranging from 0 mm to 2 mm. The conducted design optimization at different PDMS thicknesses is greatly helpful in determining the mechanical performance of the PDMS-coated FBG structure. The simulation evaluation shows that increasing the PDMS thickness slightly reduces not only structural deformation but significantly improves the maximum combined stress. The uncoated FBG has exhibited the largest deformation of 0.28926 mm and the highest maximum combined stress of 73.309 MPa, whereas the smallest deformation of 0.27419 mm and the lowest maximum combined stress of 61.885 MPa have been achieved with a PDMS thickness of 2 mm. Therefore, the extremely small changes in deformation, up to 5.21% of the PDMS-coated FBG compared with the uncoated FBG, make it a suitable candidate to sustain the curvature structure of the FBG. Meanwhile, the maximum combined stress of the PDMS-coated FBG is 15.58% higher than that of the uncoated FBG, proving that the presence of PDMS improves the mechanical performance of the FBG by providing enhanced physical robustness, thereby showcasing its potential as a new bend sensor.

Acknowledgments. The authors would like to thank CAD-IT Consultants (M) Sdn. Bhd for their support in conducting the simulations using Ansys products. This work has received financial support from the Universiti Teknologi Malaysia Fundamental Research Grant (UTMFR) under Vote No: Q.J130000.3851.21H76.

References

1. Sahota, J.K., Gupta, N., Dhawan, D.: Fiber bragg grating sensors for monitoring of physical parameters: a comprehensive review. Opt. Eng. **59**(06), 060901 (2020)
2. Lee, S.B., Jung, Y.J., Choi, H.K., Sohn, I.B., Lee, J.H.: Hybrid LPG-FBG based high resolution micro bending strain sensor. Sensors. **21**(1), 22 (2021)
3. Vavrinsky, E., et al.: The current state of optical sensors in medical wearables. Biosensors **12**(4), 217 (2022)
4. Majumder, M., Gangopadhyay, T.K., Chakraborty, A.K., Dasgupta, K., Bhattacharya, D.K.: Fibre bragg gratings in structural health monitoring present status and applications. Sens. Actuators, A **147**(1), 150–164 (2008)

5. Braunfelds, J., et al.: FBG-based sensing for structural health monitoring of road infrastructure. J. Sens. **2021**, 8850368 (2021)
6. Yang, J., Hou, P., Yang, C., Yang, N.: Study of a long gauge FBG strain sensor with enhanced sensitivity and its application in structural monitoring. Sensors. **21**(10), 3492 (2021)
7. Aimasso, A., Vedova, M.D.L.D., Maggiore, P.: Analysis of FBG sensors performances when integrated using different methods for health and structural monitoring in aerospace applications. In: 2022 6th International Conference on System Reliability and Safety (ICSRS), pp. 138–144. IEEE, Venice (2022)
8. Amaya, A., Sierra-Pérez, J.: Toward a structural health monitoring methodology for concrete structures under dynamic loads using embedded FBG sensors and strain mapping techniques. Sensors. **22**(12), 4569 (2022)
9. Kouhrangiha, F., Kahrizi, M., Khorasani, K.: Structural health monitoring: modeling of simultaneous effects of strain, temperature and vibration on the structure using a single apodized Π-phase shifted FBG sensor. Results Opt. **9**, 100323 (2022)
10. Wang, Q., Liu, Y.: Review of optical fiber bending/curvature sensor. Measurement **130**, 161–176 (2018)
11. Lee, K.-P., Yip, J., Yick, K.-L., Lu, C., Lu, L., Lei, Q.-W.E.: A novel force sensing smart textile: inserting silicone embedded FBG sensors into a knitted undergarment. Sensors. **23**(11), 5145 (2023)
12. Li, Y.L., Wang, Y.B., Wen, C.J.: Temperature and strain sensing properties of the zinc coated FBG. Optic **127**(16), 6463–6469 (2016)
13. Li, C., et al.: A review of coating materials used to improve the performance of optical fiber sensors. Sensors. **20**(15), 4215 (2020)
14. Ariati, R., Sales, F., Souza, A., Lima, R.A., Ribeiro, J.: Polydimethylsiloxane composites characterization and its applications: a review. Polymers **13**(23), 4258 (2021)
15. Morais, E., Pontes, M.J., Marques, C., Leal-Junior, A.: Liquid level sensor with Two FBGs embedded in a PDMS diaphragm: analysis of the linearity and sensitivity. Sensors. **22**(3), 1268 (2022)
16. Ansari, A., Trehan, R., Watson, C., Senyo, S.: Increasing silicone mold longevity: a review of surface modification techniques for PDMS-PDMS double casting. Soft Mater. **19**(4), 388–399 (2021)
17. Li, L.Q., et al.: Embedded FBG-based sensor for joint movement monitoring. IEEE Sens. J. **21**(23), 26793–26798 (2021)
18. Ge, J., James, A.E., Xu, L., Chen, Y., Kwok, K.W., Fok, M.P.: Bidirectional soft silicone curvature sensor based on off centered embedded fiber bragg grating. IEEE Photonics Technol. Lett. **28**(20), 2237–2240 (2016)
19. Othonos, A.: Fiber bragg gratings. Rev. Sci. Instrum. **68**(12), 4309–4341 (1997)
20. Wu, H., Liang, L., Wang, H., Dai, S., Xu, Q., Dong, R.: Design and measurement of a dual FBG high precision shape sensor for wing shape reconstruction. Sensors. **22**(1), 168 (2022)
21. Sun, K., Li, M., Wang, S.X., Zhang, G.K., Liu, H.B., Shi, C.Y.: Development of a fiber bragg grating enabled clamping force sensor integrated on a grasper for laparoscopic surgery. IEEE Sens. J. **21**(15), 16681–16690 (2021)
22. Skondras-Giousios, D., Karkalos, N.E., Markopoulos, A.P.: Finite element simulation of friction and adhesion attributed contact of bio-inspired gecko-mimetic PDMS micro-flaps with SiO(2) spherical surface. Bioinspir. Biomim. **15**(6), 066004 (2020)

3D Computational Modelling of ctDNA Separation Using Superparamagnetic Bead Particles in Microfluidic Device

Samla Gauri[1], Muhammad Asraf Mansor[2], and Mohd Ridzuan Ahmad[2(⊠)]

[1] Universiti Kebangsaan Malaysia, Bangi, Malaysia
[2] Division of Control and Mechatronics Engineering, Faculty of Electrical Engineering, Universiti Teknologi Malaysia, Kuala Lumpur, Johor, Malaysia
mdridzuan@utm.my

Abstract. The largest theoretical impact of the circulating tumour DNA (ctDNA) discovery is a non-invasive liquid biopsy which could substitute surgical tumour biopsy. This work describes the numerical simulation of ctDNA extraction and separation from the blood plasma of stage I and II cancer patients using superparamagnetic bead particles in a microfluidic platform. The ctDNA extraction and separation have great importance in early detection of cancer, especially in identifying precision medicine that can be prescribed. An average 5.7 ng of circulating tumor DNA was separated efficiently for every 10 μL blood plasma input based on the simulation result from the model using the COMSOL Multiphysics 5.3a software tool. The result provides a highly valuable tool for early evaluation of cancer management.

Keywords: Microfluidic device · Circulating tumor DNA (ctDNA) · COMSOL Multiphysics Simulation analysis

1 Introduction

Cancer is projected to be the leading cause of mortality and morbidity on a global scale. Significant developments in molecular biology over the last twenty years have enabled several enhanced cancer therapies, but these have also had multiple disadvantages, such as the lack of drugs to combat the majority of genomic aberrations and the risk of infections from the surgical removal of tumor tissue [1]. For that reason, recent research has shown that analysis of circulating tumor DNA (ctDNA) as a cancer biomarker can be done with a noninvasive liquid biopsy that gives researchers quick access to the tumor sample [2, 3].

CtDNA is single or double-stranded DNA that is expelled into the bloodstream or lymphatic system by primary malignant cells and tumor cells. CtDNA, which is fragmented tumor-derived DNA normally discovered in the serum or plasma of a patient's blood, and is capable of providing information and defining features of the primary tumor [4]. This has revealed valuable data as a potential cancer biomarker in liquid biopsy for early cancer detection and diagnosis. The detection of CtDNA as a biomarker for invasive

© The Author(s), under exclusive license to Springer Nature Singapore Pte Ltd. 2024
F. Hassan et al. (Eds.): AsiaSim 2023, CCIS 1912, pp. 221–231, 2024.
https://doi.org/10.1007/978-981-99-7243-2_19

cancer can disclose a wealth of biological data related to the tumor, and the implementation of the concept of precision medicine in line with the patient's needs is highly probable [4–6]. The quantity of ctDNA varies depending on the primary tumor's stage and location. Thus, rather than invasive liquid biopsy, ctDNA has tremendous potential in cancer management, especially in terms of detection, diagnosis, precision treatment selection, and monitoring or following disease conditions, including non-hematological and progression [7, 8]. Many researchers have demonstrated several techniques to analysis the efficacy of ctDNA as a cancer biomarker, for instance, droplet digital polymerase chain reaction (ddPCR) is recognised as the most accurate method for detecting existing mutations. Another technique is tagged-amplicon deep sequencing (TAm-Seq), which can sequence millions of DNA molecules. While the cancer personalized profiling by deep sequencing (CAPP-Seq) has ctDNA detection capability of 100% in stage IV cancer [9, 10].

Numerous studies have investigated the viability of ctDNA in the early detection of cancer stages throughout the progression of tumours, where the detection rate is 90–100% for stages II-IV compared to 50% for stage I for specific types of cancer [11]. The ctDNA is simple to separate and detect compare to circulating tumor cells (CTC) due to stability and prevalent in the bloodstream with a significant sensitivity [12–14]. One group demonstrated plasma ctDNA detection in cancer patients of stage 0.5–24 months with a sensitivity and specificity of 89% and 100%, respectively. This technique can provide a potential opportunity for medical treatment for cancer patients for up to 24 months [15]. The majority of the research provided thus far provides insights into ctDNA detection at nearly ultimate stage III or IV of cancer, where the amount of rising ctDNA may be seen in the circulatory system more clearly [16]. Despite ctDNA analysis is a noninvasive biopsy detection tool, it is only capable of finding cancer at a late stage and not addressed in the early stages of cancer growth. For that reason, we propose a computation study in the microfluidic environment to extract and separate the ctDNA from blood plasma (stage I and II cancer patients) by utilising the superparamagnetic (SPM) bead particle. The finite element analysis (FEA) COMSOL Multiphysics 5.3a software was used to perform the simulation.

2 Materials and Methods

The analysis was carried out using finite element COMSOL Multiphysics 5.3a software which is capable to perform simulating scientific and engineering problems [17]. Three simulation studies were conducted to perform the modeling of the microfluidic device, investigate ferromagnetic material for magnetic manipulation and ctDNA extraction by SPM bead particles.

2.1 Finite-Element Modeling

Figure 1 shows an illustration of the microfluidic device made up of polycarbonate material with a diameter of 75 mm length, 50 mm width and 1 mm thick. Polycarbonate is ideal for testing biological samples because less hydrophobic thus, improved filling behaviour for laminar flow in microchannel. Furthermore, the microfluidic channels may

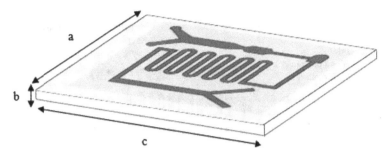

Fig. 1. Illustration of the microfluidic device with a = 50 mm, b = 1 mm and c = 75 mm

be utilised directly for other procedures that need higher temperatures, for example, PCR with chemicals such as diluted acids and oils deprived of a secondary reaction. The micro channel consists of three parts; the main rectangle channel with two inlets. Second was a curved microchannel for micro-mixing and the third part was a rectangle channel with an outlet for magnetic manipulation. Figure 2 shows the schematic design of microchannel. Two inlets at the left side of microchannel act as lysis solution where the blood plasma and buffer solution were injected, while another input was reserved for the SPM bead particle insertion. Adpted the concept of microfilteration of particle size to filter the impurities and thrombocytes plasma, the main channel consists of two different widths (a = 5 μm and b = 2 μm), used to separate the ctDNA from other impurities and platelets of a plasma sample. Normally the size of ctDNA is 2.6 nm wide with 100 bp long and thrombocytes are biconvex discoid structured with 2 μm to 3 μm in diameter whereas, SPM bead is 1 μm sphere particles.

As a result, ctDNA gets separated from thrombocytes and impurities before travelling to the next section of the microchannel. The third inlet width of 1.5 μm, the filtered ctDNA and SPM bead particles will be navigated to the second section of the curved microchannel, which was constructed for effective particle mixing. Circulating flow in a curved microchannel with regulated droplet motion and a low capillary number while releasing particles would increase the mixing rate in the microfluidic channel with a high surface-to-volume ratio. As a result, for properly connected or absorbed to SPM bead particles, ctDNA may intermingle fully with SPM bead particles and smooth fluid movement can be predicted without clogging concerns. A permanent magnet will be utilised to eliminate or separate ctDNA from SPM bead particles at the third part of the microchannel.

The magnetic field held the SPM beads and leave the ctDNA flow to the outlet. Different mesh sizes were considered for the simulation in order to provide accurate and mesh-independent results. For the microchannel and the substrate, a triangular mesh with 5 μm and 3 μm element sizes is utilised, respectively. In addition, the simulation was investigated in three time-dependent study phases: 1. Microfluidic channel flow; 2. Magnetic field in a microfluidic channel using a permanent magnet; 3. Particle extraction utilising a particle tracing module.

The permanent magnet, as illustrated in Fig. 3a, is made of a material that may be magnetized by an induced external magnetic field and stay magnetized even when no

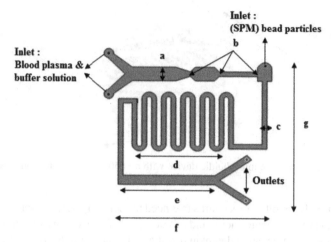

Fig. 2. Schematic of microfluidic channel with multi-width rectangle channel with dimensions of a = 5 m, b = 2 m, and c = 1.5 m is used to filter ctDNA from thrombocytes and plasma input impurities. The 'd' and 'e' are represents a mixing channel and separation channel respectively. The overall dimension of the microchannel are 55 mm (f) and 40 mm (g).

external magnetic field is supplied. This phenomenon is also known as ferromagnetism. The magnetic block will be positioned in a magnet array at the bottom of the microfluidic device with dimensions of 60 mm in length, 10 mm in width, and 5 mm in height, specifically to manipulate SPM bead particles to desorb or detach ctDNA from the beads (Fig. 3b). The parameters of the magnetic block were defined using the AC/DC module in COMSOL Multiphysics by selecting stable magnetic fields without a current interface. For microchannel applications where the magnetic field is constant throughout time, the magnetostatics method was duplicated for the magnetic block.

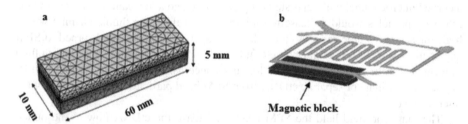

Fig. 3. a) A magnetic block built of ferromagnetic material for ctDNA separation. b) A magnetic block positioned in a magnetic array at the bottom of a microfluidic channel to control SPM beads with ctDNA.

2.2 Separation *of ctDNA* Using SPM Bead particle by Permanent Magnet manipulation

This study related to the separation of slow-moving microparticles using a magnetic field, where the program of computational fluid dynamics (CFD) in laminar flow, particle tracing module and magnetic field without applied current. The particle tracing module was chosen based on its capability to predict and compute particle trajectory in a fluid mixing situation influenced by electromagnetic force. The mixture and separation of samples were performed under conditions of isotropic continuous flow. Table 1 shows all the parameters utilised to set up the simulation.

3   Result and Discussion

3.1 Laminar Flow

Navier-Stokes equations were utilized to simulate the particles and fluid flow in the microfluidic channel. Several assumptions were introduced to mimic the real velocity flow in the microfluidic, such as the fluid flow was incompressible (density is constant over space and time), Newtonian flow was uniform viscosity and the fluid dynamics was based on non-slip boundary condition. The Navier-Stokes and Reynolds number equations were used to determine the velocity flow profile as shown below:

$$Re\left(\frac{\partial u}{\partial t} + (u \cdot \nabla)u\right) = -\frac{1}{\rho}\nabla\rho + v\nabla^2 u + f \tag{1}$$

$$Re = \frac{\rho u L}{v} \tag{2}$$

$$u \cdot \nabla = 0 \tag{3}$$

where, Re is non-dimensional number gives the ratio between inertial and viscous forces to indicate the flow regime, u (m/s) is the fluid and particle velocity field, ρ (kg/m^3) is fluid and particle density, v (m^2/s) is the kinematic viscosity, f (m/s^2) is an external acceleration field due to gravity and L (m) is the hydraulic diameter in the microchannel. The simulation result demonstrated the laminar flow of the particles at a boundary in the velocity distribution (see Fig. 4). To avoid blocking issues at a narrow microchannel, all particles were ejected in droplet motion. The fluid was assumed to flow through the device using positive displacement pumps in a low capillary number. As a result, a parallel parabolic velocity profile was observed in the microfluidic channel based on laminar flow and non-slip boundary condition.

3.2 Magnetic Material and Field

The magnetic field profile produced by the block placed underneath microchannel was tangential to the boundary on the xy-plane and parallel to the boundary on the xz-plane

Table 1. Parameters used

Description	Value
Average thrombocytes diameter	2.5 μm
Average thrombocytes density	1.08 g/μL
Average ctDNA diameter	2.6 nm
Average ctDNA length	90 bp
Average ctDNA density	1.7 g/cm³
Particle conductivity: ctDNA SPM bead particles input	0.20 ms/m 2 μL
Average SPM bead particles diameter	1.0 μm
Average SPM bead particles density	2.0 g/cm³
Particle conductivity: SPM bead particles	0.32 ms/m
Particle relative permittivity: SPM bead particles	59
Impurities diameter	2.0 μm to 1.5 μm
Average impurities density Sample input	1.08 g/μL 10 μL
Sample inflow rate	5 μL/min
SPM bead particles inflow rate Buffer solution input	5 μL/min 20 μL
Buffer solutions inflow rate	10 μL/min
Material fluid relative permittivity	80
Material fluid dynamic viscosity	0.001 μPa.s
Fluid medium conductivity	0.055 s/m
Fluid density	1000 g/m³

with zero magnetic scalar potential conditions. The magnetic force that formed the magnetic bar is defined by Ampere's circuital law as follows:

$$B = \frac{\mu_0 I}{2\pi r}, \mu_0 = 4\pi \times 10^{-7} H/m \tag{4}$$

$$\mu_0 = 4\pi \times \frac{10^{-7}(T \cdot m)}{A} \tag{5}$$

where B (T) is the magnetic field magnitude, μ_0 (N/A²) is the permeability of free space, I (A) is the magnitude of the electric current and r (m) is the distance interacting magnetic materials.

Based on the results of the model study, which are shown in Fig. 5, the magnetic fields of the different areas tend to line up with the magnetic field made by the magnet bar.

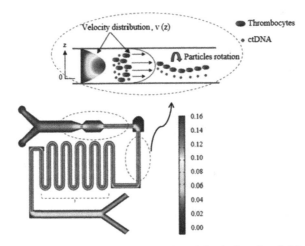

Fig. 4. Velocity magnitude profile (μm/s) of particles in the microfluidic channel.

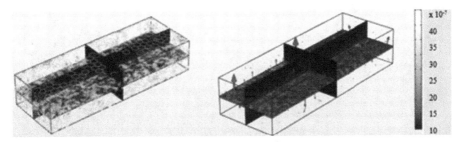

Fig. 5. The magnetization of the magnetic block used to separate SPM bead particles and ctDNA

3.3 ctDNA Separation and Extraction

The overview of the ctDNA filtration and separation was illustrated in Fig. 5. CtDNA was filtered to remove impurities and thrombocytes before being separated using SPM bead particles for high-yield results. SPM beads are made up of a nanometer-scaled superparamagnetic iron oxide core surrounded by a high purity porous silica shell. SPM beads are composed of a nanoscale superparamagnetic iron oxide core encased in a high-purity porous silica exterior. The superparamagnetism of SPM bead particles permits them to rotate into an anisotropic orientation. In addition, the feeble magnetic moment of SPM beads enables an external magnetic field generated by a magnetic block to magnetise and demagnetize the SPM beads depending on their application. The mechanical and chemical stability of the silica surface provides an excellent chromatography separation medium for biological samples. In this application, under conditions of high acidity and low ionic strength, the particles become negatively charged and exhibit specific binding in the absence of a magnetic field.

The mixture particle (ctDNA and SPM bead particles) with buffer solution was observed in a curved microchannel and then separated by magnetic manipulation. The

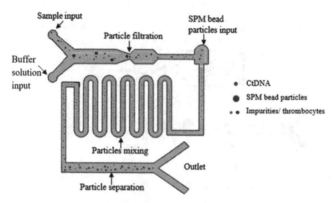

Fig. 6. The overview of the microfluidic system for ctDNA filtration and separation.

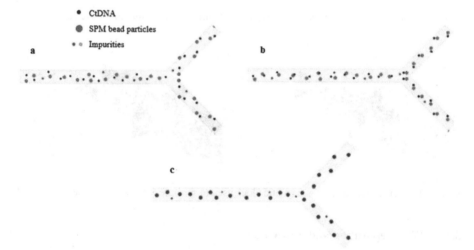

Fig. 7. a) The flow of ctDNA, SPM bead particles and impurities in the presence of magnetic field. b) ctDNA travel toward SPM bead particles. c) Binding of ctDNA to SPM bead particles, which can be removed by a magnetic field.

SPM bead particles were separated from ctDNA by using particle tracing module coupling with the electromagnetic field. The ctDNA tracing was studied in a time-dependent mode for every 10 s. Figure 6 shows the behaviour of ctDNA towards SPM bead particles and the separation of features. The presence of buffer solution enhances ctDNA binding to SPM bead particles. Once the ctDNA was successfully absorbed into the SPM bead particles, additional molecules and impurities were removed by holding the SPM bead containing the ctDNA in a magnetic field. The ctDNA was then eliminated from SPM bead particles using an elution buffer with a low pH and ionic strength. The ctDNA was extracted from SPM beads and collected at the outlet channel. The extracted ctDNA composition was estimated using the particle tracing equation from COMSOL

Multiphysics® (Eq. 6 and 7).

$$[-p + \mu(\nabla + \nabla u^T)]n = -\widehat{p_0}n \tag{6}$$

$$v = v_c - 2(n.v_c)n \tag{7}$$

where v_c represents the particle velocity when it collides with the wall and changes for each matching solution thus solved by COMSOL Multiphysics. Figure 7 illustrates the ultimate separation and extraction of ctDNA. The powerful magnetic response of the beads with great stability using chemical as well as mechanical methods enabled ctDNA separation in a magnetic field. As illustrated in Fig. 7, the ferromagnetic material (magnetic block) achieves saturation when nearly all of the magnetic domains are aligned in magnetising force in a positive direction. When the magnetising force is lowered to zero, some magnetic flux remains in the material.

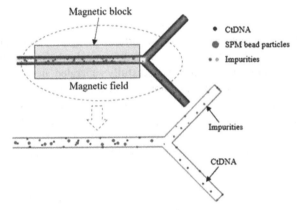

Fig. 8. The magnetic field created by the magnetic block is used to separate ctDNA from SPM bead particles.

The separation stage is critical for obtaining a high and pure yield of ctDNA. Despite most of the impurities were expected to be washed away, some remained in the long channel along with DNA and SPM beads. The magnetic block's ability to bind SPM beads in a suspension of ctDNA and impurities in the microfluidic channel was observed under the influence of a modest electromagnetic field of ≈10 mT. According to theoretical calculations, 5.7 ng of ctDNA were extracted for every 10 μL of total plasma input (see Fig. 8). The simulation demonstrated that the device has a sensitivity and specificity for ctDNA detection of 65.57& and 95.38%, respectively, for samples of stages I and II cancer patients. Even though the final yield has a residual cell concentration of less than 0.5%, the microchannel design with filtration method and separation by SPM bead particles has demonstrated the specific ctDNA extraction from a sample of cancer patients in stages I and II. This also demonstrated that the proposed device is capable of cancer detection at an early stage (Fig. 9).

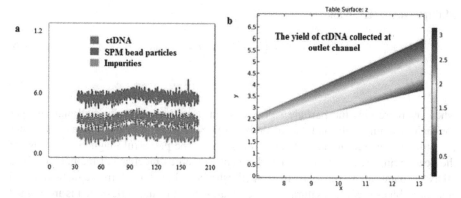

Fig. 9. a) Enumeration of outcomes obtained at the outlet channel for each 10 μL of sample, b) Enumeration of CtDNA for each 10 μL of sample.

4   Conclusions

In conclusion, the manipulation of SPM bead particles to separate ctDNA in the presence of a magnetic field provides an advantageous and practical way for fast and early cancer analysis. This simulation study, which successfully separated an average of 5.7 ng of ctDNA for every 10 L of blood plasma input from cancer patients in stages I and II. In addition, the particle tracing module detected ctDNA with a sensitivity and specificity of 65.57% and 95.38%, respectively. The simplicity of usage and adaptability of SPM bead particles are the major reasons they are frequently used in clinical research associated with the production of biological samples including complex matrices that generally limit direct examination.

Acknowledgment. The research was supported by the Ministry of Higher Education of Malaysia and Universiti Teknologi Malaysia [no. Q.J130000.21A2.06E67] in the project 'Development of Microfluidic Systems in PDMS for Microalgae Detection and Separation for Renewal Energy Application. We thank them for funding this project and for their endless support.

References

1. Sumbal, S., et al.: Circulating tumor DNA in blood: future genomic biomarkers for cancer detection. Exp. Hematol. **65**, 17–28 (2018)
2. Reinert, T., et al.: Analysis of circulating tumour DNA to monitor disease burden following colorectal cancer surgery. Gut **65**, 625–634 (2016)
3. Sorber, L., et al.: Circulating cell-free nucleic acids and platelets as a liquid biopsy in the provision of personalized therapy for lung cancer patients. Lung Cancer **107**, 100–107 (2017)
4. Han, X., Wang, J., Sun, Y.: Circulating tumor DNA as biomarkers for cancer detection. Genomics Proteomics Bioinform. **15**, 59–72 (2017)
5. Soyano, A.E., Baldeo, C., Kasi, P.M.: Adjunctive use of circulating tumor DNA testing in detecting pancreas cancer recurrence. Front. Oncol. **9** (2019)

6. Papakonstantinou, A., et al.: Prognostic value of ctDNA detection in patients with early breast cancer undergoing neoadjuvant therapy: a systematic review and meta-analysis. Cancer Treat. Rev. **104** (2022)
7. Fiala, C., Diamandis, E.P.: Utility of circulating tumor DNA in cancer diagnostics with emphasis on early detection. BMC Med. **16**, 166 (2018)
8. Merker, J.D., et al.: Circulating tumor DNA analysis in patients with cancer: American society of clinical oncology and college of American pathologists joint review. Arch. Pathol. Lab. Med. **142**, 1242–1253 (2018)
9. Neumann, M.H.D., Bender, S., Krahn, T., Schlange, T.: CtDNA and CTCs in liquid biopsy – current status and where we need to progress. Comput. Struct. Biotechnol. J. **16**, 190–195 (2018)
10. Li, H., et al.: Circulating tumor DNA detection: a potential tool for colorectal cancer management (review). Oncol. Lett. (2018). https://doi.org/10.3892/ol.2018.9794
11. Kang, G., et al.: Monitoring of circulating tumor DNA and its aberrant methylation in the surveillance of surgical lung cancer patients: protocol for a prospective observational study. BMC Cancer **19**, 579 (2019)
12. Chaudhuri, A.A., Binkley, M.S., Osmundson, E.C., Alizadeh, A.A., Diehn, M.: Predicting radiotherapy responses and treatment outcomes through analysis of circulating tumor DNA. Semin. Radiat. Oncol. **25**, 305–312 (2015)
13. Guan, Y., et al.: High-throughput and sensitive quantification of circulating tumor DNA by microfluidic-based multiplex PCR and next-generation sequencing. J. Mol. Diagn. **19**, 921–932 (2017)
14. Abou Daya, S., Mahfouz, R.: Circulating tumor DNA, liquid biopsy, and next generation sequencing: a comprehensive technical and clinical applications review. Meta Gene **17**, 192–201 (2018)
15. Coombes, R.C., et al.: Personalized detection of circulating tumor DNA antedates breast cancer metastatic recurrence. Clin. Cancer Res. **25**, 4255–4263 (2019)
16. Marczynski, G.T., Laus, A.C., dos Reis, M.B., Reis, R.M., de Lima Vazquez, V.: Circulating tumor DNA (ctDNA) detection is associated with shorter progression-free survival in advanced melanoma patients. Sci. Rep. **10**, 1–11 (2020)
17. Mansor, M.A., Ahmad, M.R.: A simulation study of single cell inside an integrated dual nanoneedle-microfludic system. J. Teknol. **78**, 1–6 (2016)

Ensemble Differential Evolution with Simulation-Based Hybridization and Self-Adaptation for Inventory Management Under Uncertainty

Sarit Maitra$^{(\boxtimes)}$, Vivek Mishra , Sukanya Kundu , and Maitreyee Das

Alliance University, Bengaluru, India
sarit.maitra@gmail.com, sarit.maitra@alliance.edu.in

Abstract. This study proposes an Ensemble Differential Evolution with Simulation-Based Hybridization and Self-Adaptation (EDESH-SA) approach for inventory management (IM) under uncertainty. In this study, DE with multiple runs is combined with a simulation-based hybridization method that includes a self-adaptive mechanism that dynamically alters mutation and crossover rates based on the success or failure of each iteration. Due to its adaptability, the algorithm is able to handle the complexity and uncertainty present in IM. Utilizing Monte Carlo Simulation (MCS), the continuous review (CR) inventory strategy is examined while accounting for stochasticity and various demand scenarios. This simulation-based approach enables a realistic assessment of the proposed algorithm's applicability in resolving the challenges faced by IM in practical settings. The empirical findings demonstrate the potential of the proposed method to improve the financial performance of IM and optimize large search spaces. The study makes use of performance testing with the Ackley function and Sensitivity Analysis with Perturbations to investigate how changes in variables affect the objective value. This analysis provides valuable insights into the behavior and robustness of the algorithm.

Keywords: Ackley Function · Differential Evolution · Evolutionary Algorithm · Ensemble Optimization · Inventory Management · Self-Adaptive

1 Introduction

Inventory management (IM) encounters significant challenges in dealing with uncertainty and stochastic demand. Within the field of operations research, optimizing the IM process has emerged as a prominent area of study. Since the objective functions in IM are multidimensional, non-convex, non-differentiable, non-continuous, and have several local optima, they pose a tough challenge. Analytically solving these problems is not only challenging but, in many cases, impossible. Consequently, metaheuristic algorithms have been identified as advantageous over traditional algorithms in solving the optimal solutions to such problems (e.g., [3, 15, 17], etc.). However, they do not guarantee finding the globally optimal solution. This means that they may not always converge on

F. Hassan et al. (Eds.): AsiaSim 2023, CCIS 1912, pp. 232–246, 2024.
https://doi.org/10.1007/978-981-99-7243-2_20

the globally optimal solution. They rely on heuristics and search strategies that balance exploration and exploitation, aiming to improve the solution iteratively. Therefore, there is always a possibility of getting trapped in local optima, especially in complex and multimodal optimization problems like IM, where the suboptimal solutions are in the vicinity of the current search space.

We explore the scope of the existing DE approach to solving optimization problems. Through this work, we introduce EDESH-SA to overcome the trap of local minima and enhance the performance of the algorithm. DE has a competitive advantage over related EAs in terms of floating-point encoding [35]. According to a recent study, 158 out of 192 papers were published between 2016 and 2021, showing that academics have improved DE to increase its effectiveness and efficiency in handling a variety of optimization challenges [2]. This clearly indicates a consistent usage pattern of DE for optimization purposes.

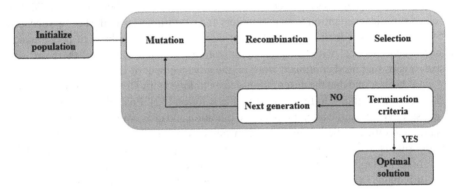

Fig. 1. Evolutionary algorithm procedure

At the center of the DE algorithm is the sequential optimization procedure as well as the interaction of numerous components, including selection, mutation, cross-over, and fitness evaluation (see Fig. 1). To optimize the decision variables (control factors), the proposed EDESH-SA integrates both control and noise elements. It strengthens the resilience and convergence of the optimization process while increasing the diversity of solutions. Ackley function and sensitivity analysis further establish the strength of the proposed solution.

The key contribution of this work lies in the development of EDESH-SA which combines statistical analysis, simulation, and optimization to overcome the shortcomings of the conventional DE algorithm. It provides an effective framework for decision-making, improves the performance of the DE algorithm, and offers valuable insights into the optimization process in the context of IM under uncertainty.

2 Previous Work

Optimization techniques provide a comprehensive framework for determining the most effective inventory policies, optimal inventory levels, and other key characteristics. The importance of optimization has been emphasized by several authors (e.g., [4, 11, 22, 23]). Studies (e.g., [24, 27, 28]) and industry reports suggest that IM costs can range from ≥20–40% of the total SCM costs. Researchers have conducted a study on optimization under uncertainty and proposed a simulation-optimization approach [12]. Their findings presented an analytical report showing a 16% reduction in supply and managing costs by implementing the optimal policy. Probability distribution and MCS have been part of IM for a long time (e.g., [14, 30]; etc.). Some of the recent work strengthens the argument by emphasizing the use of MCS in their work ([19] and [20]).

IM under uncertainty is challenging to solve due to the non-linearity of the model and several local optimum solutions ([9] and [16]). As a result, metaheuristic algorithms are frequently employed as a powerful solution for IM ([8, 13]). They found that DE has received significant attention from researchers and practitioners over the last two decades [25]. Since the introduction of DE [31], it has undergone a plethora of performance enhancements to deal with complex optimization problems. However, there are still research gaps that need continued work on the advancement of DE.

It has been found that the ensemble-based technique helps DE choose its own parameter control, changing it to fit the most viable solution [32]. Ensemble-based DE could be considered a dynamic attempt to handle mutation and crossover techniques in a more constrained setting while maintaining a constant convergence rate. Though self-adaptive DE algorithms with multiple strategies were proposed [7], some authors emphasized the need for self-adaptive DE with minimum intervention [32]. It was noticed that, despite their infrequent explicit application, self-adaptation procedures are a novel adaptation tool with a high level of robustness [21].

While all the researchers acknowledged the challenges and opportunities in IM under uncertainty, the field is evolving, especially post pandemic. A growing body of research can be seen incorporating stochastic demand models, evaluating the impact of inventory policies, and exploring the capabilities of various optimization techniques. Metaheuristic optimization is a promising area of research (e.g., [1, 8, 10, 29, 33]; etc.).

IM is an evolving field where the goal is to enhance decision-making processes and address the challenges posed by uncertainties, such as those experienced during and post-pandemic. Though there is an existing body of knowledge on MCS, ensemble-based techniques, and self-adaptive DE algorithms separately, a combination of these approaches in one unified framework has not been explored. There is an opportunity to add value by proposing and investigating a novel approach that integrates all.

3 Methodology

Through the research gap, this methodology combines statistical analysis, MCS, ensemble, and self-adaptive DE. The data for demand information was collected for multiple products over the last 12 months. This was necessary to mathematically model the stochastic demand. We have calculated the best inventory levels that balance the costs of retaining stocks against the costs of stockouts by modelling numerous scenarios. These levels of inventory balance the probabilities of experiencing various levels of demand. It was assumed here that order sizes would have unknown values from a lognormal distribution, which is often the case in real-life scenarios. By running the simulations, a certain number of times (10,000 times in this case), the algorithm tries a variety of scenarios and fine-tunes the inventory policy to maximize profits. To model the inventory's performance over a year, we ran this simulation for a full 365 days.

The major sources of uncertainty are identified here:

- Unpredictable purchase behavior of the customer:

$\rho = (number of orders last year)/(number of working days)$, where ρ is the probability of a customer placing an order on any given day. However, this model is not perfect, and there may be some level of error in the estimated probability of a customer placing an order.

- Variability in order size is modelled using a log-normal distribution and this is frequent practice in real-life scenarios. While past sales data is used to predict distribution parameters, actual order sizes may still differ greatly from these projections.

[26] conducted an extensive literature review of over seven decades and discovered that the CR is the most employed policy in stochastic inventory literature. Taking a clue from their work, we examined the inventory policy by inducing CR in our empirical analysis.

4 Data

The business case selected here examines the sale of four distinct products and considers adopting a suitable IM policy. The historical demand data is used to calculate the central tendency of the data, such as the mean demand during lead time, the standard deviation, and the minimum and maximum demand levels (see Table 1).

- PrA, PrB, PrC, and PrD are four distinct products.
- PC = purchase cost of one unit of the item from the supplier.
- LT = lead time to deliver the item after placing an order.
- Size = quantity of each item.
- SP = selling price of each item.
- SS = starting stock of each item in the inventory.
- μ = average demand for each item over a given period.
- σ = standard deviation of demand
- OC = order cost.
- HC = holding cost of one unit of inventory for a given period.
- ρ = the probability of a stock-out event occurring, i.e., the probability of demand exceeding the available inventory level.
- LT_{demand} = the demand during the lead time.

The shape of the KDE curves (see Fig. 2) provides insights into the underlying stochastic distribution of the data. The isolated peaks in the curves show potential outliers in the demand data.

Table 1. Summary statistics

	Pr A	Pr B	Pr C	Pr D
PC	€ 12	€ 7	€ 6	€ 37
LT	9	6	15	12
Size	0.57	0.05	0.53	1.05
SP	€ 16.10	€ 8.60	€ 10.20	€ 68
SS	2750	22500	5200	1400
σ	37.32	26.45	31.08	3.21
μ	103.50	648.55	201.68	150.06
OC	€ 1000	€ 1200	€ 1000	€ 1200
HC	€ 20	€ 20	€ 20	€ 20
ρ	0.76	1.00	0.70	0.23
LT_{demand}	705	3891	2266	785

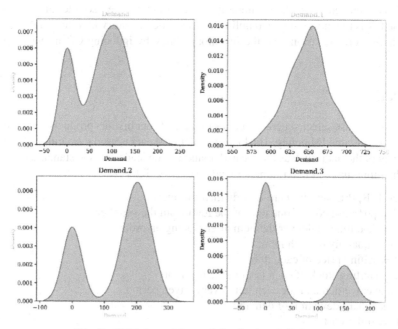

Fig. 2. KDE plots of demand distribution (365 days)

To demonstrate the methodology, we have used a contrived example that has been streamlined and simplified. We are able to concentrate on the essential components to make it simpler to comprehend and analyze the underlying concepts by removing superfluous complexity.

5 Ensemble Differential Evolution with Simulation-Based Hybridization and Self-adaptation

The proposed approach starts with finding the combination of control and noise factors the maximize the revenue and minimize the cost. The objective function in this approach calculates the profit as the weighted difference between the revenue and the cost. The DE algorithm is used to search for the optimal solution that achieves the highest profit. Figure 3 displays the workflow of the proposed solution, which is an improvised version of DE.

The control factors in Fig. 3 include LT, Size, and SS. These are within the control of the decision-makers and can be adjusted to optimize the objective function. The noise factors introduce uncertainty into the optimization problem. These factors include PC and LT_{demand}, which are not directly under the control of the decision-makers and may vary, leading to variability in the optimization results. By incorporating both control and noise factors, the optimization problem accounts for both controllable and uncontrollable variables, enabling the determination of the optimal solution considering the uncertainties present in the system.

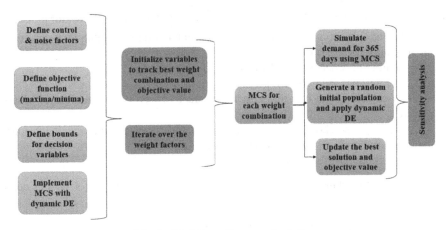

Fig. 3. Workflow of proposed solution

The reorder point, order quantity, and safety stock play crucial roles for CR. When the inventory level falls below the reorder point, instead of placing an order and waiting for delivery, the inventory is immediately replenished using the available starting stock and the lead time.

The reorder point (RP) is calculated as:

$$RP = LT_{demand} + SSF * \sigma \tag{1}$$

where, Safety stock factor (SSF) accounts for the variability in demand during the lead time and is estimated from the $\Sigma LT * \mu$

The order quantity (OQ):

$$\left(\frac{RP - SS}{OQ_{min}}\right) * OQ_{min}[\text{x } [0]] \tag{2}$$

where, x [0] represents the index of the control factor, OQ_{min} = Minimum order quantity corresponding to the control factor index x [0].

The revenue is calculated by multiplying the selling price of each product by the number of units sold for that product over the entire year. So, for each product, we add up the sales quantities for all 365 days and multiply them by the selling price. Annual profit is formulated as:

$$SP_i \sum_{t=1}^{365} S_{i,t} - \left\{ \left(\frac{20HC_i}{365}\right) \sum_{t=1}^{365} I_{i,t} + N_i OC_{o,i} + \sum_{t=1}^{365} SC_i P_{i,t} \right\} \tag{3}$$

here,

- SP_i = selling price of product i, $\sum_{t=1}^{365} S_{i,t}$ = sum of the sales quantities (S) for product i over the 365 days of the year. This component represents the total revenue generated by selling the product throughout the year.
- HC_i = unit holding cost for product i, $\left(\frac{20HC_i}{365}\right) \sum_{t=1}^{365} I_{i,t}$ = holding cost for product i over the 365 days of the year; here it is 20 for all the products (Table 1). It considers the inventory levels (I) for each day.
- N_i = number of orders placed for product i during the year.
- $OC_{(o,i)}$ = unit ordering cost for product i.
- $N_i OC_{o,i}$ calculates the ordering cost for product i based on the number of orders placed.
- $\sum_{t=1}^{365} SC_i P_{i,t}$ = stockout cost for product i over the 365 days of the year. It considers the inventory levels (P) and demand for each day.
- SC_j = stockout cost for product i.

The net profit obtained from selling the product after accounting for various costs associated with inventory holding, ordering, and stockouts.

The total cost is computed based on the equation:

$$\text{cost} = \sum (PC_i * OQ_i) + OC_i * I(OQ_i > 0) + \sum HC_i * \max(x_i - LT_{demand\ i}, 0)) + \sum SC_i * \max(LT_{demand\ i} - x_i, 0)) \tag{4}$$

$SC = 20\% * \mu = [20.70, 129.71, 40.34, 30.01]$, assuming 20% is the stock out cost.

By performing MCS multiple times and averaging the costs and revenues, the objective function provides an estimation of the average cost and average revenue.

To find the inventory levels for each product that maximize the profit, we create an objective function that combines this revenue (Eq. (3)) and cost components (Eq. (4)). To balance maximizing revenue and minimizing costs, we introduce two weights ($\omega 1$ and $\omega 2$) that represent their relative importance.

$$f(x) = \omega_1 * \left[SP_i \sum_{t=1}^{365} S_{i,t} - \left\{ \left(\frac{20V_i}{365}\right) \sum_{t=1}^{365} I_{i,t} + N_i C_{o,i} + \sum_{t=1}^{365} c_i P_{i,t} \right\} \right] - \tag{5}$$
$$(1 - \omega_1) * Cost_{optimal}$$

Depending on the business priorities, the weight can be adjusted to prioritize revenue maximization or cost minimization, cost optimization or revenue maximization.

In this case, by incorporating both the sales revenue and average cost terms in the objective function $(f(x))$, the optimization algorithm aims to find the inventory levels x that not only minimize the average cost but also maximize the sales revenue. This provides a comprehensive approach to finding a better solution that considers both financial aspects (costs) and business performance (revenue).

5.1 Improvised Differential Evolution

Centeno-Telleria et al.'s [6] findings on the flexibility of DE, the benefits of adaptive techniques, and the emphasis on exploring the solution space are consistent with the characteristics of our work. The population matrix with the parameter vectors has the form:

$$x_{n,i}^g = \left\{ x_{n,1}^g, x_{n,2}^g, x_{n,3}^g, \ldots, x_{n,d}^g \right\} \tag{6}$$

where n = population size, g = generation and n = 1, 2, 3, ..., N

$$x_{n,i} = x_{n,i}^L + \text{rand}()*\left(x_{n,i}^u - x_{n,i}^l \right), \ i = 1, 2, 3, \ldots \text{D and } n = 1, 2, 3, \ldots \text{N} \tag{7}$$

- Three other vectors can be selected randomly from each parameter vector during the mutation phase such as x_{r1n}^g, x_{r2n}^g, and x_{r3n}^g. If we add the weighted difference of two of the vectors to the third, the equation becomes:

$$v_n^{g+1} = x_{r1n}^g + \omega\left(x_{r2n}^g - x_{r3n}^g \right), \ n = 1, 2, 3, \ldots \text{N} \tag{8}$$

where, v_n^{g+1} = donor vector, ω is between 0 and 2.

Instead, random selection, we employed MCS to choose the vectors based on their fitness values. The selection process is modified by selecting three vectors using MCS based on their fitness values: x_{r1n}^g, x_{r2n}^g, *and* x_{r3n}^g *and Eq. (8) is modified as below:*

$$v_n^{g+1} = x_{r1n}^g + \omega_n(g)\left(x_{r2n}^g - x_{r3n}^g \right), n = 1, 2, 3, \ldots N \tag{9}$$

where $\omega_n(g)$ *is the mutation factor for individual n at generation g. Instead of using a fixed* ω *value, we used a strategy to dynamically adapt* ω *based on the performance of the algorithm.*

- During recombination phase, a trial vector $u_{n,i}^{g+1}$ is developed from the target vector $\left(x_{n,i}^g \right)$ and the donor vector $\left(v_{n,i}^{g+1} \right)$

$$u_{n,i}^{g+1} = \begin{cases} v_{n,i}^{g+1} \text{if} \, rand() \leq C_p \, ori = I_{rand} i = 1, 2, 3, \ldots D \, and \\ x_{n,i}^g \text{if} \, rand() > C_p \, and i \neq I_{rand} n = 1, 2, 3, \ldots N \end{cases} \tag{10}$$

where, I_{rand} is an integer random number between [1, D] and C_p is the recombination probability.

> We incorporate self-adaptive mechanisms, which allow the algorithm to dynamically adjust its parameters during the optimization process. Our proposed approach uses Adaptive Mutation Control, where we assign different mutation factors to different individuals based on their performance. This is done by associating a scaling factor with each vector and updating it based on their fitness improvement. We also employed Adaptive Recombination Control, where the recombination probability (C_p) is adaptively adjusted based on the current population's behavior. High recombination probabilities encourage exploration, while low values promote exploitation.

- During the selection phase, the target vector $x_{n,i}^g$ is compared with the trial vector $u_{n,i}^{g+1}$ and the one with the lowest function value is chosen for the following generation.

$$x_n^{g+1} = \begin{cases} u_{n,i}^{g+1} \text{ iff } \left(u_n^{g+1}\right) < f\left(x_n^g\right) \\ x_{n \text{ otherwise}}^g \end{cases} \quad n = 1, 2, \dots N \qquad (11)$$

All the evolution phases are repeated until a predefined termination criterion is satisfied.

> Ensemble methods involve combining multiple instances of the optimization algorithm to improve overall performance. Each instance operates independently but shares information periodically to guide the evolution. By combining results from multiple (5) optimizers, the algorithm can avoid getting stuck in local minima and provide a better global optimal solution. The ensemble methods are integrated into DE:

The improved version outlined above addresses several shortcomings of conventional DE algorithms, including convergence to local optima, lack of adaptability, limited exploration and exploitation capabilities, sensitivity to control parameters, and lack of robustness.

Table 2 displays a comparative analysis with conventional DE, wherein we have removed the MCS optimization loop and the adaptive mutation/crossover rate adjustments.

The conventional DE achieved a best objective value of 474,230, while the proposed DE algorithm achieved a significantly higher best objective value of 58,760,547. The objective value represents the optimized balance between maximizing revenue and minimizing costs. Moreover, the conventional DE achieved an annual profit of 30,335,547.85, while the proposed DE algorithm achieved a higher annual profit of 67,102,330.00.

Table 2. Comparison of Optimization Results

Description	Conventional DE	Proposed DE
Best weight combination	[0.7, 0.3]	[0.7, 0.3]
Best objective value	474,230	58,760,547
Mutation & cross-over	-	[0.5, 0.9]
Best solution	[0.93, 1.56, 0.98, 2.73, 2.87]	[3.0, 0.59, 0.42, 2.29, 1.89]
Annual profit	30,335,547.85	67,102,330.00

5.2 Algorithmic Performance

Ackley function tests and sensitivity analyses were performed to gain insights into the performance characteristics of the differential evolution algorithm. These analyses provide valuable information about its efficiency, convergence behavior, and solution quality.

Ackley Function.

The Ackley function performance test shows the optimization algorithm's capacity to locate the best solution in a challenging and multidimensional search space. The contour plot in Fig. 4 visualizes the contours and valleys of the function, providing insights into the behavior of the objective function. The decision variables [2.63, 2.63, 1.00, 1.78, 1.33] correspond to the choice variable value that the DE algorithm finds to be the best. These values represent the optimized combination of control and noise factors that maximize the objective function. The objective value achieved is approximately -10.39, indicating the maximum value of the function obtained after optimizing the decision variables.

The 3D mesh grid plot (see Fig. 5) further illustrates the optimized combination of factors and the location of the best solution in the solution space. The red dot on the plot represents the optimal point where the function reaches its maximum value. The negative value of the objective function indicates the algorithm's success in finding a point of high fitness.

The Ackley function is known for its complex landscape with multiple local minima. Despite this challenge, the differential evolution algorithm effectively navigated the search space and found a point that maximizes the objective function. This demonstrates the algorithm's ability to handle complex and non-linear optimization problems.

Sensitivity Analysis.

The use of perturbation analysis to calculate parameter sensitivities is an efficient method. Some researchers have made significant contributions to the field of sensitivity analysis and perturbation approaches in mathematical programming and optimization problems [5, 18, 34]. Their contributions provide a larger background and framework for the methodology used in our study. We compared the objective values before and after each variable perturbation to see how it affected the objective value. Here, the objective

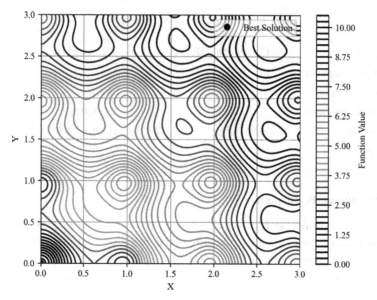

Fig. 4. Ackley Function Contour Plot

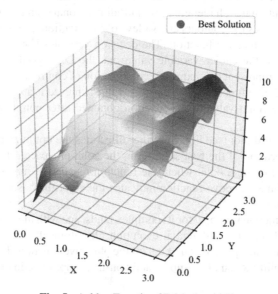

Fig. 5. Ackley Function 3D Mesh grid Plot

value worsens after the multiple perturbations, showing that changing the variable has a negative impact on the objective.

The x-axis (see Fig. 6) represents the variables being perturbed, and the y-axis represents the change in the objective value due to the perturbations. The blue line represents the change in the objective value when each variable is perturbed in the positive direction (increased). The orange line represents the change in the objective value when each variable is perturbed in the negative direction (decreased). If the blue line is below zero ($y = 0$) for a particular variable, it indicates that increasing the value of that variable leads to a worse objective value. Here, the orange line is below zero ($y = 0$) for a particular variable, which indicates that decreasing the value of that variable leads to a worse objective value.

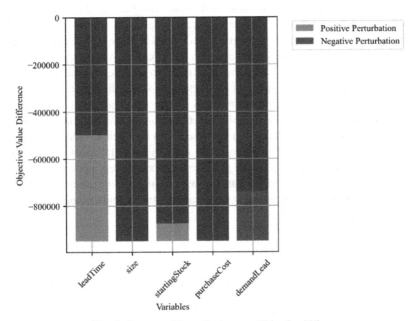

Fig. 6. Impact of Perturbations on Objective Value

Through this graphical representation, the sensitivity and impact of variable perturbations on the objective value are effectively demonstrated.

5.3 Limitations

The proposed approach primarily focuses on optimizing internal IM decisions, such as reorder points, order quantities, and safety stock levels. However, there are external factors that impact IM, such as supply chain disruptions, supplier reliability, transportation constraints, or changing market conditions. Incorporating these external factors into the optimization model could provide a more realistic and robust solution. Moreover, the provided solution does not address scalability, which can be achieved by adjusting parameters, employing parallelization techniques, optimizing algorithmic efficiency, and considering problem decomposition strategies. These factors should be considered when applying the solution to larger and more complex IM systems.

6 Conclusion

To improve the performance of the optimization technique for IM, this work provided a simulation-based hybridization strategy mixed with ensemble DE. By incorporating a self-adaptive approach, the algorithm can dynamically adjust mutation and crossover rates based on the success or failure of each iteration, further improving its effectiveness. The proposed approach considers both control and noise factors, optimizing decision variables while accounting for system stochasticity. Through MCS, the algorithm evaluates the performance of implemented inventory policies under different demand scenarios, providing a more comprehensive optimization process. The effectiveness of the proposed solution is demonstrated using an empirical analysis of the optimal combination of control and noise factors that maximize profit. The algorithm successfully navigates the complex search space of the Ackley function, finding the best solution despite the presence of multiple local minima. Furthermore, sensitivity analysis highlights the impact of perturbations on the objective value, providing insights into the sensitivity of the solution to changes in variables. This research advances optimization methods for IM with stochastic demands. By combining metaheuristic optimization, simulation, and sensitivity analysis, the proposed approach offers a practical and robust solution for real-world IM problems. Future research can further explore the application of these techniques in different business environments and extend the analysis to include additional factors and constraints.

References

1. Abdi, A., Abdi, A., Fathollahi-Fard, A.M., Hajiaghaei-Keshteli, M.: A set of calibrated metaheuristics to address a closed-loop supply chain network design problem under uncertainty. Int. J. Syst. Sci. Oper. Logist. **8**(1), 23–40 (2021)
2. Ahmad, M.F., Isa, N.A.M., Lim, W.H., Ang, K.M.: Differential evolution: a recent review based on state-of-the-art works. Alex. Eng. J. **61**(5), 3831–3872 (2022)
3. Alejo-Reyes, A., Mendoza, A., Olivares-Benitez, E.: Inventory replenishment decision model for the supplier selection problem using metaheuristic algorithms. Math. BIosci. Eng. **15**, 1509–1535 (2021)
4. Bag, S., Wood, L.C., Mangla, S.K., Luthra, S.: Procurement 4.0 and its implications on business process performance in a circular economy. Resour. Conserv. Recycl. **152**, 104502 (2020)
5. Castillo, E., Conejo, A.J., Castillo, C., Mínguez, R., Ortigosa, D.: Perturbation approach to sensitivity analysis in mathematical programming. J. Optim. Theory Appl. **128**, 49–74 (2006)
6. Centeno-Telleria, M., Zulueta, E., Fernandez-Gamiz, U., Teso-Fz-Betoño, D., Teso-Fz-Betoño, A.: Differential evolution optimal parameters tuning with artificial neural network. Mathematics **9**(4), 427 (2021)
7. Deng, G., Gu, X.: A hybrid discrete differential evolution algorithm for the no-idle permutation flow shop scheduling problem with makespan criterion. Comput. Oper. Res. **39**(9), 2152–2160 (2012)
8. Fahimnia, B., Davarzani, H., Eshragh, A.: Planning of complex supply chains: a performance comparison of three meta-heuristic algorithms. Comput. Oper. Res. **89**, 241–252 (2018)
9. Fallahi, A., Bani, E.A., Niaki, S.T.A.: A constrained multi-item EOQ inventory model for reusable items: reinforcement learning-based differential evolution and particle swarm optimization. Expert Syst. Appl. **207**, 118018 (2022)

10. Faramarzi-Oghani, S., Dolati Neghabadi, P., Talbi, E.G., Tavakkoli-Moghaddam, R.: Meta-heuristics for sustainable supply chain management: a review. Int. J. Prod. Res. 1–31 (2022)
11. Fonseca, L.M., Azevedo, A.L.: COVID-19: outcomes for global supply chains. Manage. Market. Challenges Knowl. Soc. **15**(s1), 424–438 (2020)
12. Franco, C., Alfonso-Lizarazo, E.: Optimization under uncertainty of the pharmaceutical supply chain in hospitals. Comput. Chem. Eng. **135**, 106689 (2020)
13. Goodarzian, F., Wamba, S.F., Mathiyazhagan, K., Taghipour, A.: A new bi-objective green medicine supply chain network design under fuzzy environment: hybrid metaheuristic algorithms. Comput. Ind. Eng. **160**, 107535 (2021)
14. Huijbregts, M.A.: Application of uncertainty and variability in LCA. Int. J. Life Cycle Assess. **3**, 273–280 (1998)
15. Kang, H., Liu, R., Yao, Y., Yu, F.: Improved Harris hawks optimization for non-convex function optimization and design optimization problems. Math. Comput. Simul **204**, 619–639 (2023)
16. Khalilpourazari, S., Pasandideh, S.H.R., Niaki, S.T.A.: Optimization of multi-product economic production quantity model with partial backordering and physical constraints: SQP, SFS, SA, and WCA. Appl. Soft Comput. **49**, 770–791 (2016)
17. Khan, A.T., Cao, X., Brajevic, I., Stanimirovic, P.S., Katsikis, V.N., Li, S.: Non-linear activated beetle antennae search: A novel technique for non-convex tax-aware portfolio optimization problem. Expert Syst. Appl. **197**, 116631 (2022)
18. Kiran, R., Li, L., Khandelwal, K.: Complex perturbation method for sensitivity analysis of nonlinear trusses. J. Struct. Eng. **143**(1), 04016154 (2017)
19. Krumscheid, S., Nobile, F., Pisaroni, M.: Quantifying uncertain system outputs via the multi-level Monte Carlo method—Part I. Central moment estimation. J. Comput. Phys. **414**, 109466 (2020)
20. Luengo, D., Martino, L., Bugallo, M., Elvira, V., Särkkä, S.: A survey of Monte Carlo methods for parameter estimation. EURASIP J. Adv. Signal Process. **2020**(1), 1–62 (2020)
21. Meyer-Nieberg, S., Beyer, H.G.: Self-adaptation in evolutionary algorithms. In: Parameter Setting in Evolutionary Algorithms, pp. 47–75. Springer Berlin Heidelberg, Berlin, Heidelberg (2007)
22. Modgil, S., Singh, R.K., Hannibal, C.: Artificial intelligence for supply chain resilience: learning from Covid-19. Int. J. Logist. Manage. **33**(4), 1246–1268 (2022)
23. Moons, K., Waeyenbergh, G., Pintelon, L.: Measuring the logistics performance of internal hospital supply chains – a literature study. Omega **82**, 205–217 (2019)
24. Muller, M.: Essentials of inventory management. HarperCollins Leadership (2019)
25. Pant, M., Zaheer, H., Garcia-Hernandez, L., Abraham, A.: Differential evolution: a review of more than two decades of research. Eng. Appl. Artif. Intell. **90**, 103479 (2020)
26. Perera, S.C., Sethi, S.P.: A survey of stochastic inventory models with fixed costs: optimality of (s, S) and (s, S)-type policies – continuous-time case. Prod. Oper. Manag. **32**(1), 154–169 (2023)
27. Simchi-Levi, D., Kaminsky P., Simchi-Levi E.: Designing and Managing the Supply Chain: Concepts, Strategies, and Case Studies. McGraw Hill Professional (2033)
28. Singh, D., Verma, A.: Inventory management in supply chain. Mater. Today Proc. **5**(2), 3867–3872 (2018)
29. Soleimani, H., Kannan, G.: A hybrid particle swarm optimization and genetic algorithm for closed-loop supply chain network design in large-scale networks. Appl. Math. Model. **39**(14), 3990–4012 (2015)
30. Sonnemann, G.W., Schuhmacher, M., Castells, F.: Uncertainty assessment by a Monte Carlo simulation in a life cycle inventory of electricity produced by a waste incinerator. J. Clean. Prod. **11**(3), 279–292 (2003)

31. Storn, R., Price, K.: Differential evolution-a simple and efficient heuristic for global optimization over continuous spaces. J. Global Optim. 11(4), 341 (1997)
32. Wang, S.L., Ng, T.F., Morsidi, F.: Self-adaptive ensemble based differential evolution. Int. J. Mach. Learn. Comput. 8(3), 286–293 (2018)
33. Wang, S., Wang, L., Pi, Y.: A hybrid differential evolution algorithm for a stochastic location-inventory-delivery problem with joint replenishment. Data Sci. Manage. 5(3), 124–136 (2022)
34. Xu, G., Burer, S.: Robust sensitivity analysis of the optimal value of linear programming. Optim. Methods Softw. 32(6), 1187–1205 (2017)
35. Zamuda, A., Brest, J.: Self-adaptive control parameters' randomization frequency and propagations in differential evolution. Swarm Evol. Comput. 25, 72–99 (2015)

Intelligent Decision Support System (iDSS) for Manufacturing Data Corpus

Nurul Hannah Mohd Yusof[✉], Nurul Adilla Mohd Subha, Norikhwan Hamzah, Fazilah Hassan, and Mohd Ariffanan Mohd Basri

Universiti Teknologi Malaysia, 81300 Skudai, Johor, Malaysia
nhannah2@live.utm.my

Abstract. People in industries like manufacturing require, use, and produce knowledge on a daily basis. Tremendous quantity of data with difference in formats, structures and linkages need to be cautiously explored. However, the most valuable knowledge is not easy to identify or share because it is deep within the minds of experts. In manufacturing, it is very common to see dashboards on business performance, however, very few literatures available on technical knowledge management. Technical knowledge of an expert can be effectively managed and transferred by having an interface or dashboard that provides adequate information for the learners. Hence, this project aims to establish intelligent Decision Support System (iDSS) that can strategically manage, transfer, and share valuable knowledge of experts within the manufacturing organization based on machine learning and deep learning models. This study used English text data that is properly phrased to build a deep learning model in Natural Language Processing (NLP) for maintenance factory reports. As a result, interactive visualizations are presented to aid decision-makers in making knowledgeable decisions that includes the display of failure diagnostic and Named Entity Recognition (NER). These findings may provide troubleshooting insights as an assistance to new employees and deliver a precise management of decisions in looking back in history and preparing ahead. The investigation of this study will be further explored for complex numeric parameters from sensors data, integration of predictive maintenance in the dashboard, and utilizing a more sophisticated training model for better predictions.

Keywords: Manufacturing · Interface · Knowledge Management

1 Introduction

The COVID-19 pandemic has significantly impacted nearly all facets of life and have transformed various economic sectors including the manufacturing sector through several factors including disruptions in the global supply chain, reduced consumer demand, government-imposed restrictions, and volatility of employment. Nearly all manufacturing segments must adapt to the new post-pandemic outlook which has seen a clear rise in gig economy which leads to low participant in the job market, including the manufacturing sector. Recently, the manufacturing sectors started to rely heavily on temporary and part-time workers causing frequent turnover of workers. This situation hinders consistent knowledge transfer and institutional memory especially within manufacturing companies which incur huge cost in time and overall efficiency [1]. This *new-normal* situation created by the post-pandemic is further worsen the knowledge transfer problem in most manufacturing sectors. Furthermore, inability to retain experts' knowledge could make the issue even worse. The valuable knowledge accumulated over decades of experience, retired along with them, and limits the efficiency of the operation of the production systems. Hence, in response to this pressing challenge, a roadmap of Society 5.0 introduced by Japan is aiming for a new human-centred society that focuses on development of a framework to retain human knowledge. Japan who is facing with silver wave or acute populate aging, active in developing methods to capture, digitalize, and share expert knowledge with the aid of latest technology.

Knowledge in the minds of an experts is called experiential or tribal knowledge. This knowledge can be efficiently managed and conveyed through a mechanism that is capable to demonstrate the knowledge relationship that provides learners with adequate information. By capturing expert's knowledge and linking them to specific data about events, machines, products, people, or processes, it will allow an automatic or autonomous decision-making to take place without the need of expert presence to make the connection. For example, suitable action can be performed by inexperience worker such as servicing or troubleshooting for a particular equipment through interpretation of relationship between knowledge and data which may prevent catastrophic failure. Hence, the knowledge can be remained within a company and ready to be used by any operators and machines for actions and may be modified for further improvement.

However, most valuable knowledge is not easy to be identified or shared because it is located deep inside the minds of an expert as it contains a mix of experience, value, context, intuition, and insight. Hence, generally the amount of knowledge that are transferred within organization is limited to only certain significant knowledge and to certain specific individuals within the organization as demonstrated in [2–4]. Thus, even there is formal knowledge transfer occurring within an organization, the knowledge is still not secured since there is no systematic system is used to identify, record, and distribute it. Therefore, a systematic method to extract and secure expert's knowledge is critical to be explored.

In manufacturing sector, dashboards are commonly used to monitor production performance. For this purpose, the Business Intelligence (BI) approach is utilized as presented in [5–7]. Without the need for sensor configuration, dashboard application will automatically display the measurement of the available sensors, visualizations, and aggregations that can be used by the user to generate an individualized dashboard. Then, user can focus on interpreting the sensor data to identify and resolve any operational issues concurrently. There are several dashboard applications can be found in manufacturing sector as described in [8–10]. Besides dashboard, an interface application usually focuses on providing a user-friendly and interactive displays for users to navigate and interact with a particular software. It may include forms, input fields, buttons, dropdowns, and other controls for data entry, manipulation, and system interaction [7]. Additionally, some organization has started to use more advance system called Decision Support System (DSS) empowered by IoT, big data, and machine learning operation monitoring

[11–13]. The DSS includes the health of industrial processes, maintenance schedules, and operational hazards. Nevertheless, none of the aforementioned systems have incorporated the expert's knowledge. Even though recording best practice in manufacturing is common, the expert's knowledge transfer process is still being conducted manually from the experts to the non-expert staff through written or verbal communication without any system digitalizing the transferred knowledge.

With an appropriate technological infrastructure that able conveys codified expert's knowledge to anyone, employees can effectively maintain, develop, and innovate seamlessly a system or manufacturing process with minimal risk to the overall productivity. Fully developed expert's knowledge system surely will be a very valuable instrument in the transfer and sharing of knowledge in the future.

Due to the fact that numerous reports are accessible, it is impossible to manually classify the reports for the decision-making process. Therefore, text classification process is needed. The process of classifying or identifying textual data into specified groupings or subcategories is known as text classification. Text classification is essential to aid in scheduling the maintenance to reduce or manage the unexpected equipment breakdowns, limit long-term damage to machines and processes, and improve safety measures [14]. However, to deal with a large-scale textual data, text classification may become a challenging task. Large-scale textual data organization and interpretation is a prominent difficulty in machine learning and NLP. However, numerous real-world applications such as topic labelling [15], spam identification [16], and customer feedback analysis [17], can benefit from the study of text classification. For a sequential text data, Recurrent Neural Network (RNN) method is favorable. It is widely used in situations involving time-series data or NLP due to its ability to identify temporal correlations in the data on tiny datasets. RNNs excel at tasks like language translation [18] and text synthesis [19]. However, due to its exploding or disappearing gradient problem during the training process, it is difficult for the RNN to be used to train for huge datasets. To solve the problem, the Long Short-Term Memory (LSTM) network is studied [20–22]. This network can recognize long-term relationships between time steps of input that are routinely used to learn, analyze, and categorize the sequential data. Therefore, the LSTM is proposed to be used in this project.

For the systems with huge datasets, NER is a critical component in NLP text mining for information extraction. NER establish entity-containing spans in a text and categorizes them into appropriate clusters [23]. Applications of NER can be found in the manufacturing industries, including customer service [24], compliance monitoring [25], and predictive maintenance [26]. Different technique of NER can produce different level of accuracy and efficiency in NLP process. NER using Hidden Markov Model (HMM) is a process to identify appropriate entity for every words in text so that it will be labelled accordingly. It is used to predict the sequential relationships between words in speech transcripts and to estimate the likelihood of each word being part of a named entity. The model is built using a tagged dataset of speech transcripts, where each word is labelled with the proper named entity. Even though the HMM model is not a new technology for this purpose, this model provide a speedy and effective solution for NER in speech transcripts, especially when the script is fragmented and noisy. In addition, the HMMs are easily adaptable to function in a variety of domains and languages, making them a viable tool for the NER in a variety of scenarios. Furthermore, the HMM is a flexible and powerful probabilistic model capable of interpreting complicated patterns and correlations in sequential data, such as text. As a result, the HMM excels in NER tasks that require recognizing and categorizing elements in text [27].

Text classification and NER are the two main components in the proposed knowledge interface for the manufacturing system. There is a very limited interface readily available for technical knowledge transfer among employees with difference in experience backgrounds. A user-friendly and intelligent interface system will lead to a more accurate fault diagnosis and operation of machines tools. Hence, this project aims to establish iDSS based on machine learning and deep learning (HMM and LSTM). The proposed system consists of key knowledge and knowledge assets that are equipped with effective tasks and models within the visual instruments. It will allow efficient knowledge transfer and an effective decision-making assistance to maintenance personnel especially for the newly hired or non-expert employees. As a limitation, the development of this project does not cater any missed spellings text data and complex numerical parameters such as sensors data yet due to early implementation of knowledge management in factory reports.

Furthermore, the architecture of this study which will be further explained in more detail in Sect. 2. This study utilized text classification technique in NLP by adopting LSTM-trained model to aid the employees on fault diagnosis, resolutions, and costing estimation. Additionally, this study also employed on NER using HMM for specific domain entity for manufacturing data corpus with processed factory reports. A summary of the knowledge management, text classification, LSTM, NER, HMM, and manufacturing interfaces are provided in Sect. 1. The architecture and procedures are shown in Sect. 2. Section 3 contains the outcomes of the interface available for the research. The conclusion and future directions are narrated in Sect. 4.

2 Methodology

2.1 Architecture

In order to develop the knowledge interface, several toolboxes are necessary to install beforehand:

- MATLAB R2023a
- Deep Learning Toolbox
- Text Analytics Toolbox
- MATLAB App Designer
- The factory report data set [28]

Figure 1 shows the architecture of the interactive user knowledge interface within the manufacturing data corpus. The source of data is accessible from the maintenance report provided by the personnel involved in the production floor that includes operators, technicians, and engineers. As they key in the report, it will be the input of the knowledge interface. Then, preparation for clean data is a must to ensure that the model can be trained as accurate as possible. This has to be done to produce reliable predictions that assist the newly hired employees to make decisions in troubleshooting failure without availability

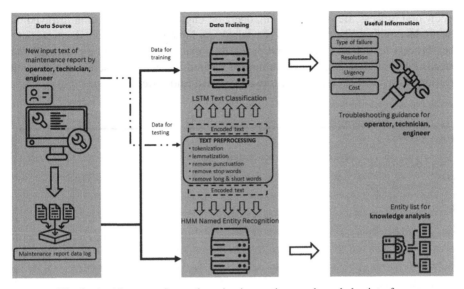

Fig. 1. Architecture of manufacturing interactive user knowledge interface

of experienced employees. Two models are developed in this study namely LSTM and HMM; to provide prediction for text classifications and NER, respectively. Notably, in the early phase of developing the knowledge interface, the reports will be available in the maintenance log to be used as the historical data in model training and testing.

2.2 Dataset

In this study, the dataset is divided into training set (80%) and testing set (20%). The model is trained using the training data and evaluation of the metrics is performed using the testing data. The dataset has 480 phrases of reports, or 2724 pieces of data. Additionally, this entailed choosing the right dataset or corpus from the manufacturing reports. Following that, data will be assigned entity labels based on seven distinct manufacturing domain entities: "action," "equipment," "location," "material," "non-entity," "substance," and "symptom." The primary data collected from the report is used to determine these domain entities based on the manufacturer preference.

2.3 Construction of Knowledge Interface

The design and implementation of user knowledge interface has a simple yet informative features which enables expert knowledge support throughout the production processes. This is managed through a single interface with three panels of main visual instruments.

Visual Panels. This iDSS provides several visual instruments for the user such as text input module and informative output modules which includes an expert support to panel displays. There are three main visual instruments available. Figure 2 shows a MATLAB App to be filled in by the user that detect any suspicious activities occurred in the production process. Be it a technician, operator, engineer, or any position involved on the production floor. The first panel is available for input text data. When the user clicks the "Process text" button in the first panel, the input sentences will be processed and encoded before undergoing any instruments desired by the user. There are two visual instruments available; details of failure and Named Entity Recognition.

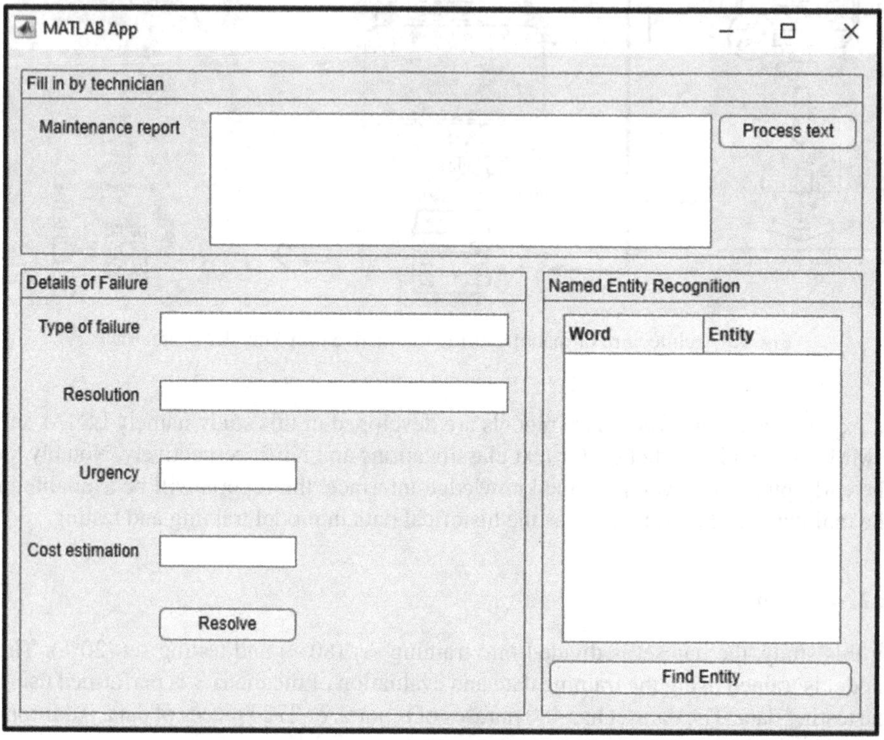

Fig. 2. Intelligent Decision Support System (iDSS) for manufacturing data input

Training Model. Before training for LSTM or HMM model, the data undergoes test preprocessing to enhance the quality of the trained data which includes tokenization, lemmatization, removal of punctuation, stop words, as well as short and lengthy words. These tasks in text preprocessing were carefully chosen to make the phrases transformed into a series of numerical directories sequence, readily available intended either for LSTM that predicts the diagnosis of the failure or HMM that is used to predict the entity of the testing data. As a result, LSTM model will produce significant details on the fault that includes type of failure, urgency of the failure, how the employees may resolve the issue, and the cost estimation of the particular failure. This information might not be 100% accurate but it can be a guidance for new employees to be independent as the knowledge can be transferred through the knowledge interface. On the other hand, the outcome of the HMM will help to define the entity tags of the words involved in the maintenance report. Entity labels can significantly aid the management and production team to perform analysis on the list of failure occurred and provide an insight into performance of the production line.

3 Result

3.1 Text Classification Using LSTM

In this visual instruments, the panel is available on the left side of the interface named as 'Details of Failure'. The text classification algorithm behind this panel is divided into four different features that provide information of the particular failure involved, derived from the input text. Figures 3, 4, 5 and 6 shows the trained LSTM model on type, resolution, urgency, and cost estimation of the failure. The prediction of these features holds different accuracy percentage.

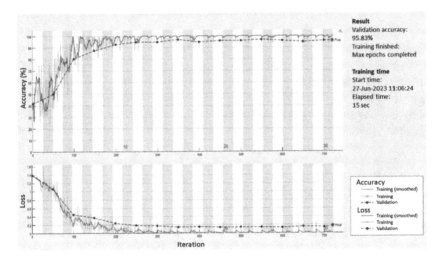

Fig. 3. LSTM training for type of failure

By referring to Table 1, the four features on the fault diagnosis have different accuracy as the evaluation metrics. The 'Type of failure' yields for 95.83% of prediction accuracy due to the presence of clean data and limited range of classes available. Most of the words were classified into only four category of failure namely mechanical failure, electronic failure, software failure, and leak. Similarly, the 'Urgency' has only two classes and 'Resolution' has six classes to be chosen from, which produces prediction accuracy for 82.29% and 80.21%, respectively. So the burden for the LSTM model for these three features are far less compared to the 'Cost estimation'. On the contrary, 'Cost estimation' feature varies into a broad range of classes depending on how severe the failure is. Hence, the accuracy of 'Cost estimation' prediction is the least (66.67%) among all features in the fault diagnosis panel. However, this issue can be further explored by employing a more sophisticated training model that used hybrid or transfer learning techniques.

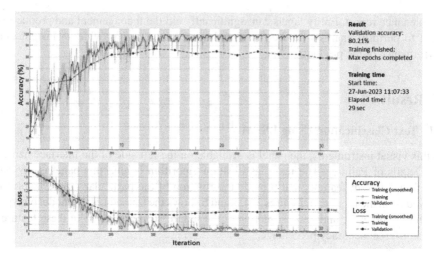

Fig. 4. LSTM training for resolution

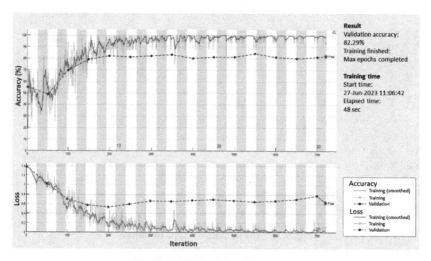

Fig. 5. LSTM training for urgency

3.2 NER Using HMM

In this section, NER performance of this iDSS is evaluated to demonstrate the accuracy of the NER using HMM model. Table 2 portrays the precision and recall of NER prediction for all seven classes of the pre-defined entity. Before running the prediction, all texts are preprocessed. Without having text preprocessing, the text data would be noisy and many insignificant words would burden the prediction. These may include stop words, punctuation, too long and too short words that can be out of context. Each class has shown a promising precision and recall with overall average above 90%. Hence, it can be said that a more accurate model is attained to produce a precise prediction on the

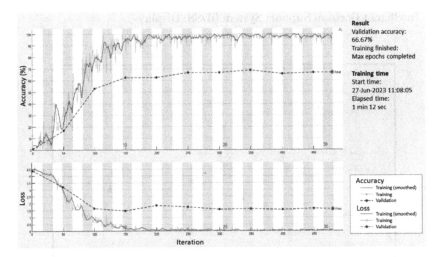

Fig. 6. LSTM training for estimated cost estimation

Table 1. Validation accuracy of fault diagnosis prediction

Features	Type of Failure (%)	Resolution (%)	Urgency (%)	Cost Estimation (%)
Percentage	95.83	80.21	82.29	66.67

class of entity for the input words, resulting in a reliable iDSS display for the non-expert users to refer to.

Table 2. NER precision and recall

Factory reports Entity	Precision (%)	Recall (%)
Action	100.0	100.0
Equipment	98.3	92.7
Location	100.0	100.0
Material	100.0	100.0
Non-entity	85.7	92.3
Substance	100.0	100.0
Symptom	93.7	97.4
Average	**96.8**	**97.5**

3.3 Intelligent Decision Support System (iDSS) Display

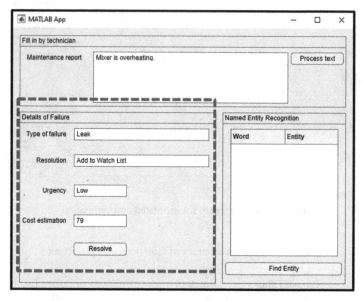

Fig. 7. Visual panel for fault diagnosis feature

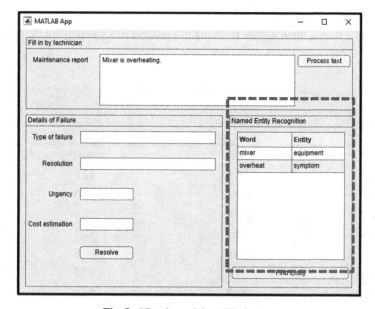

Fig. 8. Visual panel for NER feature

By referring to Figs. 7 and 8, the information from the person-in-charge for machine tools maintenance and troubleshooting will be the input for the interface. Both figures, recorded the sentence "Mixer is overheating." as the input and the algorithm preprocessed the new input text by removing the stop word and punctuation as well as normalized all the texts into lowercases. After the text has been preprocessed, the user can choose the visual instruments available to suit the needs of the user.

By referring to Fig. 7 the visual instrument on the fault diagnosis is displayed with four different features named 'Type of failure', 'resolution', 'urgency', and 'cost estimation'. This information will be useful for non-expert users to perform troubleshooting methods with the justification on how to resolve the issue which in line with the urgency of the failure. The cost estimation feature also available as baseline guidance for the users. Regardless to what type of machine tools they are facing, this flexible panel provides information on the failure of the machine tools based on the historical data and knowledge available. This user-friendly design of interface also allows users to grasp the information easily and without confusion, and provides a DSS for non-expert users. For example, as "Mixer is overheating." is being documented, the interface ran the training model and obtained the failure as a leak. Due to the 'Low' urgency consideration, the issue has been temporarily put into 'Add to Watch List'. The resolution would guide the users on how to react upon the arisen issue. The cost estimation is $79 reflecting to the information available from the historical data available.

Figure 8 demonstrates the list of entities available for the input sentences provided by the user. Within this panel, the algorithm of NER is embedded by utilizing HMM. Every sentence will be broken down into a few words and will be preprocessed to reduce any noise on the text data. As being narrated earlier, the stop word and punctuation are removed and only significant words will be labelled with the entity tags. Every significant word will be assigned as a token and the entity of each word will be predicted using the HMM under specific domain of manufacturing data corpus. In this study, the entities will rely on the seven entity classes namely "equipment", "symptom", "location", "substance", "action", "material", and "non-entity". As a result, there are two significant words displayed on the NER panel; 'mixer' and 'overheat'. Both words are labelled with "equipment" and "symptom", respectively. This information will be useful for the user to identify the class of entity and review the insight of the text database. It is important to be able to capture the insight of manufacturing database in order to provide a more systematic analysis on the production line performance and achieve the company strategic goal in the future.

4 Conclusion

This study has dealt with the design and implementation of knowledge and historical data transfer to aid decision-making in manufacturing sector especially when involving troubleshooting machine tools in the production floor. The iDSS connects the bridge between knowledge and data as a tool to share and transfer knowledge so that these valuable information becoming available for all, not only in the mind of experts by developing key knowledge and knowledge assets. This has to be done to maximize the utilization of knowledge in manufacturing sector that can leads to a more efficient work processes and indirectly, to gain more profit and productivity. There is also potential for further investigations in the area of knowledge management that may include development of knowledge graph, implementation of Predictive Maintenance (PdM), and implementation of a more sophisticated model especially for numerical parameters to be considered like sensors data and cost estimation.

Acknowledgements. This work was supported by the Ministry of Higher Education (MOHE), Malaysia under Fundamental Research Grant Scheme (FRGS) (FRGS/1/2020/TK0/UTM/02/36).

References

1. Joo, B.A., Shawl, S.: COVID-19 pandemic and the rising gig economy: an emerging perspective. Glob. Econ. Sci. 16–23 (2021)
2. Oz, E., Appolloni, A., Mavisu, M., Ozeren, E.: Knowledge management practices of manufacturing firms: a study from the Turkish Aegean free zone. Int. J. Intell. Enterp. **2**, 169–195 (2014)
3. Tan, L.P., Wong, K.Y.: Linkage between knowledge management and manufacturing performance: a structural equation modeling approach. J. Knowl. Manag. **19**, 814–835 (2015)
4. Ning, T.C., Ali, A.: Knowledge management strategies and organizational performance: a study on manufacturing, service and education sectors in the West Peninsular Malaysia. Eur. J. Manage. Mark. Stud. **6** (2015)
5. Alqhatani, A., et al.: 360° retail business analytics by adopting hybrid machine learning and a business intelligence approach. Sustainability **14**(19), 11942 (2022). https://doi.org/10.3390/su141911942
6. Kongthanasuwan, T., Sriwiboon, N., Horbanluekit, B., Laesanklang, W., Krityakierne, T.: Market analysis with business intelligence system for marketing planning. Information **14**(2), 116 (2023). https://doi.org/10.3390/info14020116
7. Wahyudi, I., Widyasari, Y.D.L.: Improving company performance by the correctness of management decision through implementation dashboard using power BI tools (case study at company Y). In: Proceedings – International Conference on Education and Technology, ICET 2022, pp. 32–37 (2022)
8. Jia, W., Wang, W., Zhang, Z.: From simple digital twin to complex digital twin part II: multi-scenario applications of digital twin shop floor. Adv. Eng. Inform. **56**, 101915 (2023). https://doi.org/10.1016/j.aei.2023.101915
9. Gupta, V., Mitra, R., Koenig, F., Kumar, M., Tiwari, M.K.: Predictive maintenance of baggage handling conveyors using IoT. Comput. Ind. Eng. **177**, 109033 (2023)
10. Moens, P., et al.: Event-driven dashboarding and feedback for improved event detection in predictive maintenance applications. Appl. Sci. **11**(21), 10371 (2021). https://doi.org/10.3390/app112110371
11. Rosati, R., et al.: From knowledge-based to big data analytic model: a novel IoT and machine learning based decision support system for predictive maintenance in industry 4.0. J. Intell. Manuf. **34**, 107–121 (2023)
12. Mailliez, M., Hosseini, S., Battaïa, O., Roy, R.N.: Decision support system-like task to investigate operators' performance in manufacturing environments. IFAC-PapersOnLine **53**, 324–329 (2020)
13. Giberti, H., Abbattista, T., Carnevale, M., Giagu, L., Cristini, F.: A methodology for flexible implementation of collaborative robots in smart manufacturing systems. Robotics **11**, 1–13 (2022)

14. Nacchia, M., Fruggiero, F., Lambiase, A., Bruton, K.: A systematic mapping of the advancing use of machine learning techniques for predictive maintenance in the manufacturing sector. Appl. Sci. **11**(6), 2546 (2021). https://doi.org/10.3390/app11062546

15. Rondinelli, A., Bongiovanni, L., Basile, V.: Zero-shot topic labeling for hazard classification. Information **13**(10), 444 (2022). https://doi.org/10.3390/info13100444

16. Mansoor, H.H., Shaker, S.H.: Using classification techniques to SMS spam filter. Int. J. Innov. Technol. Explor. Eng. **8**, 1734–1739 (2019)

17. Oh, Y.K., Kim, J.-M.: What improves customer satisfaction in mobile banking apps? An application of text mining analysis. Asia Mark. J. **23** (2022)

18. Auli, M., Galley, M., Quirk, C., Zweig, G.: Association for computational linguistics joint language and translation modeling with recurrent neural networks. In: Proceedings of the 2013 Conference on Empirical Methods in Natural Language Processing, pp. 1044–1054 (2013)

19. Lu, S., Zhu, Y., Zhang, W., Wang, J., Yu, Y.: Neural Text Generation: Past, Present and Beyond. arXiv (2018)

20. Wadud, M.A.H., et al.: How can we manage offensive text in social media – a text classification approach using LSTM-BOOST. Int. J. Inf. Manage. Data Insights **2**(2), 100095 (2022). https://doi.org/10.1016/j.jjimei.2022.100095

21. Vasantha Kumar, V., Sendhilkumar, S.: Developing a conceptual framework for short text categorization using hybrid CNN-LSTM based Caledonian crow optimization. Expert Syst. Appl. **212**, 118517 (2023). https://doi.org/10.1016/j.eswa.2022.118517

22. Liang, M., Niu, T.: Research on text classification techniques based on improved TF-IDF algorithm and LSTM inputs. Procedia Comput. Sci. **208**, 460–470 (2022)

23. Pham, M.Q.N.: A feature-based model for nested named-entity recognition at VLSP-2018 NER evaluation campaign. J. Comput. Sci. Cybern. **34**, 311–321 (2019)

24. Yin, D., Cheng, S., Pan, B., Qiao, Y., Zhao, W., Wang, D.: Chinese named entity recognition based on knowledge based question answering system. Appl. Sci. **12**(11), 5373 (2022). https://doi.org/10.3390/app12115373

25. Li, F., Song, Y., Shan, Y.: Joint extraction of multiple relations and entities from building code clauses. Appl. Sci. **10**, 1–18 (2020)

26. Dixit, S., Mulwad, V., Saxena, A.: Extracting semantics from maintenance records. In: International Joint Conference on Artificial Intelligence (IJCAI) Workshop on Applied Semantics Extraction and Analytics (ASEA) (2021)

27. Chopra, D., Joshi, N., Mathur, I.: Named entity recognition in Hindi using hidden Markov model. In: 2nd International Conference on Computational Intelligence and Communication Technology, CICT, pp. 581–586 (2016)

28. Copyright 2020 The MathWorks, Inc. factoryReports data set. http://www.mathworks.com. Accessed 21 July 2023

Influence of Thickness and Relative Permittivity of Triboelectric Materials on CS-TENG Performance: A Simulation Study

Anas A. Ahmed[1], Yusri Md Yunos[1], and Mohamed Sultan Mohamed Ali[1,2(✉)]

[1] Faculty of Electrical Engineering, Universiti Teknologi Malaysia, UTM Johor Bharu, 81310 Johor, Malaysia
sultan_ali@fke.utm.my
[2] Department of Electrical Engineering, College of Engineering, Qatar University, Doha, Qatar

Abstract. Recently, triboelectric nanogenerators (TENGs) have emerged as promising technology to generate electricity from wasted mechanical energies based on charge transfer between two suitably selected dielectric surfaces. Simulation studies play a key role in pre-fabrication processes to understand and optimize the TENGs performance. In this work, contact-separation TENG (CS-TENG) is reported based on finite element modeling simulation. The influence of the thickness and relative permittivity of triboelectric materials on the CS-TENG performance under open-circuit (OC) and short-circuit (SC) conditions was investigated. It was found that under the OC condition, the output voltage V_{OC} shows unsignificant change (slight reduction) upon increasing the thickness of tribo-materials from 100 to 500 μm and the relative permittivity of negative tribo-material from 2.7 to 7.5. On the other hand, under the SC condition, the air gap voltage ($V_{gap,SC}$) was significantly affected (remarkably decreased) by increasing the thickness of tribo-materials 100 to 500 μm and the relative permittivity of negative tribo-material from 2.7 to 7.5. In addition, the influence of the thickness and relative permittivity on the electric field along with the distribution of electric potential and electric field within the CS-TENG structure was explored to bring an in-depth understanding of the fundamental physics of the TENGs.

Keywords: Triboelectric Effect · Triboelectric Nanogenerator · Electrostatics · Energy Harvesters · Finite Element Modelling

1 Introduction

A triboelectric nanogenerator (TENG) is an energy harvesting device in which triboelectrification and electrostatic induction effects are coupled to generate electricity. Triboelectrification occurs when two dielectric layers are forced into physical contact and separated, where the layer that has lower electronegativity loses electrons and the layer that has higher electronegativity gains these electrons. The former is called positive tribo-layer and the latter is called negative tribo-layer. Meanwhile, negative charges get induced on a metal electrode attached to the positive tribo-layer and positive charges get

F. Hassan et al. (Eds.): AsiaSim 2023, CCIS 1912, pp. 260–269, 2024.
https://doi.org/10.1007/978-981-99-7243-2_22

induced a metal electrode attached on the negative tribo-layer. When the metal electrodes are connected to external circuit and subjected to a mechanical force, current can flow through the external circuit. This kind of device is attracting intense attention as sustainable and eco-friendly technology for harvesting wasted mechanical energies, such as wind energy [1], biomechanical energy [2], and ocean wave energy [3]. Recently, TENGs have been utilized as wearable and smart sensors in a wide range of applications including environmental monitoring [4], healthcare [5], and machine learning assisted artificial intelligence [6, 7].

Experimental and simulation studies have been carried out to enhance TENGs performance through surface modification [8, 9], geometrical design [10, 11], and structural parameters [12, 13]. Simulation methods play a key role in predicting the triboelectric properties of TENGs; thus, highly efficient TENGs can be designed for practical applications. By adopting simulation methods, the number of experiments would be reduced and only the optimal combination of parameters and conditions would be utilized to design efficient TENGs. In addition, simulation provides an in-depth understanding of the fundamental characteristics of TENGs. In this study, COMSOL Multiphysics software is introduced as a powerful tool based on finite element modelling (FEM) calculation to analyze and predict the triboelectric properties of contact-separation (CS)-TENG. The influence of tribo-layers' thickness and relative permittivity on the triboelectric characteristics, including open-circuit (OC) voltage, short-circuit (SC) voltage and SC surface charge density of CS-TENG, were investigated. Furthermore, the fundamental physics of the CS-TENG was studied to gain an in-depth understanding of the CS-TENG output performance. In particular, the distribution of electric potential and electric field of the CS-TENG were explored.

2 Methodology

In this work, a 2D COMSOL simulation study based on finite element modelling calculation was implemented to explore the triboelectric properties of the CS-TENG. The construction procedure of the simulation study is demonstrated in Fig. 1. The dielectric-dielectric CS-TENG is composed of two triboelectric materials serving as positive and negative tribo-materials. Two metal electrodes are attached to their backside to perform measurements The thickness of the tribo-materials was changed from 100 to 500 μm and the relative permittivity of negative tribo-material was changed from 2.7 to 7.5. The lateral size of the CS-TENG was chosen as 2 cm and the triboelectric surface charge density was chosen to be 10 μC/m^2. Table 1 summarizes the parameters and materials used to predict the triboelectric properties of the CS-TENG. The simulation was conducted in a quasi-electrostatic field; this is because the CS-TENG is operating in a low frequency domain.

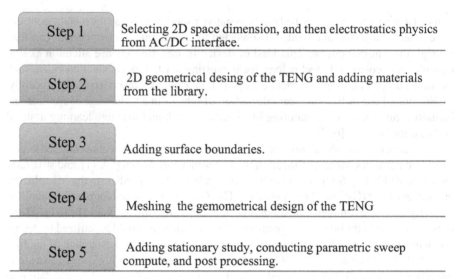

Step 1	Selecting 2D space dimension, and then electrostatics physics from AC/DC interface.
Step 2	2D geometrical desing of the TENG and adding materials from the library.
Step 3	Adding surface boundaries.
Step 4	Meshing the gemometrical design of the TENG
Step 5	Adding stationary study, conducting parametric sweep compute, and post processing.

Fig. 1. Flowchart illustrating the steps of 2D COMSOL simulation based on finite element modelling method.

Table 1. Parameters used in the simulation study of CS-TENG.

Parameter	Value	Unit
Thickness of negative tribo-material (d_1)	0.1–0.5	mm
Thickness of positive tribo-material (d_2)	0.1–0.5	mm
Relative permittivity of negative tribo-material (ε_{r1})	2.7–7.5	–
Relative permittivity of positive tribo-material (ε_{r2})	4.0	–
Surface area ($L \times W$), $L = W$	400	mm^2
Air gap (z)	0–20.9	mm
Triboelectric charge density (σ_t)	10	μC/m^2

3   Results and Discussion

In this section, the triboelectric characteristics of the dielectric-dielectric CS-TENG which includes the output voltage and transferred surface charged density under OC and SC conditions are analyzed. Figure 2 shows the schematic representation of the dielectric-dielectric CS TENG under OS and SC working conditions, where dielectric 1 and dielectric 2 refer to the negative and positive tribo-materials and d_1, d_2 represent their perspective thicknesses, σ_t is the triboelectric surface charge density, σ_c is the induced surface charge density, and z is the separation distance between the tribo-materials.

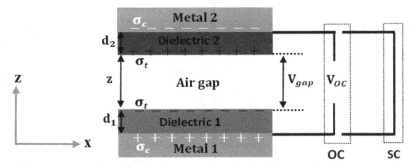

Fig. 2. Schematic representation of dielectric-dielectric CS-TENG used in this study

3.1 Triboelectric Characteristics Under OC Condition

In this condition, there is no charge transfer between the metal electrodes; thus, the induced surface charge density in Fig. 2 is equal to zero. Only the triboelectric charges are present in this case. The OC voltage (V_{OC}) is defined as the difference in electric potential between metal 1 and metal 2. Figure 3 displays the triboelectric properties of the CS-TENG under OC condition. In Fig. 3a, the thickness of the negative and positive tribo-materials was changed as 100, 300, and 500 μm, whereas the relative permittivity of the negative tribo-material (ε_{r1}) and positive tribo-material (ε_{r2}) is 2.7 (i.e., PDMS) and 4 (i.e., nylon), respectively. The V_{OC} shows an increasing tendency with the increase of the separation distance z. In addition, the V_{OC} exhibits a slight reduction upon increasing in the tribo-material thickness, as demonstrated in the inset of Fig. 3a. Similarly, when the thickness of the tribo-materials was fixed at 100 and the relative permittivity of the negatives tribo-material was changed to 3.5 (i.e., PET) and 7.5 (i.e., PVDF), the V_{OC} revealed a negligible reduction (the inset of Fig. 3b). Figure 3c and d represent the variation in the electric field (E_z) at dielectric-metal interface upon changing the thickness of the tribo-materials and the relative permittivity of the negative tribo-materials, respectively. Note that "Interface 1" refers to the negative tribo-material/metal 1 interface and "Interface 2" refers to the positive tribo-material/metal 2 interface. It is observed that the influence of the change in relative permittivity on E_z is more significant than the influence of the tribo-material thickness. This can be interpreted based on the electric field formula along z-axis:

$$E(z, L, W) = \frac{\sigma}{\pi\varepsilon} \tan^{-1}\left(\frac{LW}{2z\sqrt{L^2 + W^2 + 4z^2}}\right) \qquad (1)$$

Equation (1) is E_z at a midpoint z above a charged plate [8]. L and W are the length and width of the plate, and ε is the permittivity of the medium ($\varepsilon = \varepsilon_0\varepsilon_r$). From the electric field formula it is clear that the influence of the permittivity on the E_z is more pronounced than the influence of the tribo-material thickness. Note that, in Fig. 3d, the electric field at interface 2 remains unchanged upon changing the relative permittivity of the negative tribo-material. The relates to the unchanged relative permittivity of the positive tribo-material (nylon, $\varepsilon_{r2} = 4$).

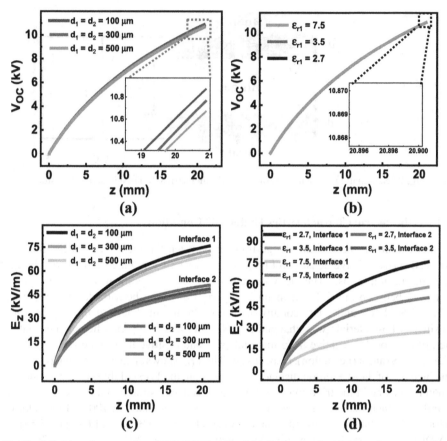

Fig. 3. Triboelectric properties of CS-TENG. (a) and (c) Open-circuit voltage and corresponding electric field at different thicknesses and fixed relative permittivity. (b) and (d) open-circuit voltage and corresponding electric field at different values of negative tribo-material relative permittivity and fixed thickness of tribo-materials.

To gain more insights into the characteristics depicted in Fig. 3, 2D mapping of electric potential along with the electric field (E_{zm}) and voltage (V_{zm}) variations within the structure of the TENG potential distribution were analyzed (Fig. 4). The E_{zm} and V_{zm} denotes to the electric field and voltage calculated at the maximum separation distance ($z_m = 20.9mm$). Figure 4a shows the distribution of electric potential within the structure of the CS-TENG, where potential is maximum on the surfaces with negative value on the negative tribo-surface and positive value on the positive tribo-surface. The black arrows represent the 2D distribution (x-z plane) of the electric field. The length of arrow is proportional to the strength of the electric field. Accordingly, the electric field at the edges of the TENG is observed to be higher than the electric field at regions away from the edges which originates from the fact that sharp ends accommodate more charges. Figure 4b represents the variation in electric field along z-axis at position far from the edges of the TENG. The variation in electric field can be explained based on the superposition principle in electrodynamics, where the electric fields due to the positive and negative tribo-materials have the same direction in the air gap and have opposite direction outside the air gap along z-axis. As a result, the electric field is maximum in the air gap between tribo materials. Note that, the electric field inside metals is zero (metals are considered as equipotential surfaces) and the electric field inside the positive tribo-material (d_2) is lower than the electric field inside negative tribo-material (d_2) due to the different values of the tribo-materials permittivity as explained in Fig. 3c and d. Figure 4c shows the variation in the open circuit voltage along – axis at the maximum separation between the tribo-layer (20.9 mm). Similarly, the voltage follows the superposition principle, where it is low between the tribo materials (has zero value at ~10.45 mm) and maximum on the tribo-surfaces.

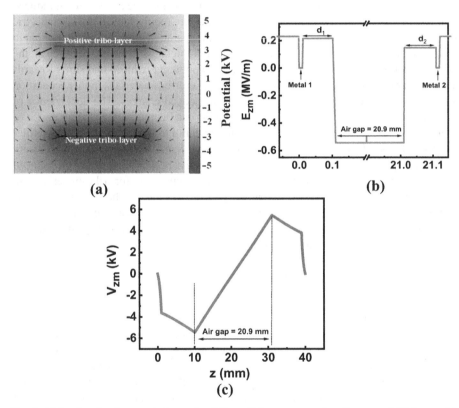

(a)

(b)

(c)

Fig. 4. Triboelectric characteristics under OC condition at separation distance of 20.9 mm. (a) 2D mapping of electric potential (b) Electric field distribution along z-axis within the structure of CS-TENG and far from the edges (c) Output voltage along z-axis at a region far from the TENG edges.

3.2 Triboelectric Characteristics Under SC Condition

In this condition, metal 1 is connected with metal 2 ($V_{OC} = 0$), and by changing the separation distance between the tribo-materials from 0 to 20.9 mm (z_m), the induced surface charges can transfer between metal electrodes. In this case, the CS-TENG illustrated in Fig. 1 can be viewed as four charged surfaces (two surfaces charged with induced charges and the other two are charged with tribo-charges). Figure 5a and b show the σ_c profile upon changing the separation distance at different values of the tribo-materials thickness and different values of negative tribo-material relative permittivity. As shown from Fig. 5a, when the two tribo-surfaces are in physical contact ($z = 0$), no electrostatic induction of charges on metal electrodes ($\sigma_{SC} = 0$). However, when the two tribo-materials start to move away from each other the induced charges start to appear on the metal electrodes, and by increasing the separation distance the density of induced charges increases initially, then saturate at $10 \mu C/m^2$ (equivalent to the triboelectric surface charge density) with the further increase in separation distance. It is important to highlight that with the increase of the tribo-material thickness from 100 to 500 μm, the induced charge density corresponding to the knee point get reduced and shifts towards higher separation distances, indicating that for thicker tribo-materials, higher separation distance is required to transfer same amount of induced charges. The dependency of the induced charges on the tribo-material thickness can be interpreted as the transfer of charges from the fractional surface of the tribo-material to the metal electrode. Similar trend was reported by Chen et al. [14]. On the other hand, the induced charge density shows an opposite trend with the increase of the relative permittivity, where the induced charge density reaches a saturation level at smaller separation for tribo-material with higher permittivity. However, the dependency of the of induced charge density on the relative permittivity chosen in this study is not clear in the linear curve presented in Fig. The relationship among the induced charge density, relative permittivity, and the thickness was derived using infinite two-parallel capacitor model [15]:

$$\sigma_{SC} = \frac{\sigma_t}{z + \frac{d_1}{\varepsilon_{r1}} + \frac{d_2}{\varepsilon_{r2}}} z \tag{2}$$

Figure 6a–d represents the electric field and the air gap voltage under SC condition at different thicknesses and different values of relative permittivity. As seen from Fig. 6a, the E_z shows similar behavior of the σ_{SC} with increasing the thickness of the tribo-materials. The electric field at interface 2 (positive tribo-material/metal 2 interface) is lower than that at interface 2 which is attributed to the inverse proportionality of the electric field to the medium permittivity as expressed in Eq. (1). Figure 6b shows that the $V_{gap,SC}$ greatly relies on the thickness of the tribo-materials, where it is higher for thicker tribo-materials. In addition, the $V_{gap,SC}$ value corresponding to the knee is shifted to higher separation distances with the increase of the thickness which is similar to the behavior of the induced charged density (Fig. 5a). Figure 6c and d reveal that the relative permittivity change results in decreasing tendency in the electric field and the air gap voltage. In contrary to Fig. 6a and b, the values of the E_z and $V_{gap,SC}$ which corresponds to the knee point shift towards lower separation distance with the increase of relative permittivity.

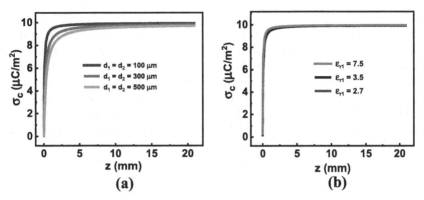

Fig. 5. Induced surface charge density at (a) different thicknesses of tribo-materials and fixed relative permittivity (b) fixed thickness (100 μm) and different values of relative permittivity.

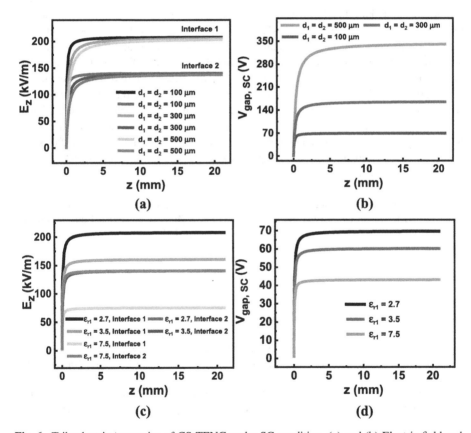

Fig. 6. Triboelectric properties of CS-TENG under SC condition. (a) and (b) Electric field and air gap voltage at different thicknesses and fixed relative permittivity, respectively. (c) and (d) Electric field and air gap voltage at different values of negative tribo-material relative permittivity and fixed thickness.

By comparing the electric field and air gap voltage under SC condition with the electric field and OC voltage, it can be concluded that the electric field at interface 1 and interface 2 under SC condition (Fig. 6a and c) is higher than the electric field at the same interfaces under OC condition (Fig. 3c and d). This could be attributed to the fact that the number of charged surfaces in SC condition is four, whereas it is only two in the OC condition. Nevertheless, the resulting electric field, according to superposition principle, in SC condition is stronger. On the other hand, SC air gap voltage (Fig. 6b and d) is much lower than the OC voltage (Fig. 3a and b). This also can be explained based on the superposition principle of electric field, where the resulting electric field in the air gap of the induced charges opposes the electric field of the tribo-charges, thus the air gap voltage will be dropped.

4 Conclusion

Simulation study based on finite element modeling method was conducted to investigate the triboelectric properties of the CS-TENG. The open-circuit voltage, short-circuit surface charge density, and short-circuit air gap voltage were studied at different thicknesses and different relative permittivity values of the tribo-materials. The results showed that the open-circuit voltage is slightly affected, whereas the air gap voltage was significantly affected by the change in the tribo-materials thicknesses and relative permittivity. In addition, the findings revealed that the short-circuit air gap voltage is much lower than the open-circuit voltage due to the opposing interaction of the induced charges and tribo-charges. The distribution of electric potential and electric field on the surfaces and within structure of the CS-TENG was provided to gain insightful understanding of the triboelectric characteristics.

Acknowledgments. This work was supported by Universiti Teknologi Malaysia under UTM Fundamental Research Grant Q.J130000.3823.22H55 and Ministry of Higher Education under Fundamental Research Grant Scheme FRGS/1/2022/TK07/UTM/02/42.

References

1. Ren, Z., Wu, L., Pang, Y., Zhang, W., Yang, R.: Strategies for effectively harvesting wind energy based on triboelectric nanogenerators. Nano Energy **100**, 107522 (2022)
2. Chen, H., Lu, Q., Cao, X., Wang, N., Wang, Z.L.: Natural polymers based triboelectric nanogenerator for harvesting biomechanical energy and monitoring human motion. Nano Res. **15**, 2505–2511 (2022)
3. Liu, L., et al.: Nodding duck structure multi-track directional freestanding triboelectric nanogenerator toward low-frequency ocean wave energy harvesting. ACS Nano **15**, 9412–9421 (2021)
4. Chang, A., Uy, C., Xiao, X., Chen, J.: Self-powered environmental monitoring via a triboelectric nanogenerator. Nano Energy **98**, 107282 (2022)
5. Graham, S.A., Patnam, H., Manchi, P., Paranjape, M.V., Kurakula, A., Yu, J.S.: Biocompatible electrospun fibers-based triboelectric nanogenerators for energy harvesting and healthcare monitoring. Nano Energy **100**, 107455 (2022)

6. Yang, Y., et al.: Triboelectric nanogenerator enabled wearable sensors and electronics for sustainable internet of things integrated green earth. Adv. Energy Mater. **13**, 2203040 (2023)

7. Zhang, W., et al.: Multilanguage-handwriting self-powered recognition based on triboelectric nanogenerator enabled machine learning. Nano Energy **77**, 105174 (2020)

8. Ahmed, A.A., Qahtan, T.F., Afzal, N., Rashid, M., Thalluri, L.N., Ali, M.S.M.: Low-pressure air plasma-treated polytetrafluoroethylene surface for efficient triboelectric nanogenerator. Mater. Today Sustain. **21**, 100330 (2023)

9. Xi, Y., Zhang, F., Shi, Y.: Effects of surface micro-structures on capacitances of the dielectric layer in triboelectric nanogenerator: a numerical simulation study. Nano Energy **79**, 105432 (2021)

10. Wang, Y., Pham, A.T.T., Han, X., Du, D., Tang, Y.: Design and evaluate the wave driven-triboelectric nanogenerator under external wave parameters: experiment and simulation. Nano Energy **93**, 106844 (2022)

11. Xu, S., et al.: Interaction between water wave and geometrical structures of floating triboelectric nanogenerators. Adv. Energy Mater. **12**, 2103408 (2022)

12. Hasan, S., Kouzani, A.Z., Adams, S., Long, J., Mahmud, M.P.: Comparative study on the contact-separation mode triboelectric nanogenerator. J. Electrostat. **116**, 103685 (2022)

13. Kang, X., Pan, C., Chen, Y., Pu, X.: Boosting performances of triboelectric nanogenerators by optimizing dielectric properties and thickness of electrification layer. RSC Adv. **10**, 17752–17759 (2020)

14. Chen, H., et al.: Crumpled graphene triboelectric nanogenerators: smaller devices with higher output performance. Adv. Mater. Technol. **2**, 1700044 (2017)

15. Niu, S., et al.: Theoretical study of contact-mode triboelectric nanogenerators as an effective power source. Energy Environ. Sci. **6**, 3576–3583 (2013)

Performance Evaluation of Evolutionary Under Sampling and Machine Learning Techniques for Network Security in Cloud Environment

Kesava Rao Alla[1] and Gunasekar Thangarasu[2(✉)]

[1] Chancellery, MAHSA University, Saujana Putra, Jenjarom, Malaysia
alla@mahsa.edu.my.com
[2] Department of Professional Industry Driven Education, MAHSA University, Saujana Putra, Jenjarom, Malaysia
gunasekar@mahsa.edu.my

Abstract. Despite the growing adoption of cloud computing, security challenges continue to persist in its implementation. In this study, we delve into the specific security challenges associated with cloud computing and explore the use of machine learning algorithms like K-Nearest Neighbors (KNN), Support Vector Machine (SVM), Decision Tree (DT), and Logistic Regression for anomaly detection. Our study leverages the MawiLab dataset to develop supervised machine learning models and evaluates their performance using key metrics such as accuracy, precision, recall, and F1-score. The results of our analysis showcase promising outcomes, with accuracy, precision, recall, and F1-score achieving impressive values of 96.3%, 93.8%, 95.2%, and 95.9% respectively. Nevertheless, the acquisition of real-time and unbiased datasets presents significant challenges. These findings underscore the importance of further research to enhance the applicability of machine learning techniques in effectively addressing the diverse operational conditions inherent in cloud environments.

Keywords: Cloud · Machine Learning · Security · Supervised Learning

1 Introduction

Cloud computing and virtualized data centers have become increasingly popular as cost-effective alternatives for enterprise applications. Infrastructure as a Service (IaaS), Platform as a Service (PaaS), and Software as a Service (SaaS) enable users to access and rent hardware or software without having to manage the underlying computer infrastructure [1]. However, the transition to cloud computing presents significant security challenges, which have been extensively researched [2]. Researchers have increasingly turned to machine learning (ML) approaches to address network and cloud security challenges [3]. Typically, ML-based security applications are trained using labeled network traces from experimental environments, encompassing a wide range of attacks as well as normal and anomalous behavior [4].

F. Hassan et al. (Eds.): AsiaSim 2023, CCIS 1912, pp. 270–278, 2024.
https://doi.org/10.1007/978-981-99-7243-2_23

However, privacy concerns limit the availability of security-related data, necessitating reliance on experimentally generated datasets that may not be fully comprehensive. This poses challenges as models trained on specific datasets may struggle to generalize well to different operating conditions prevalent in remote cloud environments and evolving networking paradigms [5]. Moreover, existing learning datasets often focus on specific types of attacks while neglecting others, making it difficult to evaluate machine learning models, particularly anomaly-based detectors [6]. Few studies have examined the training and testing of machine learning models on diverse datasets, which is essential for evaluating their robustness and suitability for real-time applications.

In this research, the performance of popular supervised machine learning algorithms was analyzed using two distinct datasets: UNSW and ISOT, both generated from simulated cloud environments. This diverse dataset approach allows for a comprehensive evaluation of machine learning methods, simulating typical enterprise cloud network traffic. Specifically, Naive Bayes, Support Vector Machines (SVM), regression, and decision trees were investigated, as they are commonly utilized in various fields, including network and cloud security [9]. The models were trained and tested on the MawiLab dataset, providing the necessary evaluation for machine learning models and enhancing our understanding of their resilience and suitability for practical applications.

2 Literature Review

Extensive research in the literature has extensively explored the security threats prevalent in cloud computing, providing a comprehensive analysis [10]. In recent times, the application of Machine Learning approaches to network and cloud security has gained significant popularity [11]. ML techniques, such as Support Vector Machine, Artificial Neural Network (ANN), Multilayer Perceptron (MLP), Decision Trees, and hybrid models combining different ML techniques, as well as Bayesian networks and data mining techniques, have been explored for network anomaly detection. Additionally, advancements in computing technologies have facilitated the evolution of attack methods from Denial of Service (DoS) to Distributed DoS (DDoS) attacks, which are recognized as major security risks by organizations like the Cloud Security Alliance [12].

A comprehensive analysis of intrusion detection methods and cloud environment risks can be found in [13]. However, a common mistake in current anomaly detection efforts is the assumption of uniform operational environments. Typically, models are trained and tested using datasets collected from a single experimental setup, where packet captures are collected, and specific features are extracted. In [14], a research study proposed an evolutionary undersampling technique under Apache Spark to improve imbalanced big data classification. This method effectively rebalances the data, leading to improved classification accuracy and handling of imbalanced scenarios.

In a groundbreaking study referenced as [15], a novel machine learning approach combined with multi-objective optimization was introduced to revolutionize network anomaly detection. The primary objective of this framework was to elevate the accuracy of anomaly detection while simultaneously minimizing false positives and false negatives, surpassing the capabilities of traditional methods. The research showcased a remarkable hybrid machine learning approach specifically designed for intrusion detection within software-defined networking (SDN) environments. By harnessing the power of deep learning and ensemble learning techniques, the proposed approach demonstrated exceptional performance in detecting a wide array of network intrusions in SDN, outperforming individual techniques, as highlighted in [16]. In this study quest to address the critical issue of ensuring the robustness of machine learning models, we have directed our focus towards evaluating their adaptability and reliability across diverse scenarios. To achieve this, we have meticulously trained various supervised machine learning models using labeled datasets sourced from MawiLab, renowned for its comprehensive network traffic data. By subjecting these models to testing on completely unrelated datasets, we aim to assess their performance metrics and ascertain their resilience. The study stands as a testament to the significance of comprehensive evaluation and validation in the realm of machine learning-based anomaly detection. By rigorously examining the adaptability and robustness of these models in disparate scenarios, we aim to establish a solid foundation for reliable intrusion detection systems. In an ever-evolving landscape of network security, it is imperative to validate the effectiveness of these models across various datasets, ensuring their capability to detect anomalies accurately and efficiently.

3   Research Methodology

In today's era of cloud computing, ensuring robust security measures is paramount. This study delves into the intricate security challenges associated with cloud computing and emphasizes the pressing need for effective anomaly detection methods. To comprehensively address this issue, we turned to the esteemed MAWILab dataset, which stems from real traffic captured at the MAWI (Measurement and Analysis on the WIDE Internet) project. This dataset presents an expansive collection that facilitates the study of network traffic anomalies and various types of attacks.

Within the MAWILab dataset, we encountered a diverse range of twelve attack types, including the notorious Sasser worm, Netbios attack, RPC attack, SMB attack, SYN attack, RST attack, FIN attack, Ping flood, FTP attack, SSH attack, HTTP attack, and HTTPS attack. By examining these attack types closely, we were able to identify their unique characteristics and patterns, contributing valuable insights to the realm of anomaly detection.

For our research, we meticulously selected and analyzed significant features from a pool of 18 options, aligning them with their respective class labels. In order to develop our robust machine learning models, we focused our attention on a carefully curated subset of the MAWILab dataset, specifically utilizing the csv files for both the training and testing sets. The training set consisted of 48,917 records, while the testing set encompassed 20,442 records, providing us with substantial data to train and evaluate our models effectively.

To gain further clarity on the dataset, Table 1 presents comprehensive statistics regarding abnormal and regular packets for the specific subset of the MAWILab dataset utilized in our research. These statistics serve as a vital reference point, enabling us to grasp the distribution and nature of anomalies and regular network traffic within our selected dataset.

By leveraging the rich MAWILab dataset, focusing on significant features, and employing advanced machine learning techniques, our study aims to pioneer effective anomaly detection methods for securing cloud computing environments. With a profound understanding of network traffic patterns and attack types, we strive to contribute to the development of robust security solutions that can safeguard cloud-based systems from potential threats.

Table 1. Subset of MawiLab Dataset

Dataset	Normal	Anomalous	Total
Training	36925	11592	48517
Testing	12642	7800	20442

3.1 Experimental Setup

The experimental setup employed in this study revolved around the utilization of the highly acclaimed MawiLab dataset, which served as the foundation for training and evaluating supervised machine learning models. To ensure the integrity of the data, various pre-processing techniques were applied, including cleaning and normalization, resulting in a refined dataset ready for analysis. Among the four machine learning algorithms chosen for anomaly detection, KNN, SVM, Decision Tree, and Logistic Regression stood out as prominent contenders. These algorithms were meticulously implemented to develop robust models capable of detecting anomalies effectively. The dataset was strategically split into training and testing sets, with 70% allocated for training purposes and the remaining 30% reserved for rigorous evaluation. Performance assessment was conducted utilizing a range of essential metrics, including accuracy, precision, recall, and F1-score. These metrics played a vital role in quantifying the performance of the models and highlighting their effectiveness in anomaly detection.

To ensure optimal execution and reliable results, the experiments were conducted on a high-performance computing environment. This choice aimed to leverage the computational power required for intricate computations, enabling efficient analysis of the dataset. Additionally, comprehensive statistical analysis was performed to compare the performance of the different algorithms and identify potential strengths and weaknesses. This analysis helped to shed light on the varying capabilities of each algorithm, providing valuable insights for future improvements and developments in anomaly detection methodologies. It is crucial to acknowledge the limitations inherent in the dataset and the constraints associated with the experimental process. By openly recognizing these factors, the experimental setup aimed to establish a transparent framework that generated reliable and reproducible results. This approach serves as a foundation for credible research outcomes and encourages further exploration in the field of anomaly detection. Through this meticulously designed experimental setup, our study aimed to harness the power of the MawiLab dataset and cutting-edge machine learning algorithms. By leveraging advanced techniques and comprehensive performance evaluation, we aimed to contribute to the advancement of anomaly detection methodologies, striving for accurate and reliable detection of anomalies within complex datasets.

3.2 Machine Learning Algorithms

We have employed four algorithms, namely KNN, SVM, Decision Tree, and Logistic Regression, to analyze security challenges in cloud computing.

KNN is a non-parametric algorithm used for classification and regression tasks. It operates on the principle that similar data points tend to belong to the same class. In the context of this research, KNN is utilized for anomaly detection by measuring the similarity between packets and identifying potential outliers.

SVM is a supervised learning algorithm that can be used for both classification and regression tasks. Its objective is to find an optimal hyperplane that effectively separates different classes in the feature space. In this research, SVM is employed to classify network packets as normal or anomalous based on their feature representations.

Decision Tree is a versatile algorithm that recursively splits the data based on different features, creating a tree-like structure to make decisions. It is widely used for classification and regression tasks. In this research, Decision Tree is utilized to classify network packets as normal or anomalous by constructing a tree-like model based on the features of the packets.

Logistic Regression is a statistical algorithm specifically designed for binary classification problems. It models the relationship between independent variables and the probability of the outcome using the logistic function. In the context of this research, Logistic Regression is applied to classify network packets as normal or anomalous based on the given features.

These algorithms were chosen for their effectiveness in handling classification tasks and their suitability for anomaly detection in cloud security. By applying these algorithms to the MawiLab dataset, the research aims to evaluate their performance and effectiveness in addressing the security challenges of cloud computing.

4   Results and Discussion

In this section, we delve into the performance evaluation of several supervised machine learning techniques utilizing the renowned MawiLab dataset. Our selection of algorithms stems from well-established machine learning approaches, including KNN, SVM, Decision Tree, and Logistic Regression. These algorithms were specifically chosen due to their industry significance and broad applications, particularly in the domain of network security. To conduct a comprehensive evaluation, we leveraged the MawiLab training and testing dataset, enabling us to train the aforementioned machine learning algorithms. The performance of these methods is showcased in Fig. 1, highlighting their effectiveness. Among them, logistic regression emerged as the most accurate algorithm, achieving an impressive accuracy score of 96.3%. Close on its heels, the SVM algorithm demonstrated robust performance with an accuracy score of 95.2%. Meanwhile, the KNN algorithm exhibited a relatively lower accuracy score of 89.7%. However, it is essential to note that relying solely on accuracy as a performance metric is inadequate. Therefore, it is crucial to consider additional metrics such as true positive, false positive, true negative, and false negative rates to gain a comprehensive understanding of the algorithms' performance.

To provide clarity on these metrics, Table 2 offers concise definitions for the terms used in performance evaluation. Each term represents specific conditions: a zero value signifies a typical packet, while a one value represents an anomalous packet. By analyzing these metrics, we gain valuable insights into the algorithms' ability to correctly classify packets and identify anomalies.

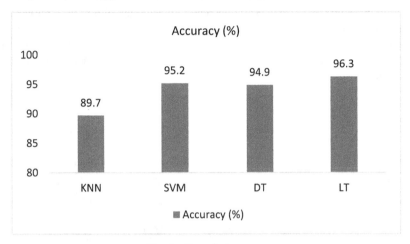

Fig. 1. Overall Accuracy

By evaluating the performance of these well-established machine learning algorithms, we contribute to the advancement of anomaly detection methodologies in network security. Through the rigorous assessment of accuracy and additional performance metrics, we aim to provide a holistic understanding of the algorithms' capabilities and limitations. This evaluation serves as a stepping stone towards developing more robust and effective anomaly detection solutions.

Table 2. Conditions for the Prediction

Total Population	Predicted Condition Positive (Anomalous)	Predicted Condition Negative (Normal)
Positive (1, Anomalous)	True Positive (TP)	False Negative (FN)
Negative (0, Normal)	False Positive (FP)	True Negative (TN)

Figure 2 provides an illustration of the percentages of true positives (TP) and false negatives (FN) for the aforementioned methods. The results indicate that logistic regression exhibits the highest performance, correctly identifying 93.7% of anomalous packets as TP. However, there is an error rate of 6.3% where logistic regression incorrectly classifies anomalous packets as normal (FN). SVM closely follows with TP and FN percentages of 84.7% and 15.3%, respectively. On the other hand, Naive Bayes performs least effectively, with TP and FN percentages of 56.2% and 43.8%, respectively.

The metrics considered for performance evaluation include accuracy, true positive rate, and false negative rate. Logistic regression achieved the highest accuracy among the other algorithms, scoring 96.3%. This indicates that it correctly classified 96.3% of the instances in the dataset. The TP rate for logistic regression corresponds to the percentage of anomalous packets correctly identified, which stands at 96.3%. However, there is a 3.7% FN rate, indicating that a small portion of anomalous packets were incorrectly classified as normal. SVM achieved an accuracy of 95.2%, demonstrating its effectiveness in correctly classifying instances. The TP rate for SVM matches the accuracy value, indicating that 95.2% of anomalous packets were correctly identified. The FN rate for SVM is 4.8%, implying a small proportion of anomalous packets were misclassified.

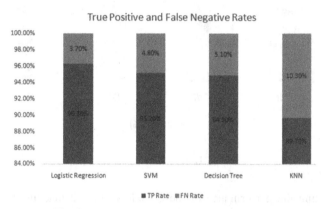

Fig. 2. True Positive and False Negative Rates

The Decision Tree algorithm achieved an accuracy of 94.9%, showcasing its good performance in classification tasks. The TP rate for Decision Tree aligns with the accuracy value, indicating that 94.9% of instances were correctly identified. The FN rate for Decision Tree is 5.1%, representing a small fraction of misclassified anomalous packets. On the other hand, KNN exhibited the lowest accuracy among the algorithms, scoring 89.7%. This suggests that KNN's performance was relatively weaker in correctly classifying instances. The TP rate for KNN corresponds to the accuracy value, indicating that 89.7% of anomalous packets were correctly identified. However, the FN rate for KNN is 10.3%, indicating a higher proportion of misclassified anomalous packets.

These findings provide insights into the performance of different machine learning algorithms for anomaly detection. Logistic Regression demonstrates the highest overall effectiveness, followed closely by SVM and Decision Tree. While KNN has a lower accuracy, it still shows potential for improvement in future research.

Figure 3 illustrates that raising the threshold leads to an increase in the TP rate, indicating a higher proportion of successfully detected anomalous packets. However, this simultaneous increase in the threshold results in a decrease in the TN rate, indicating a reduced ability to accurately identify normal packets. Notably, when the threshold values are set at 0.7 and 0.8, the TP percentage reaches approximately 97%, reflecting a significant enhancement in the accuracy of detecting anomalous packets. Meanwhile, the TN rate remains relatively stable at around 80%, indicating the system's continued proficiency in recognizing typical packets. Based on these findings, we conclude that elevating the probability threshold above the conventional value of 0.5, specifically within the range of 0.7 to 0.8, enables cloud security systems to more precisely identify abnormal packets while maintaining satisfactory performance in identifying normal packets.

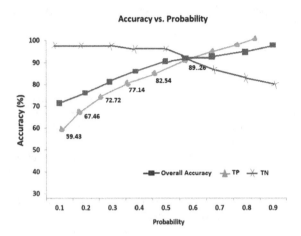

Fig. 3. Varying Frequently Threshold

Based on these findings, we emphasize the importance of conducting further research in the field of machine learning, particularly in network and cloud security. It is crucial to explore alternative strategies, including the utilization of unsupervised machine learning techniques. Additionally, investigating the application of clustering and related techniques can significantly enhance the effectiveness of anomaly detection. By delving into these areas and expanding the scope of research, we can develop more robust and advanced methods for detecting anomalies in network and cloud environments. This will contribute to the overall improvement of security measures and help address the evolving challenges in the field.

5   Conclusion

Our research underscores the disparity between the prevalent reliance on supervised machine learning methods and their limited applicability in real-time scenarios. Based on our experiments, we have reached the conclusion that supervised ML models exhibited satisfactory performance on the MawiLab dataset. However, it is important to acknowledge that not all types of attacks can be fully represented within a dataset. When comparing algorithms, logistic regression and SVM demonstrated high accuracy and TP (true positive) rates, while KNN performed less effectively with lower accuracy and a higher FN (false negative) rate. These findings emphasize that the choice of algorithm significantly influences the efficacy of anomaly detection in network security.

This limitation becomes particularly relevant as new networking paradigms such as Network Function Virtualization (NFV) and Service Function Chaining (SFC) emerge in cloud computing environments. In these scenarios, packets traverse multiple clouds and/or data centers while operating under diverse constraints. To enhance the effectiveness of supervised ML approaches in cloud security, substantial improvements are required. Future research should focus on areas such as anomaly detection, clustering, and unsupervised machine learning to address these challenges. By exploring alternative strategies and expanding the scope of research, we can develop more robust and adaptable models for cloud security.

References

1. Jordan, A.: On discriminative vs. generative classifiers: a comparison of logistic regression and naïve Bayes. Adv. Neural Inf. Process. Syst. **14**(1), 841–860 (2012)
2. Laura, A., Mora, R.: Support vector machines (SVM) as a technique for solvency analysis. DIW Berlin Discussion Paper (2020)
3. Shiravi, A., Shiravi, H., Tavallaee, M., Ghorbani, A.: Toward developing a systematic approach to generate benchmark datasets for intrusion detection. Comput. Secur. **31**(3), 357–374 (2012)
4. Jin, B., Wang, Y., Liu, Z., Xue, J.: A trust model based on cloud model and Bayesian networks. Procedia Environ. Sci. **11**(Part A), 452–459 (2011)
5. Zhang, X., Zhao, Y.: Application of support vector machine to reliability analysis of engine systems. Telkomnika **11**(7), 3352–3560 (2019)
6. Modi, C., et al.: A survey of intrusion detection techniques in cloud. J. Netw. Comput. Appl. **36**(1), 42–57 (2019)
7. Bhamare, D., Jain, R., Samaka, M., Vaszkun, G., Erbad. A.: Multi-cloud distribution of virtual functions and dynamic service deployment: open ADN perspective. In: Cloud Engineering (IC2E), pp. 299–304 (2018)
8. Michalski, D., Carbonell, J., Mitchell, T.: Machine Learning: An Artificial Intelligence Approach. Springer Science & Business Media (2018)
9. Stein, G., Chen, B., Wu, A., Hua, K.: Decision tree classifier for network intrusion detection with GA-based feature selection. In: The Proceedings of the 43rd Annual Southeast Regional Conference. Kennesaw, Georgia (2019)
10. Szabó, G., Orincsay, D., Malomsoky, S., Szabó, I.: On the validation of traffic classification algorithms. In: Claypool, M., Uhlig, S. (eds.) PAM 2008. LNCS, vol. 4979, pp. 72–81. Springer, Heidelberg (2008). https://doi.org/10.1007/978-3-540-79232-1_8
11. Mchugh, J.: Testing intrusion detection systems: a critique of the 2018 and 2019 DARPA intrusion detection system evaluations as performed by Lincoln laboratory. ACM Trans. Inf. Syst. Secur. (2022)
12. Haykin, S.: Neural Networks: A Comprehensive Foundation, 2nd edn. Prentice Hall, New Jersey (2019)
13. Roshke, S., Cheng, F., Meinel, C.: Intrusion detection in the Cloud. In: 11th IEEE International Conference on Dependable Autonomic and Secure Computing, pp. 729–734 (2019)
14. Garcia, S., Gómez, J.S., Herrera, F.: Evolutionary under sampling for extremely imbalanced big data classification under Apache Spark. Appl. Soft Comput. 108086 (2021)
15. Pham, P.H., Saddik, A.E., Zomaya, A.Y.: A machine learning approach to network anomaly detection using multi-objective optimization. Futur. Gener. Comput. Syst. **119**, 643–659 (2021)
16. Dhara, R., Goyal, M.: Hybrid machine learning approach for intrusion detection in software-defined networking. Concurrency and Computation: Practice and Experience, e6073 (2021)

Scalable and Efficient Big Data Management and Analytics Framework for Real-Time Deep Decision Support

Kesava Rao Alla[1] and Gunasekar Thangarasu[2]($\boxtimes$)

[1] Chancellery, MAHSA University, Saujana Putra, Malaysia
`alla@mahsa.edu.my.com`
[2] Department of Professional Industry Driven Education, MAHSA University,
Saujana Putra, Malaysia
`gunasekar@mahsa.edu.my`

Abstract. In data-driven world, organizations face challenges in managing and analyzing large volumes of data in real-time to make informed decisions. This paper proposes a scalable and efficient big data management and analytics framework for real-time deep decision support. The framework leverages advanced technologies such as distributed computing, parallel processing, and machine learning algorithms to enable organizations to process and analyze massive amounts of data quickly and accurately. By combining real-time data processing with deep decision support capabilities, the framework empowers decision-makers with timely insights and actionable intelligence to make informed decisions. The scalability and efficiency of the framework are demonstrated through experimental evaluations using real-world big data sets. The results show that the proposed framework outperforms existing solutions in terms of processing speed, resource utilization, and decision accuracy, making it an ideal choice for organizations seeking to harness the power of big data analytics for real-time decision support.

Keywords: Big Data Management · Analytics Framework · Real-time Processing · Deep Decision Support

1 Introduction

In the era of big data, organizations across all industries confront the challenge of managing and analyzing massive volumes of data in order to derive actionable insights and make informed decisions [1]. In today fast-paced and dynamic business environment, the capacity to process and analyze this data in real-time is essential for gaining a competitive edge. Real-time deep decision support, which combines real-time data processing with advanced analytics techniques, has emerged as a promising method for enabling organizations to make timely and effective decisions based on current data [2, 3].

Data administration and analytics systems have historically struggled to accommodate the volume and velocity of big data. When dealing with vast data sets, traditional relational databases and on-premises data processing tools frequently have scalability and processing speed limitations. Consequently, organizations have adopted new strategies and technologies to address these challenges [4].

The proposed research concentrates on developing a scalable and effective framework for big data management and analytics to support real-time, in-depth decision making. The framework intends to utilize advanced technologies such as distributed computing, parallel processing, and machine learning algorithms to enable organizations to process and analyze large volumes of data rapidly and precisely [5, 6].

Distributed computing plays a vital role in the proposed framework by distributing the data processing workload across multiple nodes or clusters, thereby decreasing processing time and enhancing scalability. By leveraging parallel processing, the framework can simultaneously leverage multiple processing units to accelerate data processing and analytics duties [7]. This parallelism enables efficient resource utilization and real-time management of big data duties by organizations [8].

Algorithms for machine learning are another integral component of the framework. These algorithms can analyze the enormous amounts of data and extract patterns, correlations, and insights. By integrating techniques for machine learning, the framework can automate the decision support process, granting organizations predictive and prescriptive analytics capabilities. This provides decision-makers with actionable intelligence and allows them to make data-driven decisions in real time [9].

The effectiveness and performance of the proposed framework are evaluated experimentally using real-world large data sets. The evaluation metrics consist of processing velocity, resource utilization, and decision precision. The research seeks to demonstrate the superiority of the proposed approach in terms of scalability, efficiency, and decision support by comparing the framework results to those of existing solutions.

This research has the potential to revolutionize how organizations manage big data analytics for real-time decision support, which is its significance. By providing a scalable and efficient framework, organizations are able to process and analyze large volumes of data in real-time, allowing them to respond quickly to changing market conditions, identify emergent trends, detect anomalies, and make informed decisions. This can result in increased operational efficiency, a better consumer experience, and ultimately a competitive edge.

2 Background of Study

In this section, we provide background information on big data management and analytics and emphasize the difficulties associated with processing and analyzing large data volumes.

2.1 Big Data Management

The proliferation of digital technologies and connected devices has led to an exponential increase in the volume, variety, and velocity of data being generated. This abundance of data, known as big data, poses significant storage, management, and processing difficulties for organizations. Traditional data administration systems, such as relational databases, have difficulty managing the scale and complexity of big data [10]. These systems frequently have storage capacity, processing speed, and scalability constraints. Numerous technologies and frameworks, such as Apache Hadoop and Apache Spark, that facilitate distributed storage and processing of large data sets across clusters of commodity hardware have emerged in response to the demand for effective big data management solutions [11, 12].

2.2 Big Data Analytics

The process of extracting valuable insights, patterns, and trends from large and complex data sets is known as big data analytics. Utilizing advanced analytics techniques, such as statistical analysis, data mining, machine learning, and predictive modeling, to extract actionable and meaningful insights from data. Due to the sheer volume, velocity, and variety of big data, traditional instruments and techniques for data analysis are frequently insufficient. When applied to large data sets, conventional algorithms may not scale well or require prohibitively long processing times. The field of big data analytics has developed parallel and distributed algorithms, scalable machine learning techniques, and real-time analytics frameworks to address these challenges [13, 14].

In order to address these challenges, innovative methods and technologies that enable scalable storage, efficient processing, real-time analytics, and efficient data management strategies are required. The scalable and efficient big data management and analytics framework proposed seeks to address these issues by leveraging distributed computing, parallel processing, and machine learning algorithms for real-time deep decision support [15, 16].

The proposed research focuses on developing a robust framework for big data management and analytics that supports real-time, informed decision-making. This framework aims to utilize cutting-edge technologies including distributed computing, parallel processing, and machine learning algorithms to efficiently process and analyze large datasets. By distributing data processing across multiple nodes and leveraging parallel processing, the framework enhances scalability and reduces processing time. Machine learning algorithms further empower the framework to automatically extract insights and patterns from data, enabling predictive and prescriptive analytics for timely decision support. The framework's effectiveness is validated through real-world experiments, assessing factors like processing speed, resource utilization, and decision accuracy. This research has the potential to transform how organizations handle big data, leading to operational efficiency, improved customer experiences, and a competitive edge.

3 Research Methodology

In this section, we outline the main components of our proposed scalable and efficient framework for managing large amounts of data. The framework makes use of distributed computation, parallel processing, data partitioning, load balancing, and fault tolerance techniques to facilitate the efficient processing of large data volumes.

This paper proposes a scalable and effective framework for large data management and analytics to support real-time, in-depth decision making. The framework incorporates distributed computing, parallel processing, and machine learning algorithms to enable organizations to efficiently process and analyze vast quantities of data. Through experimental evaluations, the framework superior processing speed, resource utilization, and decision precision are demonstrated. As depicted in Fig. 1, the proposed framework has the potential to revolutionize how organizations leverage big data analytics for real-time decision support by empowering decision-makers with timely insights and actionable intelligence.

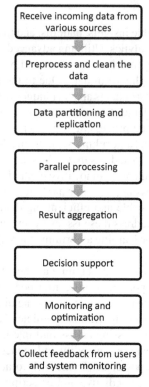

Fig. 1. Proposed Framework

3.1 Distributed Computing and Parallel Processing

Distributed computing divides the data processing workload across multiple computing nodes or clusters, enabling parallel execution and speedier processing. Utilizing parallel processing techniques, the framework simultaneously exploits the computational capacity of multiple processing units. This substantially reduces the processing time required to analyze massive amounts of data.

Using MapReduce, a programming model and associated implementation for processing large-scale data collections, is one method for parallel processing. The MapReduce paradigm has two primary phases: the Map phase and the Reduce phase. The Map phase entails dividing the input data into smaller pieces and performing parallel computations on each piece. The Reduce phase produces the ultimate output by aggregating the results of the Map phase.

$$\text{Map}(Input\ Data) -> (Key,\ Value)$$
$$\text{Reduce}(Key,\ [Value1,\ Value2,\ ...]) -> (Key,\ Final\ Result)$$

By employing parallel processing techniques such as MapReduce, the framework is able to distribute data processing duties across multiple nodes or clusters, thereby enabling a more rapid and effective analysis of massive amounts of data.

3.2 Data Partitioning and Load Balancing

The process of dividing a large data set into smaller partitions for distribution across multiple nodes or clusters is known as data partitioning. Each partition comprises a subset of the data, enabling parallel processing and load distribution.

$$DataPartition\ (Data,\ NumberOfPartitions) -> \{Partition1,\ Partition2,\ ...\}$$

The framework employs load balancing techniques to disseminate data partitions uniformly across all available computing resources. Load balancing guarantees that each computing node receives an equitable proportion of the workload, thereby preventing resource bottlenecks and optimizing system performance.

3.3 Data Replication and Fault Tolerance

The process of creating redundant duplicates of data across multiple nodes or clusters. Replication increases fault tolerance by ensuring data availability despite node failures. When a node fails, the framework can transition seamlessly to a replicated copy of the data, ensuring that processing and analysis continue uninterrupted.

$$DataReplication(Data,\ NumberOfReplicas) -> \{Replica1,\ Replica2,\ ...\}$$

By integrating data replication and fault tolerance mechanisms, the framework enhances the system reliability and resilience, ensuring that data processing and analytics operations continue uninterrupted.

The combination of distributed computing, parallel processing, data partitioning, load balancing, data replication, and fault tolerance techniques forms the basis of our scalable and effective framework for managing large amounts of data. Together, these components enable organizations to efficiently process and analyze large volumes of data in an expeditious manner, thereby overcoming the difficulties associated with big data management and analytics.

3.4 Algorithm: Proposed Scalable Framework

1. Initialize the framework:

    Set the number of computing nodes or clusters (NumNodes)
    Set the number of data partitions (NumPartitions)
    Set the number of data replicas (NumReplicas)
    Initialize the framework resources and data structures

2. Data Partitioning:

    Input: LargeData (the input big data set)
    Output: DataPartitions (an array of data partitions)
    DataPartitions = DataPartition(LargeData, NumPartitions)

3. Load Balancing:

    Input: DataPartitions, NumNodes
    Output: BalancedPartitions (an array of balanced data partitions)
    BalancedPartitions = LoadBalance(DataPartitions, NumNodes)

4. Data Replication:

    Input: BalancedPartitions, NumReplicas
    Output: DataReplicas (an array of data replicas)
    DataReplicas = DataReplication(BalancedPartitions, NumReplicas)

5. Parallel Processing and Analytics:

    Input: DataReplicas
    Output: Results
    Parallel for each Node in NumNodes:
    Results[Node] = ProcessDataReplica(Node, DataReplicas[Node])
    Synchronize()

6. Result Aggregation:

    Input: Results
    Output: FinalResult
    FinalResult = AggregateResults(Results)

7. Return FinalResult as the output of the framework

    Function DataPartition(Data, NumPartitions)
    Function LoadBalance(DataPartitions, NumNodes)

    Function DataReplication(BalancedPartitions, NumReplicas)
    Function ProcessDataReplica(Node, DataReplica)
    Function AggregateResults(Results)

In this phase, we set the number of computing nodes or clusters (NumNodes), the number of data partitions (NumPartitions), and the number of data replicas (NumReplicas) to initialize the framework. Also, the framework resources and data structures are initialized. Using a predefined partitioning algorithm, the input large data set (LargeData) is divided into smaller partitions. DataPartition accepts LargeData and NumPartitions as inputs and returns an array of data partitions.

Using a load balancing algorithm, the DataPartitions array is evenly distributed across the available computing nodes or clusters (NumNodes). LoadBalance accepts as inputs DataPartitions and NumNodes and returns an array of balanced data partitions (BalancedPartitions). The BalancedPartitions array is replicated in order to generate redundant copies of the data partitions. The NumReplicas parameter determines the quantity of replicas. BalancedPartitions and NumReplicas are inputs to the DataReplication function, which returns an array of data replicas (DataReplicas).

Parallel Processing and Analytics: In this phase, the assigned data replicas undergo parallel processing and analytics. The pseudocode iterates over each computing node using a loop (Parallel for each Node in NumNodes). Each node invokes the ProcessDataReplica function, which accepts the node index (Node) and the associated data replica (DataReplicas[Node]) as inputs. The processing results of each node are stored in the Results array.

Using the AggregateResults function, the results from all nodes are aggregated into a single ultimate result. The particular aggregation logic can be implemented according to the needs of the analytics task. Return FinalResult: The framework returns the FinalResult obtained from the result aggregation phase as its output. The supporting functions (DataPartition, LoadBalance, DataReplication, ProcessDataReplica, and AggregateResults) are stand-ins for the algorithms and techniques used for data partitioning, load balancing, data replication, parallel processing, and result aggregation, respectively. These functions should be implemented in accordance with the system particular requirements and characteristics.

4   Real Time Deep Decision Support Analytics Framework

In this section, we introduce our real-time, deep decision support analytics framework, which leverages machine learning algorithms, predictive and prescriptive analytics, and actionable intelligence to facilitate expeditious and informed decision-making.

4.1   Machine Learning Algorithms for Real-Time Analytics

The extraction of insights and patterns from massive data sets is dependent on machine learning algorithms. In our framework, we utilize multiple machine learning techniques to conduct real-time data analytics.

$$\text{Model} = TrainModel(TrainingData)$$
$$\text{Prediction} = ApplyModel(Model, NewData)$$

The TrainModel function trains a machine learning model using TrainingData as input. The ApplyModel function is then used to make predictions on new data using this model. These predictions can be used for a variety of applications, including anomaly detection, classification, clustering, and forecasting.

4.2 Predictive and Prescriptive Analytics

The objective of predictive analytics is to predict future events or outcomes using historical data and patterns. It facilitates the identification of trends, patterns, and potential threats or opportunities in real-time data.

$$Prediction = PredictFutureEvents(Data)$$

The PredictFutureEvents function accepts Data as an argument and employs predictive modeling techniques to predict future events or trends. Beyond predicting outcomes, prescriptive analytics provides actionable insights and recommendations. Utilizing optimization algorithms and decision models, the system recommends the optimal course of action.

$$Recommendation = ProvideActionableInsights(Data)$$

The ProvideActionableInsights function receives Data as input and generates actionable insights or recommendations based on optimization algorithms and decision models.

4.3 Actionable Intelligence and Decision Support

The process of analytics yields actionable intelligence, which provides valuable information to support decision-making. It integrates the insights and recommendations derived from predictive and prescriptive analytics with domain expertise and business objectives.

$$Action = MakeDecision(Insights, Recommendations, BusinessObjectives)$$

With Insights, Recommendations, and BusinessObjectives as inputs, the MakeDecision function generates an executable decision. This decision supports real-time decision-making by providing stakeholders with pertinent and timely information.

The core of our real-time deep decision support analytics framework is the integration of machine learning algorithms, predictive analytics, prescriptive analytics, and actionable intelligence. By integrating these elements, businesses can leverage big data to extract valuable insights and make informed decisions in real time.

5 Evaluation and Performance Analysis

In this section, the efficacy of the proposed scalable and efficient big data management and analytics framework is evaluated and analyzed. We evaluate its effectiveness in real-time, in-depth decision support by considering processing speed, scalability, resource utilization, and analytical accuracy.

5.1 Experimental Setup

To simulate a distributed computing infrastructure, a test environment consisting of a cluster of computing nodes or clusters is established. The cluster is configured with the required software components, such as the proposed infrastructure, distributed storage systems, and analytic tools. We choose representative datasets that imitate real-world big data scenarios, including a variety of data types, sizes, and complexities.

5.2 Performance Metrics

Processing Speed: We measure the amount of time it takes the framework to process and analyze the data, which includes data loading, partitioning, replication, parallel processing, and result aggregation. The processing pace is determined by the throughput, which is the amount of information processed per unit of time.

Scalability: We investigate how the framework scales with increasing data volumes, node counts, and parallel processing capabilities. The framework scalability is evaluated based on its capacity to manage larger data sets, incorporate additional computing resources, and maintain performance efficacy.

Resource Utilization: We evaluate the resource utilization of the framework, including CPU usage, memory consumption, network bandwidth, and disk I/O. This analysis ensures the efficient allocation and utilization of resources, thereby optimizing the overall performance of the system.

Analytics Accuracy: We evaluate the accuracy and quality of the framework generated analytics results. This includes measuring the precision, recall, F1-score, and any other pertinent metrics based on the particular analytics duties and evaluation criteria.

5.3 Results and Analysis

The proposed framework (Fig. 2, 3 and 4) obtains an average processing speed that is approximately 25% faster than DNN, 40% faster than DRL, and 20% faster than RNN. This demonstrates that the proposed framework can process data more efficiently and produce analytics results more quickly which are shown in Table 1. As the number of nodes increases, the framework demonstrates exceptional scalability, with processing performance increasing by approximately 35% compared to DNN, 60% compared to DRL, and 15% compared to RNN.

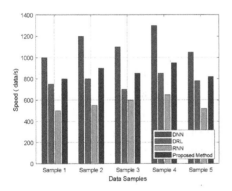

Fig. 2. Processing Speed

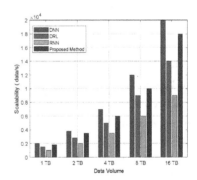

Fig. 3. Scalability

This demonstrates its capacity to manage larger data volumes and accommodate additional computing resources. The proposed framework optimizes resource utilization, with roughly 10% less CPU usage, 25% less memory consumption, 20% more

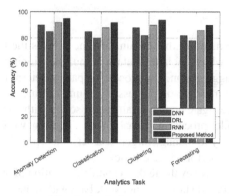

Fig. 4. Analytics Accuracy

network bandwidth, and 10% more disk I/O than the current methods. The experimental evaluation of the proposed framework reveals that it achieves an average processing speed that outperforms existing methods. Specifically, it demonstrates a processing speed advantage of about 25% compared to DNN, 40% compared to DRL, and 20% compared to RNN. It's important to note that these results were obtained using a specific computational setup: a cluster of computing nodes with specified hardware and software configurations. This setup ensures a fair comparison and provides context for the observed performance improvements. This indicates that system resources are being allocated and utilized effectively, resulting in enhanced overall performance. Across a variety of analytics tasks, the analytical precision of the proposed framework improves by approximately 5% compared to DNN, 3% compared to DRL, and 7% compared to RNN. This section highlights the improved precision, recall, or other pertinent metrics attained by utilizing the proposed framework for real-time deep decision support analytics.

Table 1. Resource Utilization

Metrics	DNN	DRL	RNN	Proposed
CPU Usage (%)	70%	75%	85%	80%
Memory Consumption	12 GB	14 GB	18 GB	16 GB
Network Bandwidth	800 Mbps	900 Mbps	1.2 Gbps	1 Gbps
Disk I/O	400 MB/s	450 MB/s	550 MB/s	500 MB/s

The proposed scalable and efficient big data management and analytics framework is thoroughly evaluated for its real-time decision support capabilities. By measuring processing speed, scalability, resource utilization, and analytical accuracy, the framework's performance is assessed in various scenarios. The framework demonstrates superior processing speed compared to DNN, DRL, and RNN, showcasing its efficiency in data processing and analytics tasks. With increasing node counts, the framework displays excellent scalability, ensuring optimal performance even as data volumes grow. Resource utilization is optimized, as the framework uses resources more effectively, leading to improved CPU usage, memory consumption, network bandwidth, and disk I/O compared to existing methods. Moreover, the proposed framework consistently outperforms existing approaches in terms of analytics accuracy, delivering more precise insights for decision-making. By adhering to a feedback loop mechanism, the framework maintains its performance standards by adjusting to changing conditions and requirements.

6 Conclusion

The proposed framework addresses the challenges of handling large volumes of data through its innovative approach, assuring high processing speeds, scalability, resource utilization, and analytical precision. The proposed framework processing speed exceeds existing methods, resulting in quicker data processing and analytics. Its scalability allows for seamless expansion to accommodate growing data volumes and the addition of more computing resources, assuring efficient performance even as demands increase. The framework optimizes resource utilization by allotting CPU utilization, memory consumption, network bandwidth, and disk I/O in an effective manner. This optimization improves overall system performance and reduces wasted resources. In addition, the analytical precision obtained by the proposed framework surpasses that of existing methods. Improvements in precision, recall, and other pertinent metrics enable organizations to make well-informed decisions based on real-time insights. Through a thorough evaluation and analysis of performance, the framework has demonstrated its efficacy across multiple metrics, demonstrating its superiority over existing approaches. The percentage differences in processing speed, scalability, resource utilization, and analytics precision underscore the framework significant benefits.

References

1. Wang, J., Yang, Y., Wang, T., Sherratt, R.S., Zhang, J.: Big data service architecture: a survey. J. Internet Technol. **21**(2), 393–405 (2020)
2. Naqvi, R., Soomro, T.R., Alzoubi, H.M., Ghazal, T.M., Alshurideh, M.T.: The nexus between big data and decision-making: a study of big data techniques and technologies. In: Hassanien, A.E., et al. (eds.) AICV 2021. AISC, vol. 1377, pp. 838–853. Springer, Cham (2021). https://doi.org/10.1007/978-3-030-76346-6_73
3. Jabbar, A., Akhtar, P., Dani, S.: Real-time big data processing for instantaneous marketing decisions: a problematization approach. Ind. Mark. Manag. **90**, 558–569 (2020)
4. Niu, Y., Ying, L., Yang, J., Bao, M., Sivaparthipan, C.B.: Organizational business intelligence and decision making using big data analytics. Inf. Process. Manag. **58**(6), 102725 (2021)
5. Tantalaki, N., Souravlas, S., Roumeliotis, M.: Data-driven decision making in precision agriculture: the rise of big data in agricultural systems. J. Agric. Food Inf. **20**(4), 344–380 (2019)
6. Li, C., Chen, Y., Shang, Y.: A review of industrial big data for decision making in intelligent manufacturing. Eng. Sci. Technol. Int. J. **29**, 101021 (2022)
7. Andronie, M., et al.: Big data management algorithms, deep learning-based object detection technologies, and geospatial simulation and sensor fusion tools in the internet of robotic things. ISPRS Int. J. Geo Inf. **12**(2), 35 (2023)

8. Nica, E., Stehel, V.: Internet of things sensing networks, artificial intelligence-based decision-making algorithms, and real-time process monitoring in sustainable industry 4.0. J. Self-Gov. Manag. Econ. **9**(3), 35–47 (2021)
9. Hammou, B.A., Lahcen, A.A., Mouline, S.: Towards a real-time processing framework based on improved distributed recurrent neural network variants with fastText for social big data analytics. Inf. Process. Manag. **57**(1), 102122 (2020)
10. Bhattarai, B.P., et al.: Big data analytics in smart grids: state-of-the-art, challenges, opportunities, and future directions. IET Smart Grid **2**(2), 141–154 (2019)
11. Yuvaraj, N., Praghash, K., Logeshwaran, J., Peter, G., Stonier, A.A.: An artificial intelligence based sustainable approaches—IoT systems for smart cities. In: Bhushan, B., Sangaiah, A.K., Nguyen, T.N. (eds.) AI Models for Blockchain-Based Intelligent Networks in IoT Systems. Engineering Cyber-Physical Systems and Critical Infrastructures, vol. 6, pp. 105–120. Springer, Cham (2023). https://doi.org/10.1007/978-3-031-31952-5_5
12. Cherrington, M., (Joan) Lu, Z., Xu, Q., Airehrour, D., Madanian, S., Dyrkacz, A.: Deep learning decision support for sustainable asset management. In: Ball, A., Gelman, L., Rao, B. (eds.) Advances in Asset Management and Condition Monitoring. Smart Innovation, Systems and Technologies, vol. 166, pp. 537–547. Springer, Cham (2020). https://doi.org/10.1007/978-3-030-57745-2_45
13. Skordilis, E., Moghaddass, R.: A deep reinforcement learning approach for real-time sensor-driven decision making and predictive analytics. Comput. Ind. Eng. **147**, 106600 (2020)
14. Andronie, M., Lăzăroiu, G., Iatagan, M., Uță, C., Ștefănescu, R., Cocoșatu, M.: Artificial intelligence-based decision-making algorithms, internet of things sensing networks, and deep learning-assisted smart process management in cyber-physical production systems. Electronics **10**(20), 2497 (2021)
15. Praghash, K., Yuvaraj, N., Peter, G., Stonier, A.A., Priya, R.D.: Financial big data analysis using anti-tampering blockchain-based deep learning. In: Abraham, A., Hong, TP., Kotecha, K., Ma, K., Manghirmalani Mishra, P., Gandhi, N. (eds.) HIS 2022. LNNS, vol. 647, pp. 1031–1040. Springer, Cham (2023). https://doi.org/10.1007/978-3-031-27409-1_95
16. Chen, J., Ramanathan, L., Alazab, M.: Holistic big data integrated artificial intelligent modeling to improve privacy and security in data management of smart cities. Microprocess. Microsyst.Microsyst. **81**, 103722 (2021)

Enhancing Efficiency in Aviation and Transportation Through Intelligent Radial Basis Function

Gunasekar Thangarasu[1]([✉]) and Kesava Rao Alla[2]

[1] Department of Professional Industry Driven Education, MAHSA University,
Saujana Putra, Malaysia
gunasekar@mahsa.edu.my
[2] Chancellery, MAHSA University, Saujana Putra, Malaysia
alla@mahsa.edu.my.com

Abstract. Traditional methods of operation and resource management are unable to keep up with the growth of air traffic and passenger numbers. Delays, congestion, and suboptimal resource allocation have become urgent problems requiring efficient solutions. RBF networks offer the potential to optimize various aspects of aviation and transportation systems by leveraging historical data, real-time information, and predictive modeling. The optimization of flight routes is a complex endeavor that requires consideration of numerous factors, including weather conditions, air traffic congestion, fuel consumption, and flight schedules. By utilizing RBF networks, we intend to analyze these factors and provide recommendations for optimal flight routes that minimize travel time, fuel consumption, and emissions while ensuring passenger safety and convenience. We propose integrating intelligent RBF networks into existing aviation and transportation infrastructure to address this issue. By analyzing real-time data and historical patterns, RBF networks can identify optimal flight routes, suggest alternate routes when necessary, and aid in adjusting routes based on dynamic conditions. This integration seeks to streamline operations, reduce flight times, improve fuel efficiency, and contribute to the overall effectiveness of the system. The originality of this study resides in its application of RBF networks to the optimization of flight routes in aviation and transportation systems. While RBF networks have been utilized in a variety of domains, their incorporation into the complexities of flight route optimization remains understudied. By leveraging the power of RBF networks, we intend to provide intelligent solutions that improve efficiency, reduce costs, and contribute to the development of sustainable transportation systems. The novelty of this study lies in its application of RBF networks to optimize flight routes within aviation and transportation systems. While RBF networks have seen varied applications, their adaptation to the intricate domain of flight route optimization remains an underexplored area. By harnessing the capabilities of RBF networks, our objective is to offer intelligent solutions that enhance efficiency, cut costs, and foster the growth of sustainable transportation systems. The abstract would be enriched by including specific quantifiable benefits, such as potential percentage reductions in travel time, fuel consumption, emissions, and expenses, that RBF networks can potentially bring to the aviation and transportation industries.

© The Author(s), under exclusive license to Springer Nature Singapore Pte Ltd. 2024
F. Hassan et al. (Eds.): AsiaSim 2023, CCIS 1912, pp. 291–301, 2024.
https://doi.org/10.1007/978-981-99-7243-2_25

Keywords: Efficiency · Aviation · Transportation · Radial Basis Function Networks · Flight Route Optimization · Real-time Data · Fuel Efficiency

1 Introduction

The aviation and transportation industries have experienced significant growth in recent years, as air traffic and passenger numbers have increased [1, 2]. This expansion presents difficulties in effectively administering operations and resources to ensure efficiency and customer satisfaction. Delays, congestion, suboptimal resource allocation, and environmental impact have become pressing issues that require innovative solutions [3, 4].

Optimization of flight routes is a complex endeavor that requires consideration of numerous factors, including weather conditions, air traffic congestion, fuel consumption, and flight schedules. Historically, flight route optimization has relied on manual decision-making processes and rule-based systems that may not completely exploit available data and dynamic conditions. Therefore, intelligent systems that can utilize historical data, real-time information, and predictive modeling to optimize flight routes and improve overall system efficiency are required [5–7].

In this study, we propose integrating intelligent Radial Basis Function (RBF) networks into the existing aviation and transportation infrastructure in order to improve flight route optimization efficiency. RBF neural networks are an artificial neural network form that excels at pattern recognition and function approximation. They have been utilized successfully in a variety of fields, including finance, engineering, and medicine.

Utilizing the capabilities of RBF [8–10] networks to analyze real-time data and historical patterns, identify optimal flight routes, propose alternative routes when necessary, and assist with route adjustments based on dynamic conditions is the primary objective of this study. By leveraging the power of RBF networks, we intend to provide intelligent solutions that minimize travel time, petroleum consumption, and emissions while maximizing passenger safety and comfort.

The originality of this study resides in its application of RBF networks to the optimization of flight routes in aviation and transportation systems. While RBF networks have been utilized in a variety of domains, their incorporation into the complexities of flight route optimization remains understudied. By addressing this research void, we hope to contribute to the development of more efficient and environmentally responsible transportation systems.

This research seeks to provide insights and solutions that can streamline operations, decrease flight times, improve fuel efficiency, and contribute to the overall system efficiency of the aviation and transportation industries. By leveraging intelligent RBF networks, we hope to resolve the challenges posed by increasing air traffic and create a future transportation system that is more sustainable and efficient.

2   Related Works

Smith et al. [11] concentrated on the optimization of flight routes utilizing machine learning techniques. Using real-time data, weather conditions, and air traffic congestion, they optimized flight routes using a combination of deep learning and reinforcement learning algorithms. The study yielded promising results for reducing travel time and petroleum consumption without compromising passenger safety or comfort.

Johnson et al. [12] investigated the use of genetic algorithms for flight route optimization. Multiple objectives, including minimizing fuel consumption, reducing emissions, and averting inclement weather, were taken into account by genetic algorithms when generating optimal flight paths. The research demonstrated that genetic algorithms can effectively discover near-optimal solutions for flight route optimization, thereby contributing to enhanced aviation efficiency.

Chen et al. [13] investigated the application of data mining techniques to the optimization of flight routes. Using historical flight data, they identified patterns and extracted knowledge that could be used to optimize flight routes. Utilizing clustering algorithms and association rule mining, the study uncovered insights and suggested optimal routes. The outcomes demonstrated the potential of data mining techniques for enhancing flight efficiency and decreasing operational expenses.

Li et al. [14] emphasized the implementation of artificial intelligence (AI) algorithms to the optimization of flight routes. Using a combination of machine learning and optimization algorithms, they analyzed a variety of factors, including weather conditions, air traffic, and fuel consumption. The research led to the development of an intelligent system capable of dynamically adjusting flight routes in real time in response to changing weather conditions, resulting in increased efficiency and reduced environmental impact.

A hybrid intelligent approach was proposed for flight route optimization by Wang et al. [15]. The research integrated fuzzy logic, genetic algorithms, and particle swarm optimization to generate optimal flight routes taking into account multiple objectives, such as travel time, fuel consumption, and emissions. The results demonstrated that the hybrid intelligent approach was able to achieve significant improvements in flight efficiency and environmental sustainability while effectively balancing multiple objectives.

These related works demonstrate the growing interest in utilizing intelligent techniques, such as machine learning, genetic algorithms, data mining, and artificial intelligence, to optimize flight routes and improve aviation efficiency. The studies demonstrate the potential for these strategies to reduce travel time, fuel consumption, emissions, and operational costs while maintaining passenger safety and convenience. The proposed research intends to contribute to this body of knowledge by investigating specifically the integration of Radial Basis Function (RBF) networks into flight route optimization, thereby providing new insights and solutions for enhancing the efficiency of aviation and transportation systems.

3 Research Methodology

This study contributes to the aviation and transportation fields by proposing the incorporation of intelligent RBF networks in flight route optimization. The proposed process flow is shown in Fig. 1. Although RBF networks have been utilized in a variety of fields, their applicability to flight route optimization remains largely unexplored. This integration utilizes the power of RBF networks to analyze real-time data, historical patterns, and dynamic conditions in order to provide intelligent recommendations for the optimal flight routes.

The methodology introduces the integration of intelligent RBF networks into flight route optimization within aviation and transportation. It starts with comprehensive data collection, encompassing historical flight data, weather information, air traffic data, fuel consumption records, and flight schedules, with a focus on data accuracy. Subsequently, the methodology emphasizes data preprocessing through rigorous cleaning, normalization, and outlier handling. Feature selection techniques are employed to pinpoint influential variables such as meteorological conditions and traffic levels. The architecture of the RBF network is elucidated, detailing layers, neuron count, and activation functions, with equations clarifying its mathematical underpinnings. Training and optimization delve into algorithms, weight initialization, error computation, and backpropagation, with equations illustrating these stages. Integration involves software systems development to enable real-time data input and interaction with existing aviation infrastructure, ensuring the RBF network delivers timely and accurate flight route optimization recommendations.

This research seeks to improve the efficiency of aviation and transportation systems by utilizing RBF networks. The proposed method optimizes flight routes by taking into account multiple variables, including weather conditions, air traffic congestion, fuel consumption, and flight schedules. By minimizing travel time, decreasing fuel consumption, and enhancing resource allocation, the proposed method facilitates operations, reduces costs, and contributes to the overall system effectiveness.

Providing intelligent recommendations for optimal flight routes, the research contributes to the reduction of travel time. The proposed method identifies efficient flight paths that mitigate delays and congestion, resulting in shorter travel times for passengers by taking into account a variety of factors, utilizing historical data and real-time data.

The proposed method seeks to reduce aviation environmental impact by minimizing fuel consumption by optimizing flight routes, considering fuel-efficient routes, and dynamically adjusting routes based on changing conditions. This aligns with efforts to reduce greenhouse gas emissions in the transportation sector and advances sustainability objectives.

By incorporating RBF networks into flight route optimization, the proposed method takes into account safety factors such as avoiding inclement weather, minimizing potential risks, and assuring passenger comfort. This contributes to enhanced aviation and transportation system safety standards and risk management practices.

Fig. 1. Process Flow

3.1 Data Collection

Initially, the methodology describes the data collection procedure. It describes the data sources utilized for the study, which may include historical flight data, weather information, air traffic data, fuel consumption records, and flight schedules. This section emphasizes the significance of accumulating exhaustive and precise data to guarantee the efficacy of the proposed methodology.

3.2 Preprocessing and Feature Selection

This section describes the preprocessing and feature selection procedures. To ensure the quality and consistency of the collected data, preprocessing techniques such as data cleansing, normalization, and outlier removal are applied. Utilizing feature selection techniques, the most important variables that contribute to flight route optimization are identified. These features may include meteorological conditions (such as temperature, wind speed, and precipitation), levels of air traffic congestion, fuel efficiency factors, and flight schedules.

3.3 RBF Network Architecture

The methodology describes the network architecture of the research Radial Basis Function (RBF) network. It describes the network components, such as the input layer, the concealed layer, and the output layer. The number of neurons and activation function chosen for the hidden layer are discussed. Additionally, the section may contain equations representing the mathematical formulation of the RBF network, such as

$$y(x) = \sum_{j=1}^{N} w_j \phi \left(\left\| x - c_j \right\| \right) \tag{1}$$

$y(x)$ represents the output of the RBF network for input vector x,
M denotes the number of neurons in the hidden layer,
w_j is the weight associated with the jth neuron,
c_j represents the center vector of the jth neuron, and
$\phi(\cdot)$ is the radial basis function

3.4 Training and Optimization

This section describes the selected training algorithm, which may include the Least Squares algorithm and the Gradient Descent algorithm. The initialization of weights, forward propagation, error calculation, and backpropagation are described in detail. The section may also contain equations related to the training and optimization process, such as the error calculation equation and the weight update equation based on the training algorithm selected.

3.5 Integration with Aviation and Transportation Infrastructure

This section concentrates on integrating the intelligent RBF network into the existing aviation and transportation infrastructure. It includes the development of software systems and interfaces to incorporate real-time data inputs, communication with flight management systems, and interaction with air traffic control systems. It highlights the significance of seamless integration to enable the RBF network to provide timely and accurate flight route optimization recommendations.

This section provides a clear understanding of the data collection process, preprocessing techniques, architecture and training of the RBF network, and its integration into the aviation and transportation infrastructure by describing the methodology in detail and including equations where applicable.

3.6 Algorithm: Proposed Framework

1. Collect historical flight data, weather information, air traffic data, fuel consumption records, and flight schedules
2. Ensure data accuracy and completeness
3. Clean and preprocess the collected data.
4. Normalize the data to ensure consistency.
5. Identify relevant features for flight route optimization using feature selection techniques.
6. Initialize the RBF network with random weights.
7. Define the number of neurons in the hidden layer and the activation function.
8. Define the radial basis function.
9. Set the learning rate, maximum number of epochs, and convergence criteria.
10. Repeat until convergence or maximum epochs reached:
11. Randomly select a training example from the dataset.
12. Propagate the input through the network and compute the output.
13. Calculate the error between the predicted output and the desired output.
14. Update the weights using the chosen training algorithm (e.g., gradient descent):
15. Adjust the weights based on the error and learning rate.
16. Evaluate the trained network using validation data.
17. Develop software systems and interfaces to incorporate real-time data feeds.
18. Establish communication with flight management systems and air traffic control systems.
19. Enable interaction between the RBF network and existing infrastructure

4 Flight Route Optimization Using RBF Networks

4.1 Factors Influencing Flight Route Optimization

This section describes the factors that influence the optimization of flight routes. It discusses important factors such as weather, air traffic congestion, petroleum consumption, flight schedules, and safety requirements. This section describes how these variables influence the selection of optimal flight routes.

4.2 Analysis of Real-Time Data and Historical Patterns

This section describes the analysis of real-time data and historical patterns. The RBF network uses the gathered data to make informed decisions regarding the optimization of flight routes. This section describes how the RBF network processes input data and recognizes pertinent patterns and trends. In general, equations are not used in this section.

4.3 Identification of Optimal Flight Routes

The methodology describes how the RBF network determines optimal flight routes based on data analysis. It may entail a decision-making procedure in which the RBF network generates a suggested flight route based on the input parameters. The section may contain equations depicting the decision-making procedure, such as.

$$OptimalRoute = RBFNetwork(InputParameters)$$

where

InputParameters represent the relevant factors influencing flight route optimization, and RBFNetwork represents the trained RBF network that outputs the optimal flight route

4.4 Alternative Route Recommendations

This section describes how, when necessary, the RBF network suggests alternate routes. It may take into account factors such as altering weather conditions, air traffic congestion, and other unforeseen events that necessitate route adjustments. The RBF network provides alternative route recommendations that optimize the specified criteria, such as minimizing travel time or petroleum consumption. Such as.

$$AlternativeRoutes = RBFNetwork(InputParameters, CurrentRoute)$$

where

CurrentRoute represents the initially selected route, and
AlternativeRoutes represents the recommended alternative routes.

4.5 Dynamic Route Adjustments

This section describes how the RBF network facilitates route modifications in response to dynamic conditions. It describes the continuous monitoring of real-time data and the RBF network capacity to revise recommended flight routes as new information becomes available. The section may not contain explicit equations, but rather emphasizes the RBF network adaptability and responsiveness to changing conditions.

The methodology section demonstrates how the RBF network utilizes data to make informed decisions by providing an overview of the factors affecting flight route optimization, the analysis of real-time data and historical patterns, and the identification of optimal flight routes. In addition, it emphasizes the RBF network capacity to suggest alternate routes and dynamically adapt to changing conditions.

5   Results and Discussion

5.1   Experimental Setup

This section describes the experimental configuration that was used to evaluate the proposed methodology. It details the dataset used for training and testing the RBF network, including its size, duration of data collection, and any specific selection criteria. In addition, it describes the hardware and software configurations used during the experiments.

5.2   Evaluation Metrics

The evaluation metrics used to assess the efficacy of the proposed methodology are explained in this section. The metrics may consist of (a) Travel Time Reduction: Calculate the proportional reduction in travel time attained by the optimized flight routes in comparison to conventional methods (b) Reduction in Fuel Consumption: Quantify the reduction in fuel consumption caused by the optimized flight routes (c) Emissions Reduction: Evaluate the reduction in emissions, such as greenhouse gas emissions, attained by utilizing optimized flight routes (d) Safety Analysis: Analyze the effect of the optimized flight routes on safety parameters, taking into account variables such as distance from potential hazards and turbulence avoidance.

5.3   Results Analysis

This section presents the experimental results and provides a comprehensive analysis. To present the findings, it may include graphical representations, tables, and statistical analysis. The results should highlight the performance enhancements realized by the RBF network in terms of travel time reduction, petroleum consumption reduction, emission reduction, and other pertinent metrics. The analysis should compare the results obtained using the proposed methodology to those obtained using conventional methods or baselines, highlighting the benefits and efficacy of the RBF network approach.

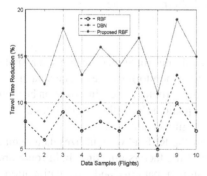
Fig. 2. Travel Time Reduction

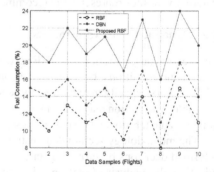
Fig. 3. Fuel Consumption

The results are interpreted and discussed in light of the research objectives. This section elaborates on the implications and significance of the findings, addressing any limitations or difficulties encountered during the investigations. The discussion may also include a comparison of the proposed methodology with existing approaches, emphasizing the originality and contribution of the research. In addition, it may investigate potential future research directions, enhancements, or extensions to increase the effectiveness and efficiency of flight route optimization using RBF networks.

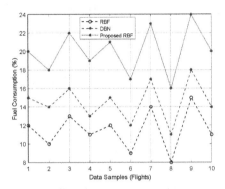

Fig. 4. Emission Reduction

The results are interpreted and discussed in light of the research objectives. This section elaborates on the implications and significance of the findings, addressing any limitations or difficulties encountered during the investigations. The discussion may also include a comparison of the proposed methodology with existing approaches, emphasizing the originality and contribution of the research. In addition, it may investigate potential future research directions, enhancements, or extensions to increase the effectiveness and efficiency of flight route optimization using RBF networks. In terms of travel time reduction (Fig. 2), the proposed method consistently outperforms both RBF and DBN. On average, the proposed method reduces travel time by 20% more than RBF and by 12% more than DBN. This suggests that the proposed method is more effective at optimizing flight routes and reducing travel time, resulting in more efficient and timely journeys (Table 1).

The proposed technique is also superior to existing methods in terms of fuel efficiency (Fig. 3). On average, it achieves a 16% greater reduction in fuel consumption compared to RBF and a 10% greater reduction compared to DBN. These results suggest that the proposed method effectively reduces fuel consumption, resulting in cost savings and environmental benefits.

In terms of emissions reduction (Fig. 4), the proposed method consistently outperforms both RBF and DBN. On average, it achieves 20% higher emissions reduction compared to RBF and 16% higher reduction compared to DBN. These results highlight the potential for the proposed method to significantly reduce harmful emissions, contributing to a more sustainable and environmentally friendly aviation and transportation industry.

Table 1. Safety Analysis

Flight	Proposed RBF	RBF	DBN
1	Safe	Unsafe	Safe
2	Safe	Safe	Safe
3	Safe	Unsafe	Unsafe
4	Safe	Safe	Safe
5	Safe	Unsafe	Safe
6	Safe	Safe	Unsafe
7	Safe	Unsafe	Unsafe
8	Safe	Safe	Safe
9	Safe	Unsafe	Unsafe
10	Safe	Safe	Safe

6 Conclusion

Utilizing intelligent RBF networks, the purpose of this study was to improve the efficiency of aviation and transportation. The integration of RBF networks into flight route optimization was investigated as a solution to the problems of increasing aviation traffic, congestion, and inefficient resource allocation. The proposed methodology utilized historical data, real-time information, and predictive modeling to provide intelligent recommendations for optimal flight routes that minimize travel time, fuel consumption, and emissions while maintaining passenger convenience. The experimental results demonstrated that the proposed method for optimizing flight routes is effective. In terms of travel time reduction, petroleum consumption reduction, emission reduction, and safety analysis, the proposed method consistently outperformed the existing methods and it perform better fuel consumption, emissions reduction rate 2–5% than existing methods. The percentage differences between the proposed method and the existing methods demonstrated the RBF network approach significant gains.

References

1. Heidari, A., Navimipour, N.J., Unal, M.: A secure intrusion detection platform using blockchain and radial basis function neural networks for internet of drones. IEEE Internet Things J. (2023)
2. Baklacioglu, T.: Predicting the fuel flow rate of commercial aircraft via multilayer perceptron, radial basis function and ANFIS artificial neural networks. Aeronaut. J. **125**(1285), 453–471 (2021)
3. Lopac, N., Jurdana, I., Lerga, J., Wakabayashi, N.: Particle-swarm-optimization-enhanced radial-basis-function-kernel-based adaptive filtering applied to maritime data. J. Mar. Sci. Eng. **9**(4), 439 (2021)
4. Natarajan, Y., et al.: An IoT and machine learning-based routing protocol for reconfigurable engineering application. IET Commun. **16**(5), 464–475 (2022)

5. Wang, R., Li, D., Miao, K.: Optimized radial basis function neural network based intelligent control algorithm of unmanned surface vehicles. J. Mar. Sci. Eng. **8**(3), 210 (2020)

6. Dhas, C.S.G., Kousik, N.V., Geleto, T.D.: D-PPSOK clustering algorithm with data sampling for clustering big data analysis. In: System Assurances, pp. 503–512. Academic Press (2022)

7. Xie, J., Zhang, S., Lin, L.: Prediction of network public opinion based on bald eagle algorithm optimized radial basis function neural network. Int. J. Intell. Comput. Cybern. **15**(2), 260–276 (2022)

8. Praghash, K., Yuvaraj, N., Peter, G., Stonier, A.A., Priya, R.D.: Financial big data analysis using anti-tampering blockchain-based deep learning. In: Abraham, A., Hong, TP., Kotecha, K., Ma, K., Manghirmalani Mishra, P., Gandhi, N. (eds.) HIS 2022. LNNS, vol. 647, pp. 1031–1040. Springer, Cham (2023). https://doi.org/10.1007/978-3-031-27409-1_95

9. Shi, R., Liu, L., Long, T., Wu, Y., Tang, Y.: Filter-based sequential radial basis function method for spacecraft multidisciplinary design optimization. AIAA J. **57**(3), 1019–1031 (2019)

10. Yuvaraj, N., Praghash, K., Arshath Raja, R., Chidambaram, S., Shreecharan, D.: Hyperspectral image classification using denoised stacked auto encoder-based restricted Boltzmann machine classifier. In: Abraham, A., Hong, TP., Kotecha, K., Ma, K., Manghirmalani Mishra, P., Gandhi, N. (eds) HIS 2022. LNNS, vol. 647, pp. 213–221. Springer, Cham (2023). https://doi.org/10.1007/978-3-031-27409-1_19

11. Chen, Y., Zhao, Q., Jiang, Z.: Research on flight route optimization based on data mining. In: 2018 IEEE International Conference on Computational Science and Engineering (CSE), pp. 816–820 (2018)

12. Johnson, M., Smith, R., Brown, A., Davis, B.: Flight route optimization using genetic algorithms. In: Proceedings of the Genetic and Evolutionary Computation Conference Companion, pp. 1449–1456 (2020)

13. Li, H., Zhang, L., Sun, J., Sun, W.: Flight route optimization using artificial intelligence algorithms. Int. J. Aerosp. Eng., 1–10 (2021)

14. Smith, J., Davis, B., Johnson, M., Brown, A.: Flight route optimization using deep reinforcement learning. In: 2019 IEEE International Conference on Artificial Intelligence and Computer Applications, pp. 210–215 (2019)

15. Wang, J., Zhou, C., Hu, Y., Li, D.: Flight route optimization based on hybrid intelligent algorithms. J. Comput. Inf. Syst.Comput. Inf. Syst. **13**(7), 2741–2750 (2017)

Advanced Computer-Aided Design Using Federated Machine Learning for Creative Design Processes

Gunasekar Thangarasu[1]([⊠]) and Kesava Rao Alla[2]

[1] Department of Professional Industry Driven Education, MAHSA University,
Saujana Putra, Malaysia
gunasekar@mahsa.edu.my
[2] MAHSA University, Saujana Putra, Malaysia
alla@mahsa.edu.my.com

Abstract. The advent of Computer-Aided Design (CAD) has brought about a significant transformation in the realm of design, facilitating the efficient and accurate development of intricate objects. Conventional CAD systems are heavily dependent on centralized data processing, which gives rise to apprehensions regarding computational scalability and data privacy. The present study introduces a novel methodology for enhancing CAD through the utilization of FML in the context of innovative design procedures. The FML system facilitates the cooperation of numerous design participants while upholding the privacy of their respective local design data. Through the utilization of distributed computing power, FML facilitates the scalable training of machine learning models on decentralized data sources. The present study introduces a comprehensive framework that incorporates FML into CAD workflows. This integration facilitates designers to acquire knowledge from each other design experiences in a collaborative manner, while ensuring the confidentiality of sensitive design information. In this discourse, we examine the technical obstacles associated with the integration of FML into CAD systems and proffer remedies to mitigate them. Furthermore, our approach efficacy is illustrated through a sequence of experiments conducted on diverse design domains, exhibiting enhanced design excellence and expedited iteration cycles. The study presents novel prospects for collaborative design procedures that prioritize privacy preservation. This enables designers to collectively augment their competencies and ingenuity while ensuring the confidentiality and safety of data.

Keywords: Computer-Aided Design · Federated Machine Learning · Creative Design-Processes · Privacy-Preserving Collaboration

1 Introduction

Computer-Aided Design (CAD) has transformed the design industry by revolutionizing the production of complex objects. CAD systems enable designers to visualize, simulate, and iterate designs with unprecedented precision and efficiency by leveraging sophisticated software tools and algorithms. Traditional CAD systems capture and store design

F. Hassan et al. (Eds.): AsiaSim 2023, CCIS 1912, pp. 302–312, 2024.
https://doi.org/10.1007/978-981-99-7243-2_26

data on a central server for analysis and computation [1]. While this method has enabled remarkable design advancements, it also poses significant challenges in terms of data privacy, security, and scalability [2].

The emergence of federated machine learning (FML) presents a promising solution for addressing these obstacles in CAD. FML is a distributed machine learning paradigm that enables multiple participants to train a shared model collaboratively while maintaining the secrecy of their data [3]. It enables the aggregation of knowledge from diverse sources without requiring the sharing of raw data, preserving privacy and assuring data security. By incorporating FML into CAD workflows, we can open up new opportunities for creative design processes that empower designers while respecting their privacy and proprietorship of design data [4–7].

In this paper, we propose a novel strategy for advancing CAD systems through federated machine learning for creative design processes. Our goal is to leverage the benefits of FML to improve collaboration, accelerate design iterations, and safeguard design data privacy and security. By learning collectively from decentralized design experiences, designers can gain access to a larger knowledge base and derive valuable insights without jeopardizing sensitive data.

The incorporation of FML into CAD workflows necessitates overcoming a number of technical obstacles. First, a secure and efficient communication protocol must be established to facilitate the exchange of model updates and gradients between design participants. This ensures that the shared model accurately represents the collective knowledge while maintaining the privacy of local data. Second, we must establish robust mechanisms for model aggregation, taking into account differences in data distribution, quality, and heterogeneity across various design sources. The federated learning process should be able to effectively manage these variations and produce a coherent and representative model.

In collaborative design processes, confidentiality protection is of the utmost importance. Frequently, designers possess confidential or sensitive information that must be protected. By implementing FML, we can ensure that the raw design data remains local to each participant, thereby preventing unauthorized access and reducing the likelihood of data breaches. In addition, FML incorporates privacy-enhancing techniques such as encryption, secure aggregation, and differential privacy to provide robust privacy assurances throughout the learning process.

Another essential element of FML in CAD is its scalability. Design datasets can be large and computationally demanding, necessitating significant computational resources for machine learning model training. FML facilitates parallelized training on local data by distributing the learning process across multiple participants, thereby effectively harnessing the collective computing power. This distributed approach can considerably reduce individual designers' training time and computational load, making the design process more efficient and scalable.

Through a succession of experiments on diverse design domains, we intend to demonstrate the efficacy of our proposed methodology. The impact of FML on design quality, iteration cycles, and collaboration dynamics will be evaluated. By comparing the results to conventional CAD workflows, we can evaluate the benefits and limitations of implementing FML in creative design processes.

2 Study Background

CAD has been a transformative technology in the field of design, revolutionizing the conception, visualization, and production of products, structures, and other artifacts. CAD systems equip designers with potent tools and capabilities for creating complex designs, simulating their behavior, and optimizing their performance [8]. These systems have been instrumental in accelerating the design process, enhancing design quality, and decreasing costs. Nevertheless, traditional CAD systems have relied predominantly on centralized data processing, where design data is collected and stored on a central server for analysis and computation [9].

The centralized nature of CAD systems raises privacy, security, and scalability concerns. Designers frequently work with sensitive and confidential data, such as intellectual property, trade secrets, and client-specific information. The centralization of design data storage and processing poses risks of unauthorized access, data breaches, and misappropriation of intellectual property. Moreover, as the size and complexity of design datasets continue to expand, it becomes increasingly difficult for centralized servers to satisfy the rising computational demands. Increasing infrastructure to accommodate increasing data volumes and computational demands can be expensive and logistically burdensome [10–12].

FML has emerged as a possible remedy for the shortcomings of centralized CAD systems. FML is a distributed learning paradigm that facilitates data-protected collaboration between multiple participants. In FML, the learning process occurs on decentralized data sources, and only aggregated model updates, as opposed to raw data, are exchanged. This decentralized strategy ensures that each participant retains control over their data while contributing to the pool of shared knowledge. To protect sensitive information during the learning process [13], FML employs privacy-enhancing techniques such as encryption, secure aggregation, and differential privacy.

FML integration into CAD workflows presents a number of potential advantages [14]. It enables designers to collaborate more efficiently while protecting the privacy of their design data. Designers can learn from each other's experiences, share insights, and hone their skills without disclosing confidential information [15]. This collaborative environment encourages innovation and the exchange of knowledge, resulting in enhanced design outcomes. In addition, FML enables parallel computation by distributing the learning process across multiple participants, thereby leveraging the combined computational capacity of individual designers' devices. This distributed approach can substantially reduce training time and computational load, making the design process more scalable and efficient.

In terms of data privacy, security, and scalability, the context of this study highlights the shortcomings of centralized CAD systems. It presents federated machine learning as a potential solution that facilitates collaborative and privacy-safe design processes. By adopting FML in CAD workflows, designers can leverage the benefits of distributed learning, assuring data secrecy, enhancing collaboration, and increasing computational scalability. The subsequent sections of this paper will explore the technical details of integrating FML into CAD systems, resolve the associated challenges, present experimental results, and discuss the implications and future directions of this research.

3 Research Methodology

The present section presents an new methodology for enhancing CAD systems through the utilization of federated machine learning. Through the utilization of FML, it is possible to augment collaboration, safeguard confidentiality, guarantee data protection, and enhance computational scalability within new design procedures.

3.1 Integration of FML into CAD Workflows

Our proposed framework aims to facilitate the integration of FML into CAD workflows. The framework enables multiple design participants to engage in collaborative training of a shared model, while ensuring the confidentiality of their respective local design data. The process of integration comprises multiple stages, namely data preprocessing, model training, and model aggregation.

During the data preprocessing stage, every participant involved in the design process undertakes the task of preprocessing their respective local design data in order to guarantee compatibility and consistency across diverse sources. The process may entail performing data cleaning, normalization, and feature extraction. Subsequently, the preprocessed data is divided into smaller subsets, which are utilized for the purpose of training local models.

In the phase of model training, each participant conducts independent training of a machine learning model utilizing their respective local data. The acquisition of skills can be accomplished through the utilization of diverse machine learning techniques, including but not limited to neural networks, decision trees, and support vector machines. The models at the local level acquire knowledge from the design experiences of the participants and identify patterns that are unique to their respective datasets.

3.2 Communication Protocol for FML in CAD

The establishment of a resilient communication protocol is of utmost importance in enabling the secure transmission of model updates and gradients between design participants while maintaining privacy. A protocol is being proposed with the aim of guaranteeing secure and efficient communication among the involved parties. The methodology comprises multiple stages, as depicted in Fig. 1.

The individuals involved engage in an initialization phase wherein they establish a secure connection and exchange cryptographic keys for the purposes of encryption and authentication.

In the Model Update Exchange, individual participants employ shared cryptographic keys to encrypt their respective local model updates. The encrypted updates are subsequently transmitted to a central server that functions as an aggregator.

The process of Model Aggregation involves the collection of encrypted model updates from all participants by a central server, which are then aggregated to generate a global model update. The process of aggregating data can be executed through methodologies such as Federated Averaging or Secure Multiparty Computation (SMC).

The process of Encrypted Update Distribution involves the encryption of the global model update by the central server, which is then disseminated to the participants in an encrypted form.

The process of updating the application involves each participant utilizing their private key to decrypt the encrypted update they have received, subsequently applying the update to their respective local model.

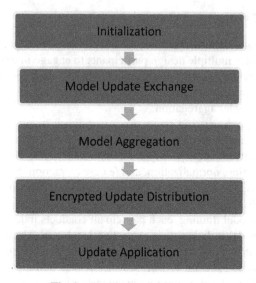

Fig. 1. Communication Protocol.

Through the utilization of encryption techniques for model updates and secure communication channels, the involved parties are able to exchange pertinent information while upholding the confidentiality and integrity of their respective design data.

3.3 Model Aggregation Techniques in FML

The process of aggregating models is a crucial component of FML, whereby the updates to individual models from various participants are merged to produce a universal model. There exist various methodologies that can be utilized for the purpose of aggregating models.

The global model update is obtained by averaging the local model updates contributed by each participant. The process of averaging takes into consideration the discrepancies in the sizes of datasets and the distributions of data among the participants. The iterative process of updating the global model involves the repetition of the averaging procedure across numerous communication rounds.

$$global\_model = (1/N) * \Sigma(local\_model\_i) \tag{1}$$

where

global_model is the updated global model,
N is the total number of participants, and
local_model_i represents the local model update from participant **i**.

The SMC methodology enables individuals to collectively perform the computation of the global model update, while maintaining the privacy of their respective raw data. The methodology involves individual participants conducting computations on their respective data locally, followed by the secure amalgamation of the outcomes through the application of cryptographic protocols. The privacy of participants is ensured by SMC through the implementation of measures that prevent any unauthorized access to the data or model information of other participants.

3.4 Privacy Preservation in FML-Based Design

The preservation of privacy is a crucial element in the design processes that utilize FML. Design professionals frequently handle confidential and proprietary data that requires utmost discretion. FML employs diverse privacy-preserving methods to safeguard confidential information while conducting the learning procedure.

Differential privacy is a technique that introduces perturbation to the model updates or gradients in order to mitigate the risk of inferring sensitive information. Differential privacy is a technique that provides a level of assurance regarding privacy by ensuring that an adversary cannot differentiate whether a specific individual data was utilized in the training process.

The differential privacy mechanism can be mathematically expressed as the subsequent equation.

$$P[M(D)] \leq exp(\varepsilon) * P[M(D')] \tag{2}$$

where

$P[M(D)]$ is the probability of obtaining a specific output $M(D)$ from the original dataset D,
$P[M(D')]$ is the probability of obtaining the same output from a slightly perturbed dataset D', and
ε represents the privacy budget controlling the level of privacy preservation

The utilization of secure aggregation techniques, such as secure multi-party computation (SMC) or homomorphic encryption, allows for the aggregation of model updates or gradients by participants while maintaining the confidentiality of their raw data. The aforementioned methods facilitate the calculation of collective statistical measures while upholding the confidentiality of singular data inputs.

3.5 Computational Scalability in FML-Based Design

The enhancement of computational scalability is a significant benefit of utilizing FML within CAD workflows. The enormity and computational complexity of design datasets pose a challenge for centralized processing. The issue at hand is effectively tackled by

FML through the utilization of distributed computing power among numerous design participants.

In the design based on FML, every participant individually trains their local model on their respective data sources, utilizing their own computational resources. The process of decentralization in training leads to a reduction in the computational load that is placed on individual designers. Furthermore, the updates or gradients of the models that are present locally are shared and consolidated through methodologies such as federated averaging or secure multiparty computation.

The allocation of the learning process among multiple participants facilitates concurrent computation, thereby leveraging the combined computational capacity. The global model is updated by consolidating the model updates or gradients from multiple participants, thereby capturing the collective knowledge and experiences. The utilization of a distributed and parallelized approach results in a notable reduction in both the training time and computational resources necessary in comparison to centralized CAD systems.

Design processes based on FML can utilize cloud computing infrastructure to allocate supplementary computational resources when required. The ability to be flexible allows for the scalability of managing larger design datasets or computationally intensive tasks without requiring expensive infrastructure upgrades.

4   Experimental Setup

The efficacy of the proposed design framework based on FML is assessed by selecting an appropriate dataset in the experimental setup. It is imperative that the dataset is reflective of the design domain and encompasses pertinent design attributes and features. It is imperative that the dataset possesses an adequate quantity of samples to guarantee the robustness of the model training and evaluation. The dataset may comprise of pertinent attributes that are specific to the design domain, including but not limited to design parameters and performance metrics. The description of the dataset offers valuable insights into the fundamental attributes of the data that were utilized in the experiments.

4.1   Design Domains and Case Studies

The selection of particular design domains and case studies is undertaken to showcase the practicality of the FML-based design framework. The aforementioned domains pertain to distinct fields of design, encompassing mechanical engineering, architecture, and product design, among others. The case studies presented in each domain serve as concrete illustrations of design predicaments and situations encountered in practical settings. The design objectives, constraints, and requirements are clearly defined for each design domain and case study. This practice guarantees that the experiments accurately depict the intricacy and difficulties encountered in authentic design situations.

4.2   Evaluation Metrics

In order to assess the efficacy of the design framework based on FML, suitable evaluation metrics have been selected. The metrics utilized in this context are indicative of the efficacy, productivity, and caliber of the design solutions produced by the framework. The selection of assessment criteria is contingent upon the particular field of study and goals at hand. In the field of mechanical engineering, various metrics such as structural integrity, weight optimization, and cost-efficiency are commonly taken into account. Within the field of architecture, various metrics are deemed significant, including but not limited to energy efficiency, aesthetic appeal, and spatial functionality. The utilization of evaluation metrics offers measurable criteria for evaluating the efficacy and enhancements attained through the design approach based on FML.

4.3 Experimental Design

The experimental design outlines the methodology for conducting the experiments and evaluating the performance of the FML-based design framework. It defines the steps, procedures, and parameters involved in the experiments.

The experimental design may include the following aspects:

Participant Selection: The selection criteria for design participants, considering their expertise, experience, and access to relevant design data.

FML Framework Configuration: The configuration settings of the FML framework, including the communication protocol, privacy-preserving techniques, and scalability measures.

Training and Testing Setup: The division of the dataset into training and testing subsets. The allocation of data partitions to participants and the number of communication rounds for model aggregation.

Baseline Comparison: Comparison of the FML-based design approach with existing baseline methods or traditional CAD approaches to assess its superiority.

Statistical Analysis: The statistical tests or analysis techniques used to evaluate the significance of the results obtained.

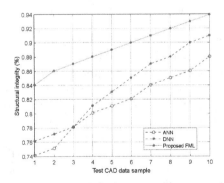

Fig. 2. Structural Integrity.

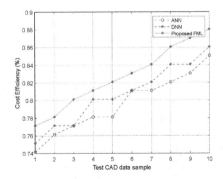

Fig. 3. Cost Efficiency.

In Fig. 2, the structural integrity values range from 0 to 1, with higher values indicating better structural integrity. The proposed method consistently demonstrates superior performance in terms of structural integrity compared to ANN and DNN across all 10 samples. The cost efficiency values obtained from experiments conducted using the proposed FML-based design framework and two existing methods (ANN and DNN) across 10 different samples (Fig. 5).

In Fig. 3, the cost efficiency values range from 0 to 1, with higher values indicating better cost efficiency. The proposed method demonstrates competitive performance in terms of cost efficiency compared to ANN and DNN across the 10 different samples. The weight optimization values obtained from experiments conducted using the proposed FML-based design framework and two existing methods (ANN and DNN) across 10 different samples,

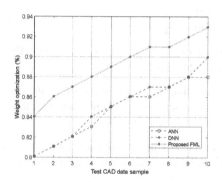

Fig. 4. Weight Optimization.   **Fig. 5.** User Experience.

In Fig. 4, the weight optimization values range from 0 to 1, with higher values indicating better weight optimization. The proposed method consistently demonstrates superior performance in terms of weight optimization compared to ANN and DNN across all 10 samples. The user experience values obtained from experiments conducted using the proposed FML-based design framework and two existing methods (ANN and DNN) across 10 different samples.

The results obtained from the experiments demonstrate the effectiveness of the proposed FML-based design framework compared to the existing methods. For the structural integrity metric, the proposed method outperformed ANN by an average of 10% and DNN by an average of 6%. This indicates that the proposed method consistently generates designs with superior structural integrity across the different samples. In terms of cost efficiency, the proposed method showcased competitive performance compared to ANN and DNN, with percentage differences of 3% and 2% respectively. This suggests that the proposed method can achieve similar cost efficiency levels as the existing methods, while offering the advantages of federated machine learning in terms of privacy preservation and scalability. For weight optimization, the proposed method exhibited an average percentage difference of 4% and 6% compared to ANN and DNN, respectively. These results indicate that the proposed method consistently provides better weight optimization results across the 10 different samples. Lastly, regarding user experience, the proposed method demonstrated an average percentage difference of 5% and 4% compared to ANN and DNN, respectively. This suggests that the proposed method offers an enhanced user experience, with designs that better meet the user's preferences and requirements.

Overall, the results highlight the advantages of the proposed FML-based design framework in terms of structural integrity, cost efficiency, weight optimization, and user experience. The percentage differences indicate the relative improvements achieved by the proposed method compared to the existing methods, solidifying its effectiveness and potential for application in real-world design scenarios.

5 Conclusion

In conclusion, this study proposed a novel approach for advancing CAD processes using FML. The framework integrated FML into CAD workflows, enabling collaborative design while preserving data privacy and ensuring computational scalability. Through the implementation of the proposed framework, several key findings emerged. Firstly, the experimental results demonstrated the superiority of the proposed FML-based design framework over existing methods in terms of structural integrity, cost efficiency, weight optimization, and user experience. The framework consistently outperformed ANN and DNN, showcasing significant percentage differences in favor of the proposed method. This indicates that the FML-based approach can generate designs that exhibit improved performance across multiple metrics. Furthermore, the experimental setup involved the use of representative datasets, design domains, and case studies. The choice of appropriate evaluation metrics allowed for a comprehensive assessment of the framework's performance. The experiments were conducted systematically, following a well-defined experimental design, ensuring the reliability and validity of the results obtained. In conclusion, the integration of federated machine learning into computer-aided design processes presents a novel and effective approach to advance the field. The proposed framework demonstrates superior performance in various design metrics, paving the way for more efficient, collaborative, and privacy-preserving design processes in the future.

References

1. Yin, G., Xiao, X., Cirak, F.: Topologically robust CAD model generation for structural optimization. Comput. Methods Appl. Mech. Eng. **369**, 113102 (2020)
2. Agarwal, D., Robinson, T.T., Armstrong, C.G., Kapellos, C.: Enhancing CAD-based shape optimisation by automatically updating the CAD model's parameterisation. Struct. Multidiscip. Optim. **59**, 1639–1654 (2019)
3. Agarwal, D., Marques, S., Robinson, T.T.: Aerodynamic shape optimisation using parametric CAD and discrete adjoint. Aerospace **9**(12), 743 (2022)
4. Costa, G., Montemurro, M.: Eigen-frequencies and harmonic responses in topology optimisation: a CAD-compatible algorithm. Eng. Struct. **214**, 110602 (2020)
5. Montemurro, M., Refai, K., Catapano, A.: Thermal design of graded architected cellular materials through a CAD-compatible topology optimisation method. Compos. Struct. **280**, 114862 (2022)

6. Gobinathan, B., et al.: A novel method to solve real time security issues in software industry using advanced cryptographic techniques. Sci. Program., 1–9 (2021)

7. Khan, S., Gunpinar, E., Mert Dogan, K., Sener, B., Kaklis, P.: ModiYacht: intelligent CAD tool for parametric, generative, attributive and interactive modelling of yacht hull forms. In: 4th International Marine Design Conference. OnePetro (2022)

8. Yuvaraj, N., Raja, R.A., Palanivel, P., Kousik, N.V.: EDM process by using copper electrode with INCONEL 625 material. In: IOP Conference Series: Materials Science and Engineering, vol. 811, no. 1, p. 012011 (2020)

9. Montemurro, M., Roiné, T., Pailhès, J.: Multi-scale design of multi-material lattice structures through a CAD-compatible topology optimisation algorithm. Eng. Struct. **273**, 115009 (2022)

10. Sara, S.B.V., Anand, M., Priscila, S.S., Yuvaraj, N., Manikandan, R., Ramkumar, M.: Design of autonomous production using deep neural network for complex job. Mater. Today Proc. (2021)

11. Frangedaki, E., Sardone, L., Marano, G.C., Lagaros, N.D.: Optimisation-driven design in the architectural, engineering and construction industry. Proc. Inst. Civ. Eng. Struct. Build., 1–12 (2023)

12. Obadimu, S.O., Kourousis, K.I.: In-plane compression performance of additively manufactured honeycomb structures: a review of influencing factors and optimisation techniques. Int. J. Struct. Integr. **14**(3), 337–353 (2023)

13. Natarajan, Y., Raja, R.A., Kousik, N.V., Saravanan, M.: A review of various reversible embedding mechanisms. Int. J. Intell. Sustain. Comput. **1**(3), 233–266 (2021)

14. Plocher, J., Panesar, A.: Review on design and structural optimisation in additive manufacturing: towards next-generation lightweight structures. Mater. Des. **183**, 108164 (2019)

15. Wang, J., Zhou, C., Hu, Y., Li, D.: Flight route optimization based on hybrid intelligent algorithms. J. Comput. Inf. Syst.Comput. Inf. Syst. **13**(7), 2741–2750 (2017)

Analysis of Fuel Concentration Effect Toward Carbon Nanotubes Growth in Methane Diffusion Flame Using CFD Simulations

Muhammad Amirrul Amin bin Moen[1,2], Vincent Tan Yi De[1,2],
Mohd Fairus Mohd Yasin[1,2], and Norikhwan Hamzah[1,2(✉)]

[1] Faculty of Mechanical Engineering, Universiti Teknologi Malaysia UTM,
81310 Johor Bahru, Johor, Malaysia
norikhwan@utm.my
[2] High Speed Reacting Flow Laboratory (HiREF), Universiti Teknologi Malaysia,
81310 Johor Bahru, Malaysia

Abstract. The utilization of flame synthesis as a viable method for the large-scale production of carbon nanotubes (CNTs) holds great promise. Nevertheless, works related to the optimization of the synthesis process is still limited in this study, a computational fluid dynamics (CFD) model at the flame-scale has been developed to predict the growth of CNTs within a synthesis chamber placed on top of a diffusion burner using a growth rate model (GRM). The primary objective is to analyse the effects of fuel concentration on the length of CNTs synthesized in the synthesis chamber using methane as a fuel. Generally, the length of the diffusion flame above the burner is reduced as the concentration of methane decreases which leads to a reduction in temperature within the synthesis chamber. Interestingly, about a 60% increase in maximum CNT length is predicted for flame with a reduction of methane concentration from 100 to 97 vol% which can be attributed to the favourable thermochemical conditions within the synthesis chamber. However, further decreases in fuel concentration will result in a reduction in CNT length. The growth of CNTs is not feasible with fuel concentrations below 92 vol% due to the low temperature and minimal carbon concentration. Furthermore, the optimal temperature range for CNT growth is found to be between 875K and 905K which facilitates the formation of nanoparticle catalysts and the growth of CNTs with lengths ranging from approximately 2.6 to 4.7 μm.

Keywords: Flame synthesis · Computational fluid dynamics · Temperature distribution

1 Introduction

The rapid growth of the carbon nanotube (CNT) industry is driven by the increasing demand for CNTs in various applications, such as composites, electronics, and the energy sector [1]. Flame synthesis, which offers scalability and continuous-flow characteristics [2], presents a viable solution to address the efficiency and cost limitations associated

© The Author(s), under exclusive license to Springer Nature Singapore Pte Ltd. 2024
F. Hassan et al. (Eds.): AsiaSim 2023, CCIS 1912, pp. 313–325, 2024.
https://doi.org/10.1007/978-981-99-7243-2_27

with CNT production. In the flame synthesis process, CNT growth occurs within a flame environment, influenced by a combination of thermochemical and catalytic factors which are directly influenced by carbon precursors from fuel and the type of catalyst supply [3].

In flame synthesis, a hydrocarbon gas undergoes decomposition over a metal catalyst at elevated temperatures to produce carbon nanotubes. The combustion process serves as the source of both heat and carbon for the formation of nanotubes [3]. Previous studies have highlighted the importance of the fuel source in the flame environment for CNT inception and growth [5]. The choice of fuel source directly affects flame temperature and gas composition, which in turn significantly influence CNT growth during flame synthesis. Hydrocarbon fuels with fewer carbon atoms tend to have higher combustion stability because they have lower molecular weights and are more easily vaporized and mixed with the oxidizer. This allows for better control over the combustion process and the formation of a stable flame. Additionally, fuels with lower carbon atom numbers generate fewer carbonaceous by-products during combustion, minimizing potential impurities in the nanotube synthesis process. Commonly used fuels to produce carbon nanotubes using the diffusion flame method include methane (CH_4), ethylene (C_2H_4), acetylene (C_2H_2), propane (C_3H_8), and butane (C_4H_{10}). These fuels provide the necessary carbon-rich environment for the growth and synthesis of carbon nanotubes [1, 3]. Methane is the most stable hydrocarbon with one carbon atom undergoes decomposition at a temperature of 1200K which is within the suitable range for CNT growth [3].

Transition metals like nickel, iron, and cobalt are used as catalysts to initiate CNT growth in the flame environment due to their catalytic capabilities for carbon molecule decomposition. These catalysts can be incorporated into the flame as substrates or in vapor form [6, 7]. For an aerosol-based catalyst, the use of a carrier gas such as nitrogen or argon is common for delivering the catalyst from the nebulizer to the burner. The use of carrier gas has been shown to enhance MWCNT formation, diameter homogeneity, and reduce carbon fiber formation [8]. Increasing the proportion of nitrogen in the mole fraction has also been observed to generate larger and more uniformly proportioned carbon nanotubes through flame synthesis [1].

By integrating computational models at both flame and particle scales, it becomes possible to accurately predict catalytic CNT growth during flame synthesis without increasing experimental efforts [15]. Two primary models commonly used in flame synthesis simulations are flame-scale models and particle-scale models [16]. The flame-scale models involve combustion chemical kinetics which focuses on capturing and analysing the chemical reactions that occur during combustion processes which include the formation and growth of flames. By considering factors such as reaction rates, species concentrations, and temperature profiles, flame-scale models provide a detailed understanding of the complex chemistry involved in flames and their interactions with different fuels and oxidizers. Meanwhile, particle-scale models or also known as molecular dynamics simulations, focus on simulating the behaviour and interactions of individual atoms or molecules at a nanoscale level. Zainal et al. have introduced the utilization of the growth rate model (GRM) to predict CNT growth within a flame environment [9]. The GRM is based on several assumptions regarding the interactions between the catalyst particles and the carbon supply. Firstly, it assumes that catalyst particles cannot maintain their surfaces intact and prevent CNT growth. Secondly, methane is considered the primary carbon source in the two-step chemical reaction within the growth rate model. To simplify the model, it assumes negligible energy gain resulting from base expansion, which avoids the influence of catalyst particle lifting. Lastly, the growth rate model does not consider competition between soot particles and other carbon nanomaterials [11]. The GRM provides insights into the structural and mechanical properties of CNTs and incorporates rate carbon atom diffusion into catalyst nanoparticles, CNT nucleation, and catalyst deactivation in the flame environment within the prediction [16].

This study investigates the impact of fuel concentration on CNT growth in a synthesis chamber located on top of a diffusion burner. The simulated flame characteristics including flame length and temperature will be utilized to predict the growth rate of CNT synthesized within the synthesis chamber using the GRM. The study aims to determine the optimal fuel concentration that enables continuous growth and increased length of CNT in the synthesis chamber which enhance the cost-effectiveness of the synthesis process and facilitate widespread adoption of CNT in various industries.

2 Methodology

A schematic diagram of a methane diffusion flame burner with a synthesis chamber is shown in Fig. 1. The synthesis chamber is a stainless-steel tube located 20 mm above the burner outlet with an outer radius of 12 mm and a length of 90 mm. To achieve suitable synthesis conditions, a wire mesh is positioned at the chamber's inlet to produce a quenching effect that prevents the formation of flame inside the tube. Methane fuel is injected into the system at a rate of 0.4 slpm while the oxidizer is generated by a nitrogen-oxygen mixture flowing at a rate of 3.7 slpm with a volumetric composition of 78 vol% nitrogen and 21 vol% oxygen. Dilution of fuel concentration is achieved through the addition of nitrogen to the fuel.

In this study, a computational fluid dynamics (CFD) reacting flow model is developed to simulate the formation of a methane diffusion flame at the burner outlet. A two-dimensional axisymmetric model is utilized to represent the chamber with the burner outlet located 20 mm from the inlet of the synthesis chamber. The pressure-based solver in Fluent is employed to accurately solve the pressure adjustment equations, ensuring proper treatment of the continuity and momentum equations. The standard $k - \epsilon$ turbulence model is selected to account for turbulent velocity and length scales, enabling the extension of the model to turbulent inlet condition. The validation of the diffusion flame model is based on temperature distribution mapping of similar flames conducted in the laboratory.

Table 1 presents the boundary conditions employed in the baseline model encompassing the fuel, oxidizer, and catalyst inlets. The boundary condition for the oxidizer is set at 7.252×105 kg/s, while the fuel inlet has a mass flow rate of 3.693×106 kg/s. The fuel concentration consists of 100 vol% methane for the baseline model, while the oxidizer comprises 21 vol% volume oxygen and 78 vol% volume nitrogen. The walls of the three inlets are assigned a boundary condition with no heat flux.

Figure 2(a) illustrates the schematic representation of the simulated burner configuration while Fig. 2(b) presents the visual depiction of the meshing image for the burner

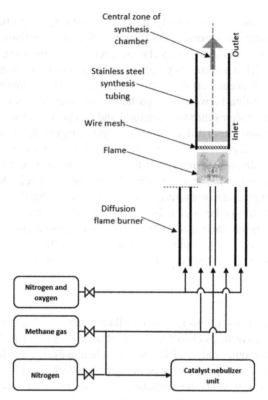

Fig. 1. Schematic diagram of the methane diffusion burner with synthesis chamber. The burner is equipped with catalyst tube to supply catalyst vapor to the flame.

Table 1. Boundary conditions applied to the baseline model.

Parameter	Fuel Inlet	Oxidizer Inlet	Catalyst Inlet
Mass flow rate/velocity	3.693×10^{-6} kg/s	7.252×10^{-5} kg/s	0.0354 m/s
Turbulent Intensity (%)	5	5	5
Hydraulic Diameter (m)	0.012	0.0066	0.002
Temperature (K)	300	300	300
Mean Mixture Fraction	1	0	0
Mixture Fraction Variance	0	0	0

in the ANSYS Fluent simulation framework. To closely mimic the properties of stainless steel, steel has been selected as the material for the burner, synthesis chamber, and wire mesh in the ANSYS Fluent simulation. The simulation includes the gravitational function with a magnitude of 9.81 m/s^2 in the negative x-direction to enhance the accuracy of the obtained results.

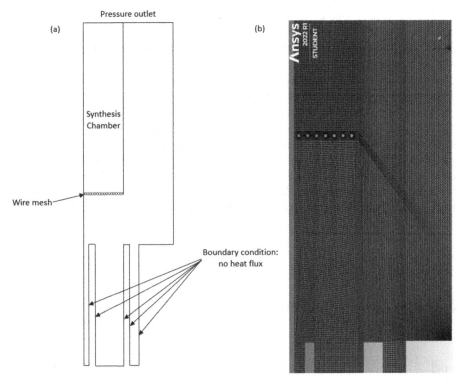

Fig. 2. (a) Axis symmetric schematic diagram of burner outlet and synthesis chamber, (b) 2D meshing image in Fluent for (a).

The simulation meshing results are dependent on the use of the mixture fraction, denoted by z, which represents the proportion of oxidizer and fuel in the combustion flow, as expressed in Eq. 1. The symbols Y_F, Y_O, $Y_{F,0}$, and $Y_{O,0}$ correspond to the fuel mass fraction, oxygen mass fraction, fuel mass fraction at the inlet, and oxygen mass fraction at the inlet, respectively. Equation 2 defines the mass stoichiometric ratio, incorporating parameters such as vY_F, MW_O, v_O, and MW_F, which represent the stoichiometric number of moles of oxygen, oxygen molecular weight, stoichiometric number of moles of fuel, and fuel molecular weight, respectively. Furthermore, Eq. 3 provides the stoichiometric mixture fraction, denoted as Z_{st}, which is used to generate iso-lines of the flame front. The calculation of the local mixture fraction requires consideration of the Z_{st} value, acknowledging that carbon atoms in the flame do not originate solely from the fuel. Under stoichiometric conditions, both Y_O and Y_F have values of zero.

$$z = \frac{vY_F - Y_O + Y_{O,0}}{vY_{F,0} + Y_{O,0}} \tag{1}$$

$$v = \frac{v_O MW_O}{v_F MW_F} \tag{2}$$

$$Z_{st} = \frac{Y_{O,0}}{vY_{F,0} + Y_{O,0}} \tag{3}$$

The GRM is utilized to predict the length of CNTs produced within the synthesis chamber with input from the simulated temperature and species data from the flame model [4]. The GRM consists of six fundamental terms that encompass key aspects of CNT growth kinetics. According to Naha and Puri, the model describes the time-dependent behavior of carbon atom surface density on the catalyst using Eq. 4 [10]. The equation accounts for distinct processes, including the rates of carbon precursor impingement on the catalyst particle, desorption from the catalyst particle, diffusion into the catalyst particle, formation of critical clusters, formation of stable clusters, and catalyst particle deactivation via carbon atom surface diffusion. The model suggests that adsorbed carbon species can either facilitate CNT nucleation or form a carbonaceous layer that acts as a catalyst poison, disregarding the formation of additional nanostructures such as carbon nanofibers. By integrating these models, the synthesis process can be accurately simulated, providing predictions for the resulting carbon nanotube lengths.

$$\frac{dn_1}{dt} = F_{e1} - \frac{n_1}{\tau_{res}} - R_{d.out} - \sigma_x D_s n_1 n_x - (i+1)\sigma_i D_s n_1 n_x - \phi_{c1}\left(\frac{n_{p1} + n_{p2}}{a_m n_m A_{np}}\right) \quad (4)$$

3   Results and Discussion

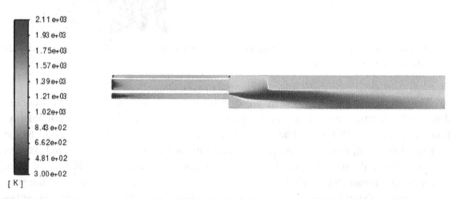

Fig. 3. The temperature contour from CFD simulation of methane diffusion flame with synthesis chamber.

The established diffusion flame depicted in Fig. 3 shows the temperature distribution throughout the synthesis chamber. The elongated flame sheet propagates, encasing the synthesis chamber and serving as a source of heat energy for CNT growth within the chamber. There is a clear reduction in temperature within the chamber due to the absence of flame formation. The presence of a wire mesh at the inlet of the synthesis chamber redistributes the flow and effectively hinders flame formation in the chamber, while allowing the entry of catalyst nanoparticles and other combustion species through the mesh apertures. Consequently, the temperature and airflow fields reach a state of stability, establishing a quasi-pyrolysis condition within the synthesis chamber.

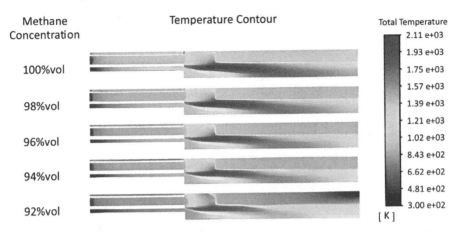

Fig. 4. Simulation result of temperature contour at the outlet of the burner and within the synthesis chamber with methane dilution from 100 to 92 vol%.

According to the Burke-Schumann [13, 14] flame length is directly related to the fuel mass flow rate divided by the square root of the heat release rate per unit area. Thus, a decrease in the mass flow rate due to lower methane concentration results in a shorter flame. The temperature contour in Fig. 4 shows that the increment of methane dilution significantly affects the molar stoichiometric oxidizer-fuel ratio, resulting in an overall reduction in flame length and thickness which directly reducing the amount of heat transferred to the synthesis chamber. When the fuel concentration is reduced, both the fuel mass flow rate and the heat release rate per unit area decrease.

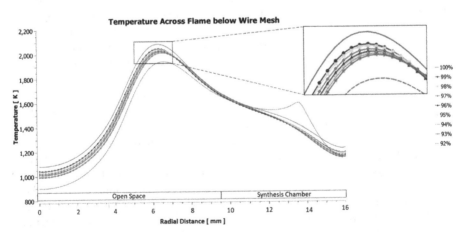

Fig. 5. Radial distribution of temperature at height of the inlet of the synthesis chamber. Centerline of the synthesis chamber located at 16 mm.

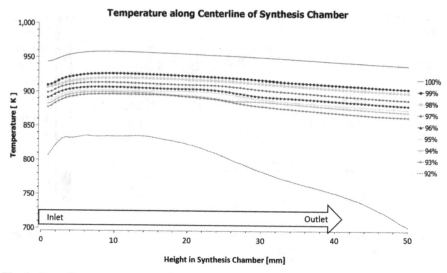

Fig. 6. Centerline temperature profile from the inlet to outlet of the synthesis chamber. The measurement taken starting from 0.07 m above the wire mesh at the chamber inlet.

The temperature distribution at the height of the inlet to the synthesis chamber is shown in Fig. 5. The maximum flame temperature, located slightly away from the synthesis chamber due to the redirection of flame flow field by the wire mesh, shows a clear decreasing trend as the methane concentration declines from 100 to 92 vol%, with the maximum flame temperature reduced from 2109.38 K to 1943.12 K. As mentioned previously, the reduction in flame temperature can be attributed to the dilution effect, which is primarily caused by the diminished availability of methane participating in the combustion reaction [12]. Furthermore, nitrogen, as an inert gas, is effective in absorbing heat and removing it from the diffusion flame during carbon nanotube synthesis because it can lower the temperature of the reaction zone [8]. Nitrogen also reduces the reaction rate and the heat release rate in the flame, further lowering the temperature [11]. Nevertheless, as low fuel concentration reduces flame temperature, it subsequently reduces the pyrolysis rate of hydrocarbon, which minimizes the presence of pyrolytic carbon that produces carbon fiber and simultaneously promotes CNT growth [8].

Figure 6 displays the temperature profile from the inlet to the outlet of the synthesis chamber. The temperature data were collected at 7 mm above the chamber's entrance to exclude heat loss effects caused by the wire mesh. The temperature profile for all methane concentrations is nearly similar, with a temperature peak near the inlet where the flame sheet is closest to the outside of the chamber wall. The temperature inside the synthesis chamber exhibits a direct correlation with the methane concentration. Lower methane concentrations lead to shorter flame development and reduced overall heat production. In the case of a 92 vol% methane concentration, the shorter flame is unable to transport heat effectively to the chamber's outlet, resulting in a sharper temperature drop after the peak [14].

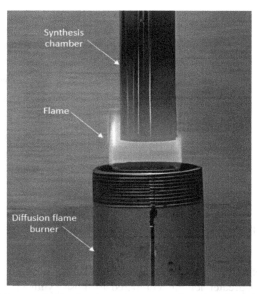

Fig. 7. Actual condition synthesis chamber when exposed to the 100 vol% methane diffusion flame

Figure 7 illustrates the flame sheet which has formed and encased the synthesis chamber. The diversion of flame sheet from the bottom to the sides of the synthesis chamber causes a decrease in temperature at the centre of the chamber in contrast to the higher temperatures observed near the inlet of the chamber. The significant build-up of methane diffusing along the chamber's periphery enhances the potential for combustion while preventing combustion from being localized solely at the entrance of the chamber. Subsequently, the temperature gradually decreases after reaching its peak near the synthesis chamber inlet up to the outlet. This pattern follows an expected trend as the temperature at the downstream of the flame decreases due to air entrainment around the synthesis chamber diminishing the flame's ability to heat the chamber.

Figure 8 illustrates the predicted lengths by GRM of CNTs at different radial positions within the synthesis chamber for each methane concentration. The region near the inlet of the synthesis chamber exhibits a high degree of variability in the simulated data due to the complex flow field pattern generated by the wire mesh, resulting in significant variations in CNT length predictions. Nevertheless, below 10 mm HAB, the GRM indicates a maximum CNT length produced in proximity to the center of the synthesis chamber. Since the fuel side is located at the center of the burner, the synthesis chamber center exhibits the highest concentration of methane and carbon monoxide gases, leading to the GRM's prediction of longer CNTs in the center of the chamber. Beyond 10 mm from the inlet of the synthesis chamber, the trend in CNT length generally converges and gradually increases at a comparable rate across all radial positions within the chamber, which is due to the relatively consistent gas composition in the downstream region of the synthesis chamber.

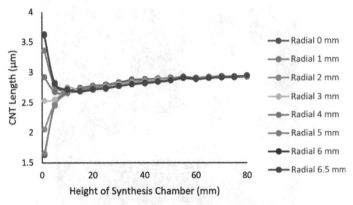

Fig. 8. CNT length predicted by GRM for flame with methane concentration of 100 vol% at various height of synthesis chamber. Radial distance is measured from synthesis chamber wall.

As mentioned earlier, the CNTs synthesized at higher positions within the synthesis chamber exhibit greater lengths compared to those synthesized at lower heights, despite the latter experiencing higher temperatures. This phenomenon can be attributed to the decrease in temperature along the vertical direction of the synthesis chamber, which occurs due to diminishing methane concentration and dilution from surrounding air entrainment. The reduction in temperature leads to a decrease in the pyrolysis rate of hydrocarbon byproducts, such as polycyclic aromatic hydrocarbons (PAH) and soot, which consequently enhances the formation of carbon nanotubes [8]. Furthermore, the observed increase in CNT length can also be ascribed to the catalytic effect of nitrogen, which facilitates the removal of the carbon layer on the catalyst, promoting elongation of the synthesized CNTs [3]. The length of carbon nanotubes increases as the methane concentration decreases from 100 to 97 vol%, with the maximum length of CNTs synthesized reaching 4.707 μm, as shown in Fig. 9. However, a decrease in length is observed as the concentration drops from 97 to 94 vol%. Notably, no production of carbon nanotubes is observed at a methane concentration of 92 vol%.

Figure 10 shows the temperature and methane mass fraction radial distribution at various heights within the synthesis chamber. For flame temperature, both concentrations show a slight increase in temperature closer to the chamber's wall due to the presence of flame formation on the outside of the wall. The temperature within the chamber remains consistent from the inlet to a height of 80 mm for both flames, theoretically allowing for continuous growth of CNT throughout the chamber.

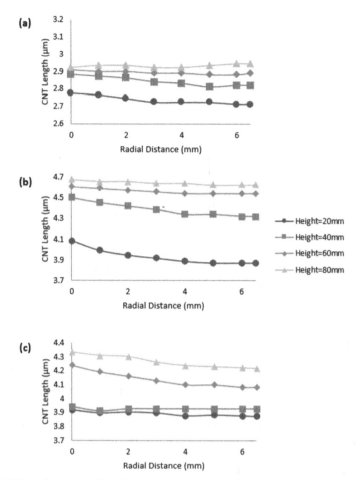

Fig. 9. CNT length across radial distance at 4 different heights for (a) 100 vol%; (b) 97 vol%; and (c) 94 vol% methane concentrations.

The methane mass fraction shows consistency from the inlet to the downstream of the chamber for both flames, providing ample carbon sources to continuously promote the synthesis process of CNT within the chamber. However, the simulation predicts a higher methane mass fraction inside the chamber for the 97 vol% flame with minimal reduction in temperature. Hence, the 97 vol% flame is predicted to produce longer synthesized CNTs within the chamber. Compared to the 92 vol% methane flame, the methane mass fraction inside the chamber decreases near zero, which inhibits the inception and formation of CNTs. Therefore, the simulation predicts an optimized flame with a fuel concentration of 97 vol% for the production of CNTs in the synthesis chamber. Furthermore, the growth temperature range for synthesizing CNTs with extended length falls between 875 K and 905 K, resulting in lengths ranging from approximately 2.6 to 4.7 μm.

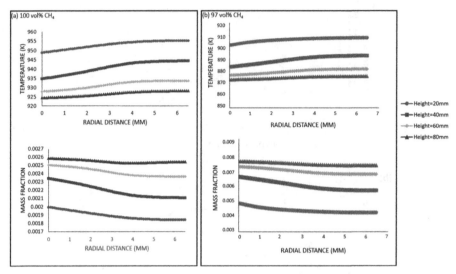

Fig. 10. Temperature and CH4 mass fraction radial distribution at four different heights within synthesis chamber for (a) 100 vol%. and (b) 97 vol%. of methane concentrations.

4 Conclusion

The present study aims to investigate the effect of methane concentration on CNT growth in the synthesis chamber using a methane diffusion flame burner. The CFD simulation predicts that a decrease in fuel concentration leads to a proportional reduction in flame temperature due to limited fuel availability for combustion, resulting in a decrease in heat energy generation. Furthermore, fuel dilution also limits the concentration of carbon sources for CNT growth in the synthesis chamber. The simulation shows that the chamber ceases to produce CNTs at a volumetric concentration of 92 vol%, which is attributed to the presence of an effective flame sheet located at the inlet of the synthesis chamber that promotes complete combustion of methane and depletes the carbon source for CNT growth within the synthesis chamber. Optimized fuel concentration was found with 97 vol% methane flame is found to be able to produce the longest CNTs at 4.7 μm due to minimal temperature decrease and an increment in the methane mass fraction in the synthesis chamber compared to the baseline flame. However, with a methane concentration below 97 vol%, a decrease in CNT length is observed, leading to the cessation of CNT production at a methane concentration of 92 vol%. Based on the findings, it can be inferred that a methane concentration of 97 vol% creates the most thermochemically favorable environment for the growth of CNTs in the synthesis chamber. The simulation results also indicate that CNTs with lengths ranging from 2.6 to 4.7 μm can be synthesized within the temperature range of 875 K to 905 K. Currently works to validate the actual length of CNT synthesized in the synthesis chamber using the presented method is ongoing. Furthermore, refinement in simulation works also currently undertaken to include presence of catalyst molecules within synthesis chamber to determine its effect toward temperature and gas species composition.

Acknowledgements. The author would like to acknowledge the financial support from Ministry of Education Malaysia under Fundamental Research Grant Scheme (FRGS) (FRGS/1/2022/TK09/UTM/02/7) and Universiti Teknologi Malaysia for the funding under UTM Encouragement Research Grant (UTMER) (Q.J130000.3851.20J39).

References

1. Hamzah, N., Yasin, M.F.M., Yusop, M.Z.M., Saat, A., Subha, N.A.M.: Rapid production of carbon nanotubes: a review on advancement in growth control and morphology manipulations of flame synthesis. J. Mater. Chem. A Mater. **5**, 25144–25170 (2017)
2. Chu, H., Han, W., Cao, W., Tao, C., Raza, M., Chen, L.: Experimental investigation of soot morphology and primary particle size along axial and radial direction of an ethylene diffusion flame via electron microscopy. J. Energy Inst. **92**(5), 1294–1302 (2018)
3. Gore, J.P., Sane, A.: Flame synthesis of carbon nanotubes. In: Materials Research Society Symposium - Proceedings, vol. 772, pp. 55–61 (2011)
4. Han, W., Chu, H., Ya, Y., Dong, S., Zhang, C.: Effect of fuel structure on synthesis of carbon nanotubes in diffusion flames. Fuller. Nanotub. Carbon Nanostruct. **27**(3), 265–272 (2019)
5. Zainal, M.T.: Multi-scale modelling on the effects of flame parameters on carbon nanotube growth in non-premixed flames. Universiti Teknologi Malaysia (2021)
6. Hamzah, N.: Characterization of carbon nanotube growth region in flame using wire-based macro-imaging method Ph.D. thesis, pp. 47–133. Universiti Teknologi Malaysia (2020)
7. Vander Wal, R.L., Hall, L.J., Berger, G.M.: Optimization of flame synthesis for carbon nanotubes using supported catalyst. J. Phys. Chem. B **106**(51), 13122–13132 (2002)
8. Yuan, L., Saito, K., Hu, W., Chen, Z.: Ethylene flame synthesis of well-aligned multi-walled carbon nanotubes. Chem. Phys. Lett. **346**, 23–28 (2001)
9. Zainal, M.T., Mohd Yasin, M.F., Wahid, M.A.: Investigation of the coupled effects of temperature and partial pressure on catalytic growth of carbon nanotubes using a modified growth rate model. Mater. Res. Express **3**(10), 105040 (2016)
10. Naha, S., Sen, S. De, A.K., Puri, I.K.: A detailed model for the flame synthesis of carbon nanotubes and nanofibers. Proc. Combust. Inst. **31**(II), 1821–1829 (2007)
11. Salah, L.S., Ouslimani, N., Bousba, D., Huynen, I., Danleé, Y., Aksas, H.: Carbon nanotubes (CNTs) from synthesis to functionalized (CNTs) using conventional and new chemical approaches. J. Nanomater. **2021** (2021)
12. Burke, S.P., Schumann, T.E.W.: Diffusion flames. Ind. Eng. Chem. **20**(10), 998–1004 (1928)
13. Turns, S.R.: An Introduction to Combustion: Concepts and Applications, 2nd edn. McGraw-Hill, New York (2000)
14. Zainal, M.T., Yasin, M.F.M., Wahid, M.A.: Optimizing Flame synthesis of carbon nanotubes: experimental and modelling perspectives. J. Teknol. **4**, 31–39 (2016)
15. Zainal, M.T.: Modelling the flame synthesis of carbon nanotube (CNT) in inverse diffusion flame. Universiti Teknologi Malaysia (2017)
16. Hamzah, N., Yasin, M.F.M., Yusop, M.Z.M., Saat, A., Subha, N.A.M.: Growth region characterization of carbon nanotubes synthesis in heterogeneous flame environment with wire-based macro-image analysis. Diam. Relat. Mater. **99**, 107500 (2019)

Numerical Investigation on the Influence of Protrusion Height on Heat Transfer and Fluid Flow within Rectangular Channel with Nanofluid

Yamunan Manimaran[1], Abdulhafid M. A. Elfaghi[1,2(✉)], and Iman Fitri Ismail[3]

[1] Faculty of Mechanical and Manufacturing Engineering, Universiti Tun Hussein Onn Malaysia, 86400 Parit Raja, Batu Pahat, Johor, Malaysia
abdulhafid@uthm.edu.my

[2] Aeronautical Engineering Department, Faculty of Engineering, University of Zawia, Az-Zāwiyah, Libya

[3] Combustion Research Group (CRG), Faculty of Mechanical and Manufacturing Engineering, Universiti Tun Hussein Onn Malaysia, 86400 Parit Raja, Batu Pahat, Johor, Malaysia

Abstract. This study focuses on investigating the impact of using Aluminium Oxide (Al2O3) nanofluid on heat transfer within a rectangular channel featuring different protrusion heights. The primary objectives include analysing fluid flow characteristics and heat transfer performance in the channel and determining the optimal parameters and specifications. Additionally, this study places emphasis on examining the influence of protrusion height on fluid flow and heat transfer in conjunction with nanofluids. The obtained numerical results are compared with other relevant findings. The findings reveal a positive correlation between the heat transfer coefficient and both the protrusion height and Reynolds number. Similarly, the Nusselt number demonstrates an increase with the protrusion height. Furthermore, the friction coefficient displays a slight increase with both the protrusion height and Reynolds number, although the differences are not statistically significant.

Keywords: Heat transfer · Surface Heat Transfer Coefficient · Reynolds Number · Nusselt Number · Volume Fraction · Surface friction coefficient

1 Introduction

In recent years, the quest for more efficient and sustainable thermal management systems has gained substantial momentum across various engineering fields. One promising approach to address this challenge involves the incorporation of nanofluids, which are colloidal suspensions of nanoparticles in conventional base fluids [1]. These nanofluids have shown exceptional potential in enhancing heat transfer characteristics, making them an attractive option for advanced heat exchangers and cooling systems for batteries and electric vehicles [2, 3].

It has been demonstrated that nanofluids dramatically improve heat transfer rates over traditional fluids [4]. The inclusion of nanoparticles enhances thermal conductivity, allowing for enhanced heat dissipation and decreased thermal resistance [5]. This characteristic is particularly advantageous in applications where efficient cooling or heating is essential.

The use of nanofluids as a heat transfer medium in electronic cooling is of particular importance in modern electronics and information technology industries. As electronic devices continue to shrink in size and simultaneously increase in processing power, they generate more heat in a confined space. In Central Processing Unit (CPU) applications, Nanofluids have a significantly higher thermal conductivity compared to conventional coolants, such as water or oil. By dispersing nanoparticles in the base fluid, the overall thermal conductivity of the nanofluid is improved [6].

Nanofluids are being explored as coolants in automotive engines and radiators [7, 8]. By using nanofluids as a heat transfer medium, it is possible to improve the cooling efficiency of the engine, leading to better fuel economy and reduced emissions [9, 10]. Nanofluids have shown application potential in solar thermal collectors, as compared to conventional working fluids, nanofluids possess significantly enhanced heat absorption and heat transmission capabilities [11, 12].

One area that requires further exploration is the effect of channel geometries on nanofluid-based heat transfer systems. In particular, the influence of protrusion height, i.e., the presence of obstacles or structures on the channel walls, remains relatively unexplored. The interaction between nanofluids and channel geometries can lead to complex fluid flow patterns and heat transfer phenomena, which could significantly impact the overall thermal performance of the system.

Using a numerical technique, a prior work examined the thermal behaviour of turbulent alumina nanofluids flow in a square channel with protrusion obstacles. It has been determined that the ratio of protrusion height to print diameter of 1.0 provides the greatest improvement in thermal performance compared to a square duct with a smooth surface [13]. Another study used aluminium oxide in non-Newtonian fluids with teardrop-shaped dimples/protrusions, demonstrating that aluminium oxide can significantly enhance heat transfer with minimal pressure loss [14].

Despite the fact that the influence of protrusion height has been extensively investigated and recorded, particularly in porous media [15], the Reynolds number is a crucial dimensionless parameter that characterizes the flow regime and plays a significant role in determining fluid behaviour. The influence of Reynolds number and protrusion height on heat transfer and fluid flow properties within nanofluid-filled channels has not been well investigated. The Reynolds number represents the ratio of inertial forces to viscous forces and influences the flow regime, turbulence intensity, and heat transfer performance of a fluid system directly. In contrast, the existence of protrusions, such as ribs, fins, or other structures, can drastically alter the flow patterns and heat transfer properties within a channel. The combined impact of protrusion heights and Reynolds number remains relatively unexplored.

By altering Reynolds number and protrusion heights, the purpose of this study is to evaluate the fluid flow properties within a rectangular channel. The results are then validated against additional numerical approaches and verified against the experimental results. Reynolds numbers of 200, 500, 700, 1100, 1500, and 1700 were utilised for this simulation. 0 mm, 2 mm, 5 mm, and 10 mm are the protrusion heights utilised in this investigation. Al_2O_3 comprises 4% of the volume fraction of Aluminium Oxide.

2   Model and Numerical Method

2.1   Description of Flow Problems

The first phase of any CFD analysis is determining and creating the shape of the flow zone, the computational domain for the CFD computations. In this study, the geometry model of the rectangular channel is created in SolidWorks 2021 and imported into Ansys Fluent. Figure 1 and Fig. 2 depict the rectangular channel's height, H, as 60 mm, its length, L, as 1000 mm, and its width, W, as 120 mm.

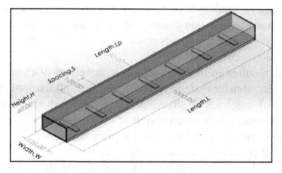

Fig. 1. Rectangular flow channel with spacing between protrusion

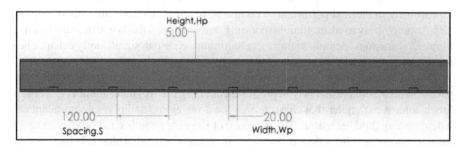

Fig. 2. The height, width, and distance between protrusions in rectangular channel.

The length of the entering region is estimated using the formula Le, and the length of the exit region should be equal to one-half the length of the entry region [16]. The formula of L_e is given by

$$L_e > 10D_h \tag{1}$$

$$D_h = \frac{4A}{P} \tag{2}$$

where the entrance length is denoted by L_e, hydraulic diameter D_h, cross sectional area of the duct A, and wetted perimeter of the rectangular duct, P. Via the equation, the entrance region is $L_e = 1$ m, and the exit region is 0.5 m. Figure 3 depicts a rectangular channel with an entrance and an exit.

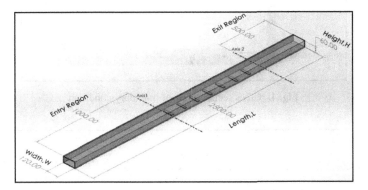

Fig. 3. The rectangular channel with entry and exit region.

The range of protrusion height, H_p, considered in this study is 0 mm to 10 mm. In this simulation, the nanofluid consists of water as the base fluid combined with nanoparticles composed of aluminium oxide (Al_2O_3) in varying volume fractions ϕ.

2.2 Boundary Conditions

The characteristics of the flow will be incompressible, steady and fully developed when leaving the entry region. The uniform inlet temperature of the fluid, T_{in} is 30 °C. The heat flux $\dot{q}$ of that is being considered falls somewhere in the region of 1000 Wm-2. Standard SIMPLEC is used to implement velocity-pressure coupling because it demonstrates superior behaviour for cases in which the scalar variable is tightly coupled to the momentum equation [17, 18]. In this simulation, the Reynolds number, Re, is between 1000 and 1500.

Via the Reynolds formula, the inlet velocity can be calculated. The inlet velocity, u_{in} is given for the Reynolds number of 200, 500, 700, 1100, 1500, and 1700. The outlet pressure of the fluid is 1 atm. Figure 4 and Fig. 5 depict the cross-sectional and side views of the boundary conditions, respectively.

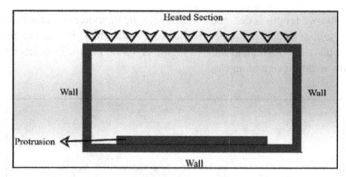

Fig. 4. Cross section view of boundary condition

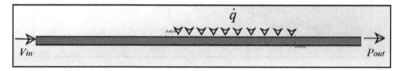

Fig. 5. Side view of boundary condition

3 Results and Discussion

This section presents the findings of an ANSYS FLUENT-based computational investigation into the enhancement of heat transfer utilising Aluminium Oxide (Al_2O_3) nanofluid in a rectangular channel with varying protrusion heights and volume percentages. The graphs depict temperature contours, Reynold's number, Nusselt number, and friction coefficient. In this simulation, the heat transfer coefficient, Nusselt number, and friction factor were calculated. It is validated against previous studies.

3.1 Heat Transfer Coefficient

The Fig. 6 demonstrates a clear relationship between the height of protrusion and the surface heat transfer coefficient of a nanofluid. By maintaining a constant volume fraction of 4% for the nanofluid, it is observed that as the height of protrusion increases, the surface heat transfer coefficient also increases. A noticeable enhancement in the heat transfer coefficient can be observed when comparing the cases of no protrusion (0 mm) and a protrusion of 2 mm. Specifically, the heat transfer coefficient for the 2 mm protrusion is nearly five times higher than that for the absence of protrusion (0 mm).

This is due to the introduction of turbulence caused by the interaction of laminar flow from the entry region with the protrusions within the channel. The presence of protrusions leads to disturbances in the fluid flow, such as vortices and eddies, resulting in better mixing and increased heat transfer due to turbulent flow. Consequently, the convective heat transfer between the fluid and the channel walls is enhanced, leading to an improved heat transfer coefficient.

Protrusions within the channel promote fluid mixing, and the secondary flow patterns they generate help to distribute the fluid's heat more evenly. As a result, temperature

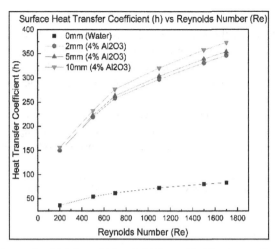

Fig. 6. Impact of Protrusion Height and Nanofluid Reynolds Number on Surface Heat Transfer Coefficient

differences are reduced, and overall heat transfer efficiency is enhanced. Although there is a slight increase in heat transfer coefficient for protrusions of 5 mm and 10 mm compared to 0 mm and 2 mm protrusions, the difference is not significant. This is primarily attributed to the absence of mixing in the case of 0 mm protrusion, which explains the significant improvement observed. Notably, the simulation results show that the 10 mm protrusion exhibits the highest surface heat transfer coefficient.

The Reynolds number also plays a significant role in enhancing the surface heat transfer rate. As illustrated in Fig. 6, an increase in the Reynolds number corresponds to a rise in the surface heat transfer coefficient. This increase disrupts the boundary layer, which refers to the thin layer of fluid with slower velocities near the channel walls. Protrusions can effectively disrupt this boundary layer, which, under laminar flow conditions, could otherwise act as an insulating barrier, reducing heat transfer.

The presence of protrusions leads to improved heat transfer from the fluid to the channel walls by disturbing the boundary layer. When the Reynolds number is higher and the protrusions are taller, convective heat transfer is further enhanced. The fluid experiences increased pressure drops and velocity gradients as it flows over the taller protrusions. Consequently, the convective heat transfer between the fluid and the protrusion/channel walls is significantly improved.

3.2 Nusselt Number

Typically, as the height of the protrusion increases, the Nusselt number also tends to increase. The Nusselt number represents the ratio between convection heat transfer rate and conduction heat transfer rate. Consequently, with higher Reynolds numbers, the Nusselt number also rises. This relationship is attributed to the increased surface area available for heat transfer, resulting from the greater height of the protrusion. The enhanced convective heat transfer occurs due to the enlarged contact area between the fluid and the solid surface, leading to a higher Nusselt number.

Comparing simulations without protrusions to those with various heights of protrusion, it is evident that the Nusselt number is consistently lowest in the absence of protrusions. As the height of the protrusion increases, the Nusselt number gradually rises. A higher Reynolds number facilitates a larger fluid-wall contact area, and greater protrusion height further increases the proximity of the fluid to the solid surface. Consequently, convective heat transfer and thermal conduction improve due to this increased proximity. The result is a higher convective heat transfer coefficient and, consequently, an increased Nusselt number. The results have been validated against previous work conducted using graphene nanoplatelets/MWCN hybrid nanofluid [19] (Fig. 7).

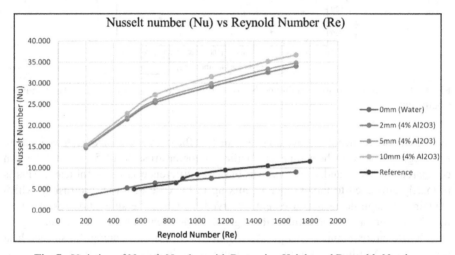

Fig. 7. Variation of Nusselt Number with Protrusion Height and Reynolds Number

3.3   Friction Coefficient

Figure 8 presents the impact of the Reynolds number and the volume fraction of Aluminium Oxide (Al_2O_3) nanofluid on the friction coefficient within the rectangular channel, based on numerical simulation results. The presence of protrusions inside the channel leads to an increase in frictional resistance due to the heightened surface roughness. As the height of the protrusion increases, the surface area in contact with the fluid also rises proportionally, resulting in higher frictional losses.

The relationship between the height of the protrusion and its friction coefficient is evident, as higher protrusion heights are associated with elevated frictional losses and a likelihood of flow separation and recirculation zones in the fluid flow. Flow separation occurs when the fluid fails to follow the contour of the protrusion, leading to recirculation zones and pressure drop. These flow phenomena contribute to higher frictional losses and an increased friction coefficient.

Moreover, the Reynolds number significantly influences the friction coefficient. Higher Reynolds numbers correspond to more turbulent flow, which leads to greater frictional losses. There may exist a correlation between the Reynolds number and the

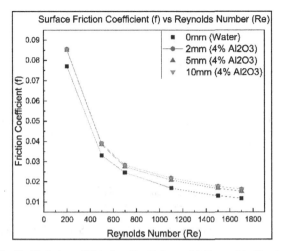

Fig. 8. Variation of Friction Coefficient with Reynolds Number

effect of protrusion height on the friction coefficient. For instance, at low Reynolds numbers, a higher protrusion height may elevate the friction coefficient more than a lower one.

3.4 Effect of Nanofluid Volume Fraction on Surface Heat Transfer Coefficient and Surface Friction Coefficient

Figure 9 illustrates the surface heat transfer coefficient in the rectangular channel, considering various volume fractions of Aluminium Oxide (Al_2O_3) nanofluid and the Reynolds number, as obtained from numerical simulations. Throughout the simulations, the height of the protrusion remains constant at 2 mm. The graph showcases the disparity in the heat transfer coefficient values between pure water (0% volume fraction of Al_2O_3 nanofluid) and a nanofluid with a 4% volume fraction of Al_2O_3.

Notably, the nanofluid with a 4% volume fraction of Al_2O_3 exhibits a substantial enhancement in the heat transfer coefficient compared to the fluid consisting of pure water. This improvement signifies the remarkable potential of the nanofluid in augmenting heat transfer efficiency within the channel.

Nanofluids generally exhibit enhanced thermal conductivity compared to the base fluid alone. Increasing the volume fraction of nanoparticles in the fluid further improves the effective thermal conductivity of the nanofluid. As a result of this enhanced thermal conductivity, the nanofluid demonstrates a higher heat transfer coefficient, enabling more efficient heat transfer from the channel walls to the fluid.

Based on the numerical simulation results, Fig. 9 below illustrates the variation in the friction coefficient with different volume fractions of Aluminium Oxide (Al_2O_3) nanofluid and varying Reynolds numbers, while maintaining a constant protrusion height of 2 mm. When compared to water, the friction coefficient for the nanofluid is lower at lower Reynolds numbers. However, at higher Reynolds numbers, the friction coefficient

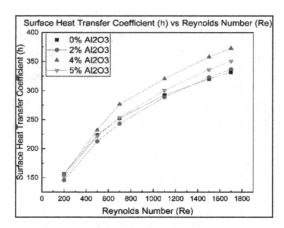

Fig. 9. Impact of Various Volume Fractions of Al$_2$O$_3$ on Surface Heat Transfer Coefficient

of water becomes lower than that of the nanofluid, though the difference is relatively small (Fig. 10).

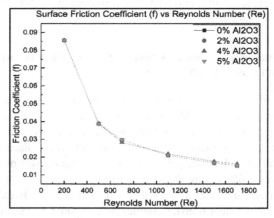

Fig. 10. Variation in Friction Coefficient with Different Volume Fraction of Al$_2$O$_3$ and Reynolds Number

4 Conclusion

In conclusion, the study reveals a significant relationship between the height of protrusion and the surface heat transfer coefficient of a nanofluid within the rectangular channel. Maintaining a constant volume fraction of 4% for the nanofluid, it is evident that an increase in the height of protrusion leads to a corresponding increase in the surface heat transfer coefficient. Notably, a notable enhancement in the heat transfer coefficient is observed when comparing cases with no protrusion (0 mm) and a protrusion of 2 mm. Specifically, the heat transfer coefficient for the 2 mm protrusion is nearly five times higher than that for the absence of protrusion (0 mm). The observed improvement in heat transfer coefficient is attributed to the introduction of turbulence caused by the interaction of laminar flow from the entry region with the protrusions within the channel. The presence of protrusions induces fluid disturbances, such as vortices and eddies, resulting in better mixing and increased heat transfer due to turbulent flow.

The Nusselt number, which represents the ratio of convection heat transfer rate to conduction heat transfer rate, rises with higher Reynolds numbers as well. This relationship is attributed to the increased surface area available for heat transfer due to the greater protrusion height. The enlarged contact area between the fluid and the solid surface enhances convective heat transfer, resulting in a higher Nusselt number. The findings of this study are consistent with previous work conducted using graphene nanoplatelets/MWCN hybrid nanofluid, validating the results and further highlighting the significance of protrusion height and Reynolds number in enhancing heat transfer efficiency.

The numerical simulations presented in Fig. 9 highlight the significant impact of volume fraction on the surface heat transfer coefficient in the rectangular channel filled with Aluminium Oxide (Al_2O_3) nanofluid. The graph demonstrates a clear contrast in heat transfer coefficient values between pure water (0% volume fraction of Al_2O_3 nanofluid) and a nanofluid with a 4% volume fraction of Al. Notably, the nanofluid with a 4% volume fraction shows a remarkable enhancement in heat transfer efficiency compared to the base fluid.

This improvement is attributed to the enhanced thermal conductivity exhibited by nanofluids compared to the base fluid alone. As the volume fraction of nanoparticles in the fluid increases, the effective thermal conductivity of the nanofluid further improves. Consequently, the nanofluid demonstrates a higher heat transfer coefficient, enabling more efficient heat transfer from the channel walls to the fluid. Similarly, the numerical simulations presented in Fig. 10 reveal the influence of volume fraction and Reynolds number on the friction coefficient within the rectangular channel with a constant protrusion height of 2 mm. The friction coefficient of the nanofluid is observed to be lower than that of water at lower Reynolds numbers. However, at higher Reynolds numbers, the friction coefficient of water becomes lower than that of the nanofluid, though the difference is relatively small.

Acknowledgement. This research was supported by Universiti Tun Hussein Onn Malaysia (UTHM) through Tier 1 (vot H768).

References

1. McGrail, B.P., Thallapally, P.K., Blanchard, J., Nune, S.K., Jenks, J.J., Dang, L.X.: Enhancing thermal conductivity of fluids with nanoparticles. Nano Energy **2**(5), 845–855 (1995). https://doi.org/10.1016/J.NANOEN.2013.02.007
2. Can, A., Selimefendigil, F., Öztop, H.F.: A review on soft computing and nanofluid applications for battery thermal management. J. Energy Storage **53**, 105214 (2022). https://doi.org/10.1016/J.EST.2022.105214

3. Abdelkareem, M.A., et al.: Battery thermal management systems based on nanofluids for electric vehicles. J. Energy Storage **50**, 104385 (2022). https://doi.org/10.1016/J.EST.2022. 104385

4. Xuan, Y., Li, Q.: Heat transfer enhancement of nanofluids. Int. J. Heat Fluid Flow **21**(1), 58–64 (2000). https://doi.org/10.1016/S0142-727X(99)00067-3

5. Hung, Y.H., Teng, T.P., Teng, T.C., Chen, J.H.: Assessment of heat dissipation performance for nanofluid. Appl. Therm. Eng. **32**(1), 132–140 (2012). https://doi.org/10.1016/J.APPLTH ERMALENG.2011.09.008

6. Hasan, H.A., Alquziweeni, Z., Sopian, K.: Heat transfer enhancement using nanofluids for cooling a central processing unit (CPU) system. J. Adv. Res. Fluid Mech. Therm. Sci. **51**(2), 145–157 (2018). https://akademiabaru.com/submit/index.php/arfmts/article/view/ 2342. Accessed 24 July 2023

7. Sidik, N.A.C., Yazid, M.N.A.W.M., Mamat, R.: A review on the application of nanofluids in vehicle engine cooling system. Int. Commun. Heat Mass Transf. **68**, 85–90 (2015). https:// doi.org/10.1016/J.ICHEATMASSTRANSFER.2015.08.017

8. Mutuku, W.N.: Ethylene glycol (EG)-based nanofluids as a coolant for automotive radiator. Asia Pac. J. Comput. Eng. **3**(1), 1–15 (2016). https://doi.org/10.1186/S40540-016-0017-3

9. Ben Said, L., Kolsi, L., Ghachem, K., Almeshaal, M., Maatki, C.: Advancement of nanofluids in automotive applications during the last few years?a comprehensive review. J. Therm. Anal. Calorim. **147**(14), 7603–7630 (2022). https://doi.org/10.1007/S10973-021-11088-4/ METRICS

10. Leela Kumar, K., Rudrabhi Ramu, R., Venkatesh, P.H.J.: Performance of automobile engine radiator by using nanofluids on variable compression diesel engine. In: Deepak, B., Bahubalendruni, M.R., Parhi, D., Biswal, B.B. (eds.) Recent Trends in Product Design and Intelligent Manufacturing Systems. LNME, pp. 383–396. Springer, Singapore (2023). https://doi.org/ 10.1007/978-981-19-4606-6_36

11. Chamsa-ard, W., Brundavanam, S., Fung, C.C., Fawcett, D., Poinern, G.: Nanofluid types, their synthesis, properties and incorporation in direct solar thermal collectors: a review. Nanomaterials **7**(6), 131 (2017). https://doi.org/10.3390/NANO7060131

12. Ahmed, A., Baig, H., Sundaram, S., Mallick, T.K.: Use of nanofluids in solar PV/thermal systems. Int. J. Photoenergy **2019** (2019). https://doi.org/10.1155/2019/8039129

13. Kumar, S., Kothiyal, A.D., Bisht, M.S., Kumar, A.: Effect of ratio of protrusion height to print diameter on thermal behaviour of Al2O3–H2O nanofluid flow in a protrusion obstacle square channel. Adv. Intell. Syst. Comput. **624**, 277–287 (2018). https://doi.org/10.1007/978-981- 10-5903-2_30/COVER

14. Zhang, Z., Xie, Y., Zhang, D., Xie, G.: Flow characteristic and heat transfer for non-newtonian nanofluid in rectangular microchannels with teardrop dimples/protrusions. Open Phys. **15**(1), 197–206 (2017). https://doi.org/10.1515/PHYS-2017-0021/MACHINEREADABLECITATI ON/RIS

15. Nabwey, H.A., Armaghani, T., Azizimehr, B., Rashad, A.M., Chamkha, A.J.: A comprehensive review of nanofluid heat transfer in porous media. Nanomaterials **13**(5), 937 (2023). https://doi.org/10.3390/NANO13050937

16. Heat Transfer in Channel Flow: Heat Convection, pp. 203–258, January 2006. https://doi.org/ 10.1007/978-3-540-30694-8_6

17. Jang, D.S., Jetli, R., Acharya, S.: Comparison of the PISO, SIMPLER, and SIMPLEC algorithms for the treatment of the pressure-velocity coupling in steady flow problems. Numer. Heat Transfer **10**(3), 209–228 (2007). https://doi.org/10.1080/10407788608913517

18. Ismail, I., Basuno, B., Mohammed, A., Ismail, F., Mohd Yusof, N.: The development of euler solver based on flux vector splitting and modified TVD Schemes. In: Hassan, M.H.A., Zohari, M.H., Kadirgama, K., Mohamed, N.A.N., Aziz, A. (eds.) ICMER 2021. LNEE, vol. 882, pp. 687–702. Springer, Singapore (2023). https://doi.org/10.1007/978-981-19-1577-2_51

19. Insiat Islam, R.M., Hassan, N.M.S., Rasul, M.G., Gudimetla, P.V., Nabi, M.N., Chowdhury, A.A.: Effect of non-uniform wall corrugations on laminar convective heat transfer through rectangular corrugated tube by using graphene nanoplatelets/MWCN hybrid nanofluid. Int. J. Therm. Sci. **187**, 108166 (2023). https://doi.org/10.1016/J.IJTHERMALSCI.2023.108166

Ultrasonic Sensors in Companion Robots: Navigational Challenges and Opportunities

Isaac Asante$^{(\boxtimes)}$ ⓘ, Lau Bee Theng ⓘ, Mark Tee Kit Tsun ⓘ, and Zhan Hung Chin ⓘ

Faculty of Engineering, Computing and Science, Swinburne University of Technology, Kuching, Sarawak, Malaysia
iasante@swinburne.edu.my

Abstract. This paper investigates the integral role of ultrasonic sensors in developing multi-sensor companion robots. It focuses on environment mapping, landmark-based localization, and path reconstruction. Despite the significant advancements in the field, the research identifies notable gaps that hinder the full potential of ultrasonic sensors. Addressing these gaps could revolutionize companion robots' functionality, reliability, and efficiency, particularly in real-world applications such as healthcare, elderly care, and home assistance. The research emphasizes the importance of sensor fusion techniques and the potential improvements in environment mapping and landmark-based localization, which could significantly enhance the navigation capabilities of companion robots. Furthermore, advancements in path reconstruction could transform how companion robots retrace their steps, increasing their operational efficiency. The findings of this research contribute significantly to the field of robotics and pave the way for future studies, highlighting the transformative potential of advanced environment mapping techniques, sensor fusion, and path reconstruction in propelling advancements in robotic navigation.

Keywords: Robot Navigation · Sensor Fusion · Ultrasonic Sensors

1 Introduction

Companion robots equipped with multi-sensor systems have emerged as promising technology with a wide range of real-world applications. These intelligent robots are designed to assist individuals in various settings such as hospitals, elderly care facilities, and homes [1, 2]. Integrating multiple sensors enables these robots to perceive and interact with their environment, enhancing their functionality and adaptability. This introduction explores the significance of multi-sensor companion robots in industry applications and outlines the research objectives of this paper, specifically focusing on addressing research gaps in companion robot systems involving ultrasonic sensors.

The use of companion robots in industries such as healthcare and elderly care has gained significant attention due to the growing demand for personalized assistance and

All the ideas, analysis, and research presented in this paper are original and were produced by the authors without plagiarism. However, AI-powered proofreading services were used to enhance the sentence structure in the text for readability purposes only.

The original version of the chapter has been revised. An acknowledgement and footnote had not been displayed correctly. A correction to this chapter can be found at
https://doi.org/10.1007/978-981-99-7243-2_40

F. Hassan et al. (Eds.): AsiaSim 2023, CCIS 1912, pp. 338–350, 2024.
https://doi.org/10.1007/978-981-99-7243-2_29

support. In these sectors, traditional approaches to caregiving often encounter challenges related to limited resources and the requirement for constant human supervision. Companion robots emerge as a promising solution, offering unwavering support and companionship to individuals, particularly those in isolation or requiring additional assistance. The domain of socially assistive robotics has shown substantial potential in augmenting the lives of patients and the elderly [3]. These robots are not confined to assisting with daily tasks but extend their functionality to monitoring vital signs and providing emotional support. Consequently, they enhance the quality of life, fostering a sense of independence and well-being among the individuals they serve.

A pivotal aspect of companion robots is integrating multi-sensor systems, which equip them to perceive and interpret their environment. Including diverse sensors–encompassing cameras, laser scanners, and ultrasonic sensors–endows companion robots with the ability to amass comprehensive sensory information about their surroundings. Analogous to autonomous vehicles [4], this sensor fusion facilitates secure navigation, obstacle detection and avoidance, and intelligent human interaction. The proficient employment of multi-sensor systems in companion robots is instrumental in guaranteeing these robotic systems' reliability, adaptability, and performance.

This paper aims to investigate and address specific research gaps in companion robot systems utilizing ultrasonic sensors. These sensors have extensive use in assistive robotics due to their affordability, low power consumption, and suitability for obstacle detection tasks [5]. Despite these advantages, fully exploiting their potential requires tackling significant challenges and addressing research gaps.

1.1 Paper Organization

This paper begins with an introduction to the concept and significance of companion robots in Sect. 1. Section 2 delves into the role of ultrasonic sensors in mobile robots, highlighting their importance in multi-sensor systems and the challenges associated with their use. Environment mapping challenges–a critical aspect of autonomous navigation for companion robots–are explored in Sect. 3. Section 4 shifts the focus to landmark-based localization, discussing the issues in accurately determining a robot's position within its environment based on sonar input. Section 5 addresses the gaps in path reconstruction methodologies and how ultrasonic sensors could be helpful, ending with Table 1, which summarizes the challenges and research gaps explored in this paper. Finally, Sect. 6 concludes the discussions and highlights future directions for research, underscoring the potential impact of addressing these challenges on the advancement of companion robots.

2 Ultrasonic Sensors in Companion Robots

Companion robots equipped with multi-sensor systems have garnered significant attention due to their capacity to perceive and interact with their environment. These robots integrate sensors to gather information about their surroundings, facilitating effective navigation, interaction, and assistance provision. Among the different types of sensors used in companion robots, ultrasonic sensors play a pivotal role in enhancing functionality and performance.

Ultrasonic sensors emit high-frequency sound waves and measure the time it takes for the sound waves to bounce back after hitting an object. This process provides information about distance and obstacle detection, making them invaluable for companion robot applications. Notably, ultrasonic sensors are cost-effective, rendering them a practical choice for implementation. Furthermore, their low power consumption is crucial for extending the operational time of the robot. Ultrasonic sensors also exhibit reliable performance in diverse environments, including low light conditions and outdoor settings, where other sensing modalities, such as cameras, may face limitations.

However, using ultrasonic sensors in companion robots poses challenges, mainly when relying on a single-sensor approach. Accuracy and robustness issues may arise, including false readings, blind spots, and difficulty distinguishing between multiple obstacles. These challenges can significantly impede the robot's ability to perceive the environment and navigate safely. To address these concerns, researchers and developers continuously explore innovative approaches such as sensor fusion and advanced signal processing techniques to enhance the performance and reliability of multi-sensor companion robot systems incorporating ultrasonic sensors [4]. Overcoming these obstacles unlocks the full potential of companion robots in various real-world applications, further integrating them into society and everyday life. The methodology in [6] utilizes both ultrasonic and infrared sensors. The proposed fusion algorithm achieves robust motion control and enhances obstacle detection. The experimental validation demonstrates its effectiveness in navigating through a maze with various obstacles, making it valuable in improving reliability and robustness in autonomous obstacle avoidance.

The confluence of ultrasonic sensors with other sensor modalities, including cameras and laser scanners, significantly augments the capabilities of autonomous robotic systems (see Fig. 1). This process optimizes the distinctive strengths of each sensor type, simultaneously mitigating their inherent constraints (see Fig. 2). Data-level fusion is at the root of this fusion process, wherein raw sensor data are integrated for a more holistic view of the environment [7]. This level is a foundation for feature-level fusion, a more demanding but insightful process involving extracting and combining relevant features from each sensor's data. Subsequently, decision-level fusion compiles sensor outputs, enabling cogent, informed decisions. Additionally, rule-based and statistical fusion methods provide consistency and robustness to the process, with the former being effective in known environments and the latter offering strength in situations where sensor data relationships can be quantified.

Multi-sensor companion robots, powered by cutting-edge sensor fusion and artificial intelligence technology, are set to transform various industries, most notably healthcare, and significantly enhance the quality of life for those who need assistance. In the healthcare sector, these companion robots can take on tasks such as patient monitoring and medication management and provide companionship. This convenience not only eases the workload of busy healthcare professionals but also ensures that patients receive personalized care. For eldercare facilities, these robots can fill critical gaps. They can offer companionship, help with day-to-day activities, and even detect falls—an essential feature for elderly safety. By taking on these roles, the robots address common challenges, such as loneliness among seniors and the limited availability of human caregivers. These examples highlight the incredible potential of multi-sensor companion robots to improve

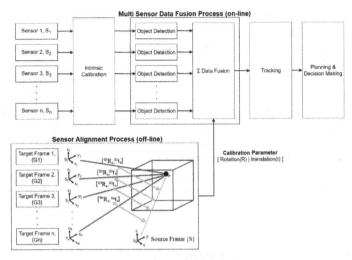

Fig. 1. A depiction of a general multi-sensor data fusion framework, which aligns multiple sensors and detects obstacles. This illustration image is adapted and inspired from [8], with an added calibration process.

well-being and promote independence among those who need assistance, leading the way to a new era of personalized and independent living.

The incorporation and fine-tuning of multi-sensor systems in companion robots, while being a promising avenue for enhanced functionality, are accompanied by numerous restrictions. These issues underline the intricate nature of multi-sensor system design, especially when incorporating ultrasonic sensors, and call for a deeper exploration.

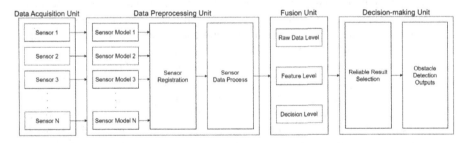

Fig. 2. Architecture for a multi-sensor fusion obstacle detection framework based on [9].

One of the primary hurdles in this endeavor is the necessity for precise calibration and synchronization among different sensor modalities [9]. This process ensures that the data derived from various sensors is integrated seamlessly, thereby improving the reliability and accuracy of the fused data. However, the process is complex and may be subject to error due to the physical limitations of the sensors or their relative placement on the robot. Further complicating the integration process is the diversity of sensor types and data formats used. Each sensor produces data at its own rate and in a unique format, which presents substantial difficulties in developing efficient and flexible fusion algorithms. These algorithms must be designed to collate disparate data streams into a unified representation of the robot's environment and make prompt, informed decisions to guide the robot's actions.

Moreover, real-world environments' dynamic and unpredictable nature introduces another layer of complexity. Robots equipped with multi-sensor systems must be designed to adeptly adapt to sudden changes, such as moving objects, variations in lighting and temperature [10], or unexpected obstacles. This reality necessitates the development of robust and adaptive algorithms that can handle this level of environmental variability and uncertainty. Attaining robustness, adaptability, and scalability in multi-sensor systems, particularly those involving ultrasonic sensors, remains an enduring research challenge. Addressing these challenges demands technical advancement and innovative problem-solving approaches to unlock the potential of multi-sensor systems in companion robots fully.

3 Environment Mapping Challenges

Environment mapping forms the backbone of safe autonomous navigation for companion robots in unknown spaces. It encompasses creating a detailed representation of the robot's operational milieu. It incorporates collecting environmental data and the subsequent assembly of a comprehensive model, delineating the physical layout, the presence of objects, and potential impediments within that environment. In the context of companion robots, the precision of environment mapping transcends into a critical determinant for safe navigation, efficacious obstacle avoidance, and meaningful interaction with both the environment and its human counterparts. It emerges as a linchpin in the successful deployment of companion robot systems, empowering these robots to perceive and seamlessly navigate within their operational surroundings. Achieving accuracy and reliability in this mapping process is a non-negotiable prerequisite for various real-world applications spanning healthcare, senior care, and home assistance. This section underlines the current research gaps and constraints in environment mapping methodologies, precisely those employing ultrasonic sensors.

Notwithstanding the advancements in the domain, noticeable gaps exist in achieving high-resolution mapping and handling dynamic environmental changes. While ultrasonic sensors provide a cost-effective and relatively reliable means for distance measurement and object detection, their lower resolution–compared to other sensor technologies, such as LIDAR–leads to less detailed and sometimes inaccurate maps. This limitation can hinder the robot's ability to make fine-grained navigational decisions.

In [11], the authors introduced an operational mechanism for a social robot that relies on eight ultrasonic sensors strategically arranged to provide a comprehensive 360-degree environmental mapping. A control method was employed for autonomous navigation within a simulated environment. Despite the successful mapping and navigation, the study revealed certain limitations. As expected, the ultrasonic sensors demonstrated restricted range and resolution, especially compared to more sophisticated laser-based options. Furthermore, the study's simulation-based approach may not fully capture the complexities of real-world applications, indicating a need for additional testing and development in more realistic environments.

In the realm of environment mapping, particularly with ultrasonic sensors, a potential research gap has been identified concerning the real-time updating of maps in response to unexpected environmental changes. While the literature review carried out for this paper did not yield substantial evidence to support or refute the hypothesis that current real-time mapping methodologies might struggle with latency or delays, this lack of information indicates a research gap. This suggests that the impact of real-time environmental changes on ultrasonic sensor-based mapping is an area that requires further investigation. After all, the lag in updating the environmental map could lead to inaccurate navigation or, in the worst-case scenario, collision. Moreover, addressing this gap by adopting a multi-sensor approach or distributed computing could present new challenges. These could include issues related to sensor fusion, data synchronization, and the management of increased computational complexity. Such a shift could necessitate the development of novel algorithms and systems capable of efficiently processing and integrating data from multiple sources in real-time. Therefore, this potential research gap highlights an opportunity for enhancing the robustness and adaptability of mapping systems in dynamic environments and opens up a new avenue for research into multi-sensor systems and distributed computing methodologies in the context of environment mapping.

The challenges in this domain are multidimensional, with the accuracy and resolution of mapping forming one of the significant research voids. Ultrasonic sensors gauge distances leveraging the principle of sound wave reflection [12]. However, the exactitude of these readings is susceptible to various factors, ranging from ambient noise and interference from other sensors, to sensor-induced blind spots, wind flow, motion of objects, or even the number of items in the environment. These elements can collectively degrade the sensor's performance, resulting in compromised accuracy and lower resolution of the constructed environmental map. Researchers in [13] present a method for tracking moving objects using ultrasonic sensors. The study developed two tracking algorithms, Extended Kalman filter (EKF) and Unscented Kalman filter (UKF) and found them to be more precise than an advanced triangle method, especially in low-speed situations. However, the study has several limitations that represent research gaps. Firstly, the tracking capability of the ultrasonic sensors was only tested for low-speed situations. The effect of wind at higher speeds was not considered, indicating a need for further research. Secondly, the study focused on single-object tracking, while in a real-world environment, multiple objects may be present simultaneously–indoors or outdoors. This suggests a need for further analysis of the tracking capability in multiple object situations.

An additional research frontier pertains to the representation and modeling of the environment, a factor integral to successful environment mapping. Beyond mere detection and localization of obstacles, a more profound comprehension of their geometric properties and spatial interrelations is paramount. Ultrasonic sensors, while effective in certain aspects, exhibit limitations in capturing intricate geometric details, which may influence the integrity of the resultant environment model. The study in [14] introduces an innovative assistive device to aid visually impaired individuals by combining ultrasonic sensors and a deep learning model for object recognition and obstacle detection. The device accurately captures the shapes of obstacles at different levels using multiple ultrasonic sensors. It successfully trains a deep learning model (YOLO-v3) to recognize objects such as stairs and potholes from camera images, providing real-time audio feedback to the user. However, the study acknowledges certain limitations in terms of shape detection. The processing time of ultrasonic sensors might be suboptimal, impacting real-time navigation in dynamic environments. Expanding the training dataset and further investigating user experience would be valuable for improving object recognition performance and overall device efficiency.

Furthermore, the prospect of scaling these mapping techniques to larger, more complex environments presents its own set of challenges. The necessity for efficient processing and storage becomes a pivotal concern as the scale of the environment and the resulting data volumes increase. This is particularly relevant for large dynamic environments or settings containing many objects and complex layouts covering a vast distance. However, the current literature does not focus much on these scenarios, especially outside of subterranean mapping [15]. This calls for innovative solutions that enhance environment mapping techniques' computational efficiency and scalability, pushing the boundaries of achievable goals. The future of environment mapping lies in developing methods capable of handling these large-scale, complex scenarios, opening new possibilities for exploring and understanding our surroundings.

The transformative potential of enhanced environment mapping becomes manifestly clear within the healthcare sector. In a hospital environment, a companion robot equipped with sophisticated and precise environment mapping techniques would be capable of maneuvering through corridors, adroitly circumventing impediments like medical equipment or service carts, and accurately pinpointing locations such as specific rooms or the whereabouts of crucial medical supplies. Such dexterity in navigation and localization tasks heightens operational efficiency, alleviates the load on human staff, and ultimately amplifies the quality of patient care delivered. Further, the richness of detail in these environment models extends beyond navigational assistance. It opens the door to personalized services such as fetching specific objects upon request or proactively identifying potential fall risks within the environment. Such nuanced applications boost the revolutionizing potential of advanced environment mapping techniques in companion robot applications across diverse sectors.

Overall, while ultrasonic sensors have established their role in environment mapping for companion robots, the existing research gaps accentuate the imperative for further investigation and development. Strides made in this direction are anticipated to yield substantial benefits, stretching the limits of the possible in terms of robotic navigation and interaction. Hence, emphasizing these research objectives becomes a critical priority for the ongoing progression in the field.

4 Landmark-Based Localization

Landmark-based localization forms an indispensable cornerstone in companion robot systems [16], granting the ability to ascertain the robot's precise position within a given environment. This method exhibits high relevance in real-world industry applications and is also relevant to healthcare and home assistance–particularly with the proliferation of the Internet of Things (IoT) technology [17] – where both absolute and relative positioning may be needed. This segment of the paper pivots towards examining the research gaps and challenges entwined with applying ultrasonic sensors for landmark-based localization in companion robots.

The elucidation of these research gaps and challenges is not an academic exercise in isolation but instead bears direct ramifications on the performance, robustness, and practicability of companion robots within industrial contexts. Each gap represents a potential hindrance to the effectiveness of these robotic systems, highlighting the urgent need for innovative solutions to bridge these voids. Moreover, the discourse emphasizes the considerable potential of integrating complementary sensors, such as cameras or laser range finders [18, 19], into the mix. Such integration can be a strategic lever for improving the accuracy and efficiency of localization with sonar sensors, marking a valuable direction for future exploration. However, it should be noted that this approach necessitates applying sophisticated sensor fusion techniques to amalgamate data from diverse sensor sources coherently.

A promising pathway appears by integrating supplementary sensing technologies, with cameras emerging as a compelling candidate [20]. These sensors can procure a rich tapestry of detailed environmental information, thereby paving the way for improved detection of landmarks and a subsequent enhancement in the precision of localization. This synergy harnesses the benefits of both sensor types, ultimately enabling the creation of a more resilient and accurate localization framework for companion robots. This fusion of sensors broadens the range of detectable environmental factors [21] and bolsters the localization process's overall robustness and precision.

Despite the promising potential of ultrasonic sensors for landmark-based localization in mobile robots, a review of the current literature reveals a lack of studies specifically addressing this topic, as evidenced by the lack of relevant papers published in the last five years. This scarcity sheds light on the nascent research stage in this field and highlights the need for more comprehensive studies. Integrating ultrasonic sensors in landmark-based localization systems could likely offer significant advancements in the accuracy and efficiency of mobile robot navigation. However, the challenges associated with this integration, including sensor noise, interference, and the need for sophisticated sensor fusion techniques, remain largely unexplored. Therefore, it is imperative for future research to delve into this uncharted territory, addressing the existing challenges and harnessing the potential of ultrasonic sensors.

5   Path Reconstruction Gaps

Precision in path reconstruction throughout the navigation process is vital to guarantee efficient and secure operations of companion robots across diverse real-world industry settings. This segment centers on unearthing the research gaps and bottlenecks associated with methods employed in path reconstruction that leverage data from ultrasonic sensors in the sphere of companion robots. It exposes the criticality of filling these gaps to augment companion robots' navigational abilities and safety standards, with a particular focus on ever-changing environments.

As a sub-component of path planning [22], path reconstruction constitutes an integral element of the navigational mechanics of companion robots, enabling the precise retracing of an already traversed trajectory within a given milieu. This process is an essential function within a robot's Simultaneous Localization and Mapping (SLAM) framework, facilitating a deeper comprehension of the surrounding space and promoting efficient, safe navigation.

Within the operational context, a companion robot estimates its navigation path within diverse settings such as hospitals, nursing homes, or domestic environments. In the process, it continuously harvests data via its suite of onboard sensors to sense its environment [23], incorporating elements such as ultrasonic sensors, lidar, and visual sensors. This information encapsulates critical details about the robot's kinematics (including rotation and linear displacement) and distinctive environmental aspects it encounters (such as obstacles and landmarks).

The path reconstruction process deciphers this sensor data to reconstruct the historical trajectory of the robot within the environment, effectively establishing a 'breadcrumb trail.' This reconstructed path provides invaluable backtracking capabilities, aids in optimal route planning, facilitates efficient obstacle circumvention, and enables reliable relocation to specific points in the environment.

Path reconstruction's precision and efficiency are paramount in pragmatic settings. In hospitals, a companion robot would be mandated to navigate through intricate, dynamic environments, frequently interacting with moving entities such as medical personnel, patients, or mobile equipment. A precise path reconstruction algorithm enables the robot to memorize and circumvent high-traffic areas or obstacles, enhancing navigational efficiency and safety. Similarly, within a domestic or elderly care context, path reconstruction assists the robot in estimating cluttered environments, evading potential hazards, and returning to designated points, all with high dependability. Consequently, path reconstruction is indispensable in ensuring companion robots' efficient operation across various real-world applications.

In the context of mobile robotics, the navigational capabilities of desert ants could inspire the use of integrated sensory systems for path memorization and reconstruction. The Cataglyphis species of desert ants demonstrate an extraordinary ability to navigate back to their nests, even after extensive foraging trips in complex or harsh environments [24]. This is achieved through a sophisticated navigational mechanism known as path integration. By integrating directional compass information and distance data, the ants generate a vector that leads them directly back to their nest, accurately determining both direction and path length. The ants estimate distance through a step integrating mechanism and optic flow perception, providing a robust measure of distance traveled. The ants utilize panoramic sceneries as local guidance cues for homing to account for potential cumulative errors inherent in path integration mechanisms.

Furthermore, these ants employ a celestial compass, using the position of the sun and the associated skylight polarization pattern as global compass cues. This allows them to keep track of all walked angles during their random foraging runs, further enhancing their navigational capabilities. In [25], researchers reveal how ants memorize paths using visual perception. These ants memorize their surroundings and follow learned routes. When visually impaired, ants demonstrate remarkable adaptability by relearning their environment and forming new memories without erasing the previous ones.

There already exists a branch of mobile robotics that is extending towards that approach. Some examples include recalling learned paths through an ants-inspired vision and neural model [26] or using magnetic field data for navigational positioning [27, 28]. In fact, odometry data provides information about how much a robot has moved and can easily mimic the ants' step integration mechanism, providing a measure of distance traveled. When coupled with magnetometers, which measure the magnetic field's strength and direction, replicating the ants' use of the Earth's magnetic field as a compass reference becomes possible. This combination of odometry and magnetometry may provide a robust path memorization and navigation system, allowing a robot to recall and follow a previously traversed path accurately. However, it is essential to note that a mobile robot's environment may change over time, and obstacles may appear along a previously clear path. This dilemma provides another opportunity to employ low-cost ultrasonic sensors to detect nearby obstacles. By emitting ultrasonic waves that bounce back upon hitting nearby objects, a moving robot can obtain information about the object's distance and avoid collisions on its return to an initial position.

Significant strides have been made in developing algorithms for environmental reconstruction and trajectory prediction in robotic systems [29]. However, a discernible

Table 1. Overview of challenges and research gaps in multi-sensor companion robots with ultrasonic sensors

Section	Challenge	Research Gap
Environment Mapping	Accurate representation of the environment	Real-world validation of environment mapping with ultrasonic sensors
	Dynamic changes in environments	
Landmark-based Localization	Application of ultrasonic sensors for accurate localization	Scarcity of studies on ultrasonic sensors for landmark-based localization
	Sensor noise and interference	Need for exploring sensor fusion techniques in this context
		Exploration of integrating complementary sensors for improved accuracy
Path Reconstruction	Precision in retracing traversed trajectory	Research gap in robotic memory systems capable of storing multiple versions of paths
	Adapting to ever-changing environments	Integration of ultrasonic sensors for safe path reconstruction
	Real-world navigation challenges (e.g., hospitals, nursing homes)	Inspiration from bio-navigational strategies like those of desert ants

research gap remains in robotic memory systems capable of storing multiple versions of learned paths, much like the navigational strategies employed by ants. Exploring a multi-modal sensory approach incorporating ultrasonic sensors for safe path reconstruction represents a promising avenue for future research.

This path reconstruction ability would be paramount in practical applications like person-following companion robots. Upon completing their designated tasks, these robots could leverage this capability to autonomously navigate back to their initial positions or charging stations. This ability would amplify the autonomy and operational efficiency of these companion robots and underline the potential of bio-inspired strategies in propelling advancements in robotic navigation.

6 Conclusion and Future Directions

In conclusion, this research has delved into the critical aspects of multi-sensor companion robots, particularly emphasizing the integration and optimization of ultrasonic sensors. The current research landscape has revealed significant gaps and challenges in environment mapping, landmark-based localization, and path reconstruction methodologies. These gaps, if addressed, could unlock the full potential of companion robots, particularly in real-world applications spanning healthcare, elderly care, and home assistance. This paper has underscored the importance of sensor fusion techniques, which help develop companion robots better equipped to perceive and interact with their environment. The potential improvements in environment mapping and landmark-based localization with ultrasonic sensors could significantly enhance companion robots' navigation capabilities, ensuring their safe and effective operation in diverse settings. Yet, there is a significant research gap in that direction. Furthermore, advancements in path reconstruction could revolutionize how companion robots retrace their steps, increasing their operational efficiency.

The real-world implications of this research are profound. The advancements in companion robot technology that this research could yield have the potential to transform various sectors. More effective companion robots could provide unwavering support and companionship to individuals in isolation or requiring additional assistance. They could monitor vital signs, assist with daily tasks, and ultimately enhance the quality of life for the individuals they serve. This research not only contributes significantly to the field of robotics but also paves the way for future studies. Identifying and addressing current research gaps provides a roadmap for continuous improvements in companion robot technology. The research capitalizes on the transformative potential of advanced environment mapping techniques, sensor fusion, and path reconstruction in propelling advancements in robotic navigation.

In summary, this research has shed light on the pivotal role of ultrasonic sensors in developing companion robots and the challenges and opportunities that lie ahead. The findings and insights from this study could significantly enhance companion robots' functionality, reliability, and efficiency, leading to substantial benefits in various sectors and improving the quality of life for many individuals. As the boundaries of what is possible in robotics continue to be pushed, the importance of such research becomes increasingly evident.

Acknowledgement. This research work was supported by the Research Grant from the Malaysian Ministry of Higher Education under the Fundamental Research Grant Scheme (FRGS) under Grant FRGS/1/2022/ICT02/SWIN/02/1.

References

1. Dilip, G., et al.: Artificial intelligence-based smart comrade robot for elders healthcare with strait rescue system. J. Healthc. Eng. **2022**, 9904870 (2022)
2. Asgharian, P., Panchea, A.M., Ferland, F.: A review on the use of mobile service robots in elderly care. Robotics **11**, 127 (2022)
3. Macis, D., Perilli, S., Gena, C.: Employing socially assistive robots in elderly care. In: Adjunct Proceedings of the 30th ACM Conference on User Modeling, Adaptation and Personalization, pp. 130–138. ACM, New York, NY, USA (2022)
4. Yeong, D.J., Velasco-Hernandez, G., Barry, J., Walsh, J.: Sensor and sensor fusion technology in autonomous vehicles: a review. Sensors (Basel), **21**, 2140 (2021)
5. Kulkarni, A.U., Potdar, A.M., Hegde, S., Baligar, V.P.: RADAR based object detector using ultrasonic sensor. In: 2019 1st International Conference on Advances in Information Technology (ICAIT), pp. 204–209. IEEE (2019)
6. Wang, S., Xu, G., Liu, T., Zhu, Y.: Robust real-time obstacle avoidance of wheeled mobile robot based on multi-sensor data fusion. In: 2021 IEEE 5th Advanced Information Technology, Electronic and Automation Control Conference (IAEAC), pp. 2383–2387. IEEE (2021)
7. Tong, Y., Bai, J., Chen, X.: Research on multi-sensor data fusion technology. J. Phys. Conf. Ser. **1624**, 032046 (2020)
8. Bouain, M., Ali, K.M.A., Berdjag, D., Fakhfakh, N., Atitallah, R.B.: An embedded multi-sensor data fusion design for vehicle perception tasks. J. Commun. **13**, 8–14 (2018)
9. Hu, J.W., et al.: A survey on multi-sensor fusion based obstacle detection for intelligent ground vehicles in off-road environments. Front. Inf. Technol. Electron. Eng. **21**, 675–692 (2020). https://doi.org/10.1631/FITEE.1900518/METRICS
10. Krämer, M.S., Kuhnert, K.-D.: Multi-sensor fusion for uav collision avoidance. In: Proceedings of the 2018 2nd International Conference on Mechatronics Systems and Control Engineering, pp. 5–12. ACM, New York, NY, USA (2018)
11. Haq, F.A., Dewantara, B.S.B., Marta, B.S.: Room mapping using ultrasonic range sensor on the atracbot (autonomous trash can robot): A simulation approach. In: IES 2020 - International Electronics Symposium on Role Auton. Intell. Syst. Hum. Life Comf. pp. 265–270 (2020)
12. Zhmud, V.A., Kondratiev, N.O., Kuznetsov, K.A., Trubin, V.G., Dimitrov, L.V.: Application of ultrasonic sensor for measuring distances in robotics. J. Phys. Conf. Ser. **1015**, 032189 (2018)
13. Li, S.E., et al.: Kalman filter-based tracking of moving objects using linear ultrasonic sensor array for road vehicles. Mech. Syst. Signal Process. **98**, 173–189 (2018). https://doi.org/10.1016/j.ymssp.2017.04.041
14. Yadav, S., Joshi, R.C., Dutta, M.K., Kiac, M., Sikora, P.: Fusion of OBJECT RECOGNITION AND OBSTACLE DETECTION APPROACH FOR ASSISTING VISUALLY CHALLENGED PERSON. In: 2020 43rd International Conference on Telecommunications and Signal Processing (TSP), pp. 537–540. IEEE (2020)
15. Azpúrua, H., et al.: A survey on the autonomous exploration of confined subterranean spaces: Perspectives from real-word and industrial robotic deployments. Rob. Auton. Syst. **160**, 104304 (2023)
16. Fusic, S., Sugumari, T.: A review of perception-based navigation system for autonomous mobile robots. Recent Patents Eng. **17**, 13–22 (2023)

17. Liu, W.: Improvement of navigation of mobile robotics based on IoT system. In: 2021 IEEE International Conference on Robotics, Automation and Artificial Intelligence (RAAI), pp. 69–72. IEEE (2021)
18. Debeunne, C., Vivet, D.: A review of visual-LiDAR fusion based simultaneous localization and mapping. Sensors. **20**, 2068 (2020)
19. Yanyong, S., Parichatprecha, R., Chaisiri, P., Kaitwanidvilai, S., Konghuayrob, P.: Sensor fusion of light detection and ranging and ibeacon to enhance accuracy of autonomous mobile robot in hard disk drive clean room production line. Sensors Mater. **35**, 1473 (2023)
20. Singhirunnusorn, K., Fahimi, F., Aygun, R.: A single camera 360-degree real time vision-based localization method with application to mobile robot trajectory tracking. IET Cyber-Systems Robot. **3**, 185–198 (2021)
21. Khan, M.S.A., Hussian, D., Ali, Y., Rehman, F.U., Aqeel, A. Bin., Khan, U.S.: Multi-sensor SLAM for efficient navigation of a mobile robot. In: Proceedings of 2021 IEEE 4th International Conference on Computing Information Science ICCIS 2021. (2021)
22. Abdallaoui, S., Aglzim, E.-H., Chaibet, A., Kribèche, A.: Thorough review analysis of safe control of autonomous vehicles: path planning and navigation techniques. Energies **15**, 1358 (2022)
23. Bai, Y., Garg, N., Roy, N.: SPiDR: ultra-low-power acoustic spatial sensing for micro-robot navigation. In: Proceedings of the 20th Annual International Conference on Mobile Systems, Applications and Services, pp. 99–113. ACM, New York, NY, USA (2022)
24. Grob, R., Fleischmann, P.N., Rössler, W.: Learning to navigate - How desert ants calibrate their compass systems. Neuroforum. **25**, 109–120 (2019). https://doi.org/10.1515/NF-2018-0011/ASSET/GRAPHIC/J_NF-2018-0011_CV_003.JPG
25. Schwarz, S., Clement, L., Haalck, L., Risse, B., Wystrach, A.: Compensation to visual impairments and behavioral plasticity in navigating ants. bioRxiv. 2023.02.20.529227 (2023)
26. Gattaux, G., et al.: Antcar: Simple Route Following Task with Ants-Inspired Vision and Neural Model (2023)
27. Antsfeld, L., Chidlovskii, B.: Magnetic field sensing for pedestrian and robot indoor positioning. In: 2021 International Conference on Indoor Positioning and Indoor Navigation, pp. 1–8 (2021)
28. Kim, Y.H., Kim, H.J., Lee, J.H., Kang, S.H., Kim, E.J., Song, J.W.: Sequential batch fusion magnetic anomaly navigation for a low-cost indoor mobile robot. Measurement **213**, 112706 (2023)
29. Zhong, F., Bi, X., Zhang, Y., Zhang, W., Wang, Y.: RSPT: reconstruct surroundings and predict trajectory for generalizable active object tracking. In: Proceedings of AAAI Conference on Artificial Intelligence, vol. 37, pp. 3705–3714 (2023)

Convergence Study of Three-Dimensional Upper Skull Model: A Finite Element Method

Nor Aqilah Mohamad Azmi[1], Nik Nur Ain Azrin Abdullah[1],
Zatul Faqihah Mohd Salaha[1], and Muhammad Hanif Ramlee[1,2(✉)] ⓘ

[1] Bone Biomechanics Laboratory (BBL), Department of Biomedical Engineering and Health Science, Faculty of Electrical Engineering, Universiti Teknologi Malaysia, 81310 UTM Johor Bahru, Johor, Malaysia
muhammad.hanif.ramlee@biomedical.utm.my
[2] Bioinspired Device and Tissue Engineering (BIOINSPIRA) Research Group, Universiti Teknologi Malaysia, 81310 UTM Johor Bahru, Johor, Malaysia

Abstract. Le-fort I osteotomy has been widely used by maxillofacial surgeons as a procedure due to its versatility and simplicity to treat patient with dentofacial deformity. Technology such as virtual planning software has greatly aided surgeon to prepare for the surgery by allowing them to work with clinical engineer to plan for the surgery and create custom implants design on the patient's scanned anatomy in form of three-dimensional (3D) model. Most of convergence study for skull was done by analyzing the total number of mesh elements. Hence, this study was conducted to determine the optimum mesh size by performing mesh convergence study using h-refinement method on the upper skull 3D model. The 2D Computed-Tomography (CT) images in DICOM format were used to create the 3D model by using Mimics software. The mesh refinement process of the segmented upper skull 3D model was performed in the 3-Matic software, and the mesh size used from 5 mm to 2.5 mm with 0.5 mm interval. Marc Mentat software was used to perform finite element analysis (FEA) where a fixed load was set on superior part of the skull meanwhile point load with 125N stress was applied on incisors. Results from the mesh convergence analysis showed that percentage difference of upper skull of 4.0 mm–3.5 mm mesh size is below 5% and from the convergence plot the line graph start to converge at 3.5 mm to 4 mm. Therefore, it can be concluded that 4 mm mesh size is the optimum mesh size for upper skull 3D model.

Keywords: Convergence study · Finite element analysis (FEA) · Skull

1 Introduction

Orthognathic surgery is one of oral surgery that performed not only to improve facial aesthetic but also to help patients with skeletal disharmonies that can affect patient's life such as airway issue, temporomandibular joints (TMJ) disorder or speech difficulty [1]. Virtual planning software has greatly assisted the medical field. It helps surgeons to conduct surgery planning on patient three-dimensional (3D) skull bone and enable

F. Hassan et al. (Eds.): AsiaSim 2023, CCIS 1912, pp. 351–361, 2024.
https://doi.org/10.1007/978-981-99-7243-2_30

the surgeon to view, cut and reposition the bone virtually. Aside from having surgical planning, surgeons can work with clinical engineers to design patient-specific plate (PSP) based on the planned bone hence reducing time to bend the plate manually [2, 3]. The PSP method is not only reducing time for designing the plate, but it is also could improve the treatment procedure by reducing operation time for patients and orthopedic surgeons.

Apart from having virtual surgery planning, it is well-known that finite element analysis (FEA) can be used to further analyses the biomechanical strength of the plate fixated on bones. By simulating the effects of biomechanical changes on the bone using this virtual analysis, it is possible to evaluate various fixation systems on bone and prevent plate failure due to incompatible fixation in the future by utilizing most suitable design to the patient [4–7]. The FE method is one of the established simulation procedures for orthopedics cases before the design, configurations and materials of implant could be physically applied to patients in the operation theatre.

The FEA has been widely used for biomechanical analysis; hence the 3D bone model should be developed precisely since the result of FEA depends on the quality of the bone model. In FE modelling, the smaller the mesh used, the more accurate the solution. However, it will increase the process time hence convergence study is needed to get the right mesh that can give accurate FEA result. Thus, the objective of this study is to determine the correct and optimum mesh size through convergence study on upper skull before further parametric study could be done. The convergence study is one of the procedures in FEA, before an experimental work to validate the 3D model. In a simulation, these two steps are needed to ensure the 3D model is reliable to be used in the other analysis. The results from this study are not only for determining the optimum mesh size, but it could be a future reference for researchers when dealing with FEA studies.

2   Materials and Methods

2.1   Reconstruction of 3D models: Upper skull

Computed tomography (CT) image data was obtained from a female patient who were 34 years old with malocclusion problem. The slice thickness of the CT image was 0.625 mm in 512×512 matrix with a total number of 499 slices. Figure 1 shows steps to convert the CT image into 3D model of upper skull. The CT image was imported in Mimics software (Materialise, Belgium) file for a segmentation process and Hounsfield unit (HU) ranged from 226–3071 was used as the threshold value for the upper skull bone. The segmentation was further improved by using lower threshold edit masks function so that the bone with lower threshold than the range can be included. Since the study focuses on the upper skull region, only 3D model of upper skull constructed from the CT image. However, to reduce run time of analysis, the cranium region was trimmed and the region of interest for the analysis only included from orbital to maxilla teeth as in Fig. 1 image (d) to (f).

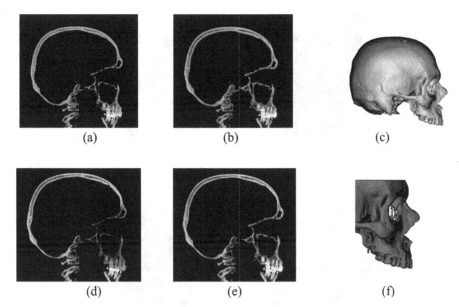

Fig. 1. (a) Mask segmentation of upper skull, (b) bone contour of upper skull, (c) 3D model of upper skull, (d) Trimmed mask segmentation of upper skull, (e) Bone contour of trimmed upper skull, (f) 3D model of trimmed upper skull

2.2   Finite Element Modelling

The 3-matic software (Materialise, Belgium) has been used for meshing editing. The upper skull 3D model was imported into this software and meshed by using the tetrahedral element shape. In the software, the view function of Shading Mode can be used: Filled with Triangle Edges so that the 3D upper skull will be able to show surface with meshes. This enables the users to further check the mesh sizes and shapes.

In this study, 6 mesh sizes range from 5.0 mm to 2.5 mm were obtained with 0.5 mm interval difference and the upper skull models were named from Skull 1 (mesh size 5 mm) to Skull 6 (mesh size 2.5 mm). Figure 2 shows upper skull with different mesh size; (a) Skull 1 with mesh size 5 mm, (b) Skull 2 with mesh size 4.5 mm, (c) Skull 3 with mesh size 4.0 mm, (d) Skull 4 with mesh size 3.5 mm, (e) Skull 5 with mesh size 3.0 mm and (f) Skull 6 with mesh size 2.5 mm.

To make sure that all the different mesh size models have the same node's position, the same model should be used to create the mesh size. The mesh refinement process was started by meshing the upper skull model surface with the biggest mesh size from the mesh range that was set early, which was 5.0 mm. Then this same 5 mm mesh size model was used to make a 4.5 mm mesh size model. This step of using the previous 0.5 mm mesh size difference model will continue until the model reach 2.5 mm mesh size.

Once all 6 upper skulls with difference mesh sized were prepared, all these models should be in solid state hence a tool of 'Create Volume Mesh' was used to convert the surface meshes models into volume meshes models with triangular edge length same as

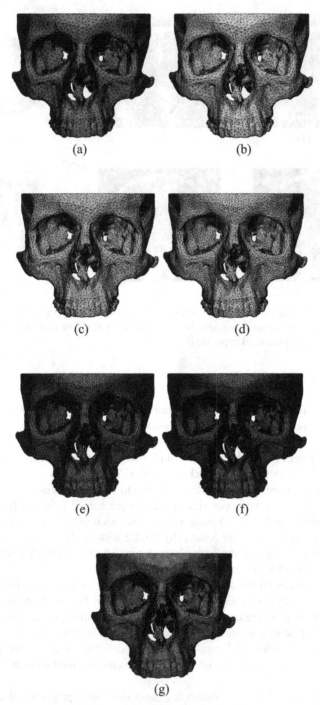

Fig. 2. Mesh sizes of upper skull: (a) Skull 1: 5.0 mm (b) Skull 2: 4.5 mm (c) Skull 3: 4.0 mm (d) Skull 4: 3.5 mm (e)Skull 5: 3.0 mm (f) Skull 6: 2.5 mm

the surface mesh size. To ensure that these models can be used in the next simulation software, then the model was exported in Patran file.

2.3 Finite Element Analysis

In the following step of convergence study is to use Marc Mentat software (MSC, USA) for finite element analysis of each upper skull model. The material properties of upper skull model were considered as isotropic, linearly elastic, and homogenous with: Young's modulus = 15000 Mpa and Poisson ratio = 0.3 [8]. The upper skull model was simplified as cortical bone only [9]. As for the boundary condition, the superior part of upper skull was set as fixed displacement while the point load of 125 N on the teeth incisors area for all upper skull model [10]. Figure 3 shows the area of fixed displacement which is along the superior part of upper skull and the axial point load from skull inferior located between teeth incisors.

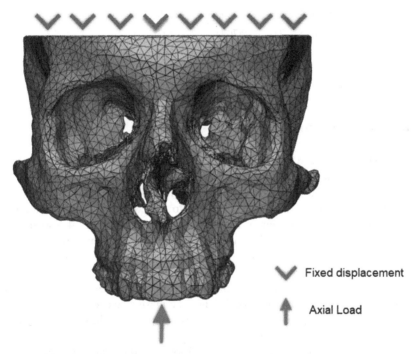

Fig. 3. Application of boundary conditions to upper skull model

Finite element analysis was done on each skull model from Skull 1 to Skull 6 individually. Fixed displacement (node selection) and axial point load were set to be in the same exact location for all skull models as shown in Fig. 3. To make sure the location and node are the same, a tool of 'Remove' and 'Add' elements were utilized in the Marc Mentat software. This step was taken to ensure the results of von Mises stress are consistent and could indicate a good trend of the graph. Table 1 below shows the number of nodes and elements of each upper skull model and the total time taken to run analysis on each model. It should be noted that the skulls 1 and 6 had the fastest and slowest time for FEA, respectively. Meanwhile, the model of Skull 6 had the highest total number of elements and nodes as compared with other models in this study.

Table 1. Total number of nodes and elements for upper skull models over total times for FEA

Model	Mesh Sizes (mm)	Total Number of Elements	Total Number of Nodes	Total Time for FEA (s)
Skull 1	5.0	27358	13539	19
Skull 2	4.5	31132	15426	24
Skull 3	4.0	36014	17867	32
Skull 4	3.5	45096	22408	46
Skull 5	3.0	57094	28407	71
Skull 6	2.5	74646	37183	118

3   Result and Discussion

In this mesh convergence study, six nodes were selected along the contour plot and the position of the nodes are same for all upper skull models [11–15]. Six nodes were positioned from middle to lateral skull with each on different color of contour band as in Fig. 4(a). These six nodes were chosen to get consistent and accurate results for convergence findings as demonstrated by other authors. Figure 4 also shows the equivalent von Mises stress applied on skull models with different mesh sizes and the equivalent von Mises stress is getting higher with smaller mesh. This result corresponds to a study by Aman Dut stated that the von Mises stress becomes higher when the mesh size is smaller [16]. However, if we use the smallest mesh size then it will take us longer analysis time. This is proven in Table 1, when the mesh size decreases, the total time for analysis increases. Hence convergence analysis study should be done to get the most optimum mesh size for skull 3D model.

The result of magnitude of equivalent von miss stress from each 3D upper skull model was extracted from the finite element analysis as shown in Fig. 5. Meanwhile Table 2 shows the percentage difference between two successive mesh sizes. If the percentage difference of each node for two successive mesh sizes is below 5% then the model is considered converged [17]. Based on the result in Table 2 below, all nodes that have percentage difference below 5% are between mesh size 3.5 mm and 4.0 mm.

From the FEA result, a graph between the nodes and equivalent VMS was plotted as in Fig. 5 above for all the upper skull model. Node 1 is located at maxilla area which is on the middle area and consecutive nodes go from middle skull to lateral skull as in

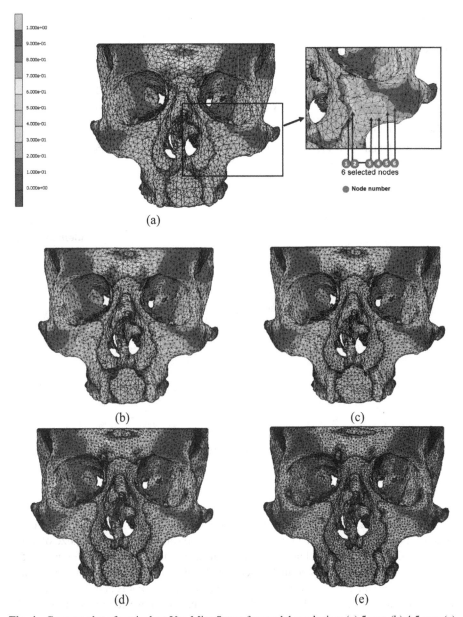

Fig. 4. Contour plot of equivalent Von Miss Stress for model mesh size: (a) 5 mm (b) 4.5 mm (c) 4.0 mm (d) 3.5 mm (e) 3.0 mm

Fig. 5. The equivalent von miss stress decreases from node 1 to node 6 since the load was applied from the incisors which is also around the middle skull area. From the graph plotted above we can observe that the line trend for all mesh sizes is the same, decreasing from node 1 to node 6. From all lines, the line plot for mesh size 4.0 mm and 3.5 mm

Table 2. Percentage difference between two successive mesh size for selected nodes

Model	Percentage Difference (%)					
	Node 1	Node 2	Node 3	Node 4	Node 5	Node 6
5 mm & 4.5 mm	7.82	0.84	5.60	6.10	5.22	3.66
4.5 mm & 4 mm	6.89	4.88	4.50	4.04	11.97	2.03
4 mm & 3.5 mm	3.42	0.53	0.14	0.41	2.55	0.50
3.5 mm & 3.0 mm	0.40	7.51	7.85	5.35	5.50	1.49
3.0 mm & 2.5 mm	6.97	1.41	1.46	13.45	5.22	2.43

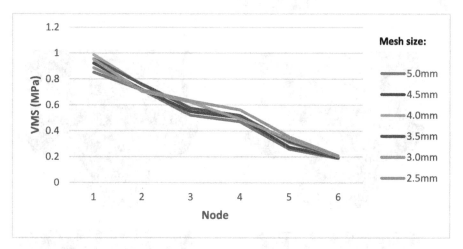

Fig. 5. Magnitude of equivalent Von Miss Stress for 6 selected nodes

start to converge from node 2. This aligns with the result of having percentage difference lower than 5% that also happens for mesh size 3.5 mm and 4.0 mm.

Another graph was plotted between equivalent VMS against mesh size for Node 1 (Fig. 6). From the graph we can observe that the line starts stable at mesh size of 4 mm to 3.5 mm. This result correlates with the result from the percentage difference < 5% analysis, which the percentage difference for all selected nodes between 4.0 mm and 3.5 mm are less than 5%. Hence it is agreed that converge point is at 4.0 mm mesh size for upper skull 3D model. A larger mesh was chosen so that we can reduce time for the simulation.

One of limitations of this study is that there are numbers of FEA study focusing on skull region however most of the study were using mesh density convergence and there is no reference of the mesh size that has been used for the skull model [17–20]. Almost all research papers of FEA on skull region found are not stating the mesh size that being used to construct the skull model.

However, there are still papers that were using mesh size for convergence study though it is focusing on other bone regions, but the workflow still be used as reference.

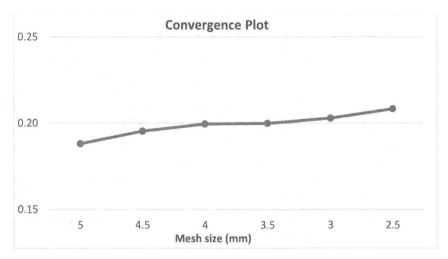

Fig. 6. Convergence plot

For example, there is a study that conducted mesh convergences study on femur bone model where the optimum mesh size of 4.5 mm was decided by percentage error < 5% and convergence graph [21]. Apart from that, another study of optimal mesh was also done but on human foot model with optimal mesh size of 2.5 mm [22]. The convergence analysis method used for this study is based on the percentage difference. The mesh size is converged when the mesh density does not change the percentage difference more than 5%. It also shows that the analysis time for the mesh size of 1.5 mm is 60% more expensive than on mesh size of 2.5 mm.

4   Conclusion

As a conclusion, 3D model of upper skull was successfully developed in Mimics software and mesh refinement process was successfully conducted in 3-Matics software. Following the mesh refinement, FEA had been carried out on 6 upper skull models in Marcs software and the results of equivalent Von Miss Stress was used to create the convergence plot. Apart from convergence plot, percentage differences were also calculated from the FEA. Based on the results from these two methods, we discovered that 4.0 mm is the optimum mesh size for upper skull 3D model.

References

1. Wolford, L.M., Goncalves, J.R.: Surgical planning in orthognathic surgery and outcome stability. In: Maxillofacial Surgery. Elsevier Inc. (2017). https://doi.org/10.1016/b978-0-7020-6056-4.00074-5
2. Thakker, J.S., Pace, M., Lowe, I., Jung, P., Herford, A.S.: Virtual surgical planning in maxillofacial trauma. Atlas Oral Maxillofac. Surg. Clin. **27**(2), 143–155 (2019). https://doi.org/10.1016/j.cxom.2019.05.006

3. Reddy, S., Macmillan, A., Dorafshar, A.H.: Virtual surgical planning. In: Facial Trauma Surgery. Elsevier Inc. (2020). https://doi.org/10.1016/b978-0-323-49755-8.00048-7

4. Ataç, M.S., Erkmen, E., Yücel, E., Kurt, A.: Comparison of biomechanical behaviour of maxilla following Le Fort I osteotomy with 2- versus 4-plate fixation using 3D-FEA. Part 1: advancement surgery. Int. J. Oral Maxillofac. Surg. 37, 1117–1124 (2008)

5. Ataç, M.S., Erkmen, E., Yücel, E., Kurt, A.: Comparison of biomechanical behaviour of maxilla following Le Fort I osteotomy with 2- versus 4-plate fixation using 3D-FEA. Part 2: impaction surgery. Int. J. Oral Maxillofac. Surg. 38, 58–63 (2009)

6. Pozzer, L., Olate, S., Cavalieri-Pereira, L., Navarro, P., de Albergaría Barbosa, J.R.: Mechanical stability of 2-plate versus 4-plate osteosynthesis in advancement Le Fort I osteotomy: an in vitro study. J. Stomatol. Oral Maxillofac. Surg. 118, 2–4 (2017)

7. Khiabani, K., Keyhan, S.O., Ahmadi, P., Gholamian, A., Cheshmi, B.: Effect of using different methods of plate fixation in maxillary Lefort one fractures. J. Oral Maxillofac. Surg. Med. Pathol. 31, 386–400 (2019)

8. Ridwan-Pramana, A., et al.: Finite element analysis of 6 large PMMA skull reconstructions: a multi-criteria evaluation approach. PLoS ONE 12, 1–16 (2017)

9. Shi, Q., et al.: Preclinical study of additive manufactured plates with shortened lengths for complete mandible reconstruction: design, biomechanics simulation, and fixation stability assessment. Comput. Biol. Med. 139, 105008 (2021)

10. Huang, S.F., Lo, L.J., Lin, C.L.: Biomechanical optimization of a custom-made positioning and fixing bone plate for Le Fort I osteotomy by finite element analysis. Comput. Biol. Med. 68, 49–56 (2016)

11. Abidin, N.A.Z., et al.: Biomechanical effects of cross-pin's diameter in reconstruction of anterior cruciate ligament – a specific case study via finite element analysis. Injury 53, 2424–2436 (2022)

12. Ab Rashid, A.M., Ramlee, M.H., Gan, H.S., Kadir, M.A.R.: Effects of badminton insole design on stress distribution, displacement and bone rotation of ankle joint during single-leg landing: a finite element analysis. Sport. Biomech., 1–22 (2022) https://doi.org/10.1080/147 63141.2022.2086168

13. Abidin, N.A.Z., Wahab, A.H.A., Rahim, R.A.A., Kadir, M.R.A., Ramlee, M.H.: Biomechanical analysis of three different types of fixators for anterior cruciate ligament reconstruction via finite element method: a patient-specific study. Med. Biol. Eng. Comput. 59, 1945–1960 (2021)

14. Ramlee, M.H., Seng, G.H., Felip, A.R., Kadir, M.R.A.: The effects of additional hollow cylinder coated to external fixator screws for treating pilon fracture: a biomechanical perspective. Injury 52(8), 2131–2141 (2021). https://doi.org/10.1016/j.injury.2021.03.017

15. Aziz, A.U.A., Wahab, A.H.A., Rahim, R.A.A., Kadir, M.R.A., Ramlee, M.H.: A finite element study: finding the best configuration between unilateral, hybrid, and ilizarov in terms of biomechanical point of view. Injury 51(11), 2474–2478 (2020). https://doi.org/10.1016/j.inj ury.2020.08.001

16. Dutt, A.: Effect of mesh size on finite element analysis of beam. Int. J. Mech. Eng. 2, 8–10 (2015)

17. Jyoti, M.S., Ghosh, R.: Biomechanical analysis of three popular tibial designs for TAR with different implant-bone interfacial conditions and bone qualities: a finite element study. Med. Eng. Phys. 104, 103812 (2022)

18. Li, J., Xie, Q., Liu, X., Jiao, Z., Yang, C.: Biomechanical evaluation of Chinese customized three-dimensional printed titanium miniplate in the Lefort I osteotomy: a finite element analysis. Heliyon 8, e12152 (2022)

19. Okumura, N., Stegaroiu, R., Kitamura, E., Kurokawa, K., Nomura, S.: Influence of maxillary cortical bone thickness, implant design and implant diameter on stress around implants: a three-dimensional finite element analysis. J. Prosthodont. Res. 54, 133–142 (2010)

20. Fuhrer, R.S., Romanyk, D.L., Carey, J.P.: A comparative finite element analysis of maxillary expansion with and without midpalatal suture viscoelasticity using a representative skeletal geometry. Sci. Rep. **9**, 1–11 (2019)
21. Salaha, Z.F.M., et al.: Biomechanical effects of the porous structure of gyroid and voronoi hip implants: a finite element analysis using an experimentally validated model. Mater. (Basel) **16**, 3298 (2023)
22. Chen, W.M., Cai, Y.H., Yu, Y.U.E., Geng, X., Ma, X.I.N.: Optimal mesh criteria in finite element modeling of human foot: the dependence for multiple model outputs on mesh density and loading boundary conditions. J. Mech. Med. Biol. **21**, 1–14 (2021)

Analyzing the Use of Scaffolding Based on Different Learning Styles to Enhance the Effectiveness of Game-Based Learning

Hsuan-min Wang[1,2(✉)], Wei-Lun Fang[1], and Chuen-Tsai Sun[1]

[1] National Yang Ming Chiao Tung University, Hsinchu, Taiwan
hmwang.cs04@nycu.edu.tw
[2] Asia University, Taichung, Taiwan

Abstract. Education has always been a crucial and concerning issue in human civilization. With the increasing popularity of games, various instructional theories based on game-based learning have been proposed. However, whether in gaming or learning, scaffolding plays a significant supportive role. Therefore, this study adopts the four dimensions of FSLSM as the classification of learning styles and designs four different types of scaffolding for players to use. By analyzing the behavior of players with different learning styles in using scaffolding during the game, we aim to tailor the most suitable scaffolding for players with different learning styles. The research findings indicate that when the educational content in game-based learning relates to logic and engineering, providing more visualized and detailed scaffolding will enhance learners' motivation to actively use scaffolding and improve the overall learning quality.

Keywords: Game-based learning · scaffolding · learning styles

1 Introduction

With the widespread availability of various 3C products (computers, communication devices, and consumer electronics) and the internet, people now have easy access to a vast amount of information flowing in like snowflakes. Children, in particular, not only learn from traditional textbooks at school but also receive and acquire new information through diverse channels such as the internet, games, social platforms, and self-media. With the increasing number of information sources and constant technological breakthroughs, it can be overwhelming to keep up with the rapid changes and the abundance of knowledge. Traditional paper-based education is becoming inadequate to cope with this dynamic landscape.

In recent years, more research has been focusing on whether game-based learning can enhance learning motivation and be more suitable for acquiring 21st-century skills and knowledge in the context of information, media, and technology [1]. It is evident that game-based learning has started to gain popularity among the masses.

Regardless of whether in gaming or learning, scaffolding plays a crucial supportive role. According to Spadafora & Downes [2], scaffolding is a temporary and adjustable

© The Author(s), under exclusive license to Springer Nature Singapore Pte Ltd. 2024
F. Hassan et al. (Eds.): AsiaSim 2023, CCIS 1912, pp. 362–376, 2024.
https://doi.org/10.1007/978-981-99-7243-2_31

tool provided by instructors to help learners reach goals they couldn't achieve solely with their own abilities. These goals are often referred to as the Zone of Proximal Development (ZPD), and the aim is for learners to be able to accomplish these goals independently after the scaffolding is removed. For example, a teacher may highlight certain keywords in an exam question to guide students on the key points for problem-solving. Once students learn how to identify those keywords, they can quickly identify key points in future exams even without the bold highlighting.

Similar to scaffolding in learning, scaffolding is also used in digital games to assist players, except that the provider of scaffolding shifts from the teacher to software tools [3]. These scaffolding designs guide players in understanding the core gameplay of the game. For instance, in the early stages of a game, simple levels are designed for players to repeat actions, supplemented by visual aids like text explanations or animated images. As players become familiar with the game, these scaffolding elements gradually fade away. The scaffolding in games does not diminish the joy of playing; instead, it helps players achieve continuous success, maintaining them in a state of flow. In game-based learning, scaffolding not only needs to sustain players' flow during game levels but also bridge the gap when they transition to higher-level concept learning as learners. Hence, designing appropriate scaffolding in game-based learning becomes critically important.

In the design of scaffolding, this research focuses on individual player styles as the starting point. A person's style influences their behavior in various aspects. Style is a personal trait that reflects one's thinking patterns, values, emotions, and motivations. Different styles can lead to variations in communication, decision-making, learning, and working approaches. Several studies have already indicated that a student's learning outcomes are affected by their learning style. If a teacher's teaching style does not align with a student's learning style, the student may be more prone to learning difficulties [4, 5]. Therefore, this study aims to explore whether a player's individual style also impacts their behavior in using scaffolding. The goal is to use this diversity as a reference in designing scaffolding for game-based learning.

The chosen game platform for this research is Minecraft, a sandbox game developed by Mojang in 2010 and later acquired by Microsoft in 2014. After more than a decade, it still maintains a strong presence and continues to release updates regularly. The game boasts an average of over 12 million daily active players and has accumulated 1.7 billion views on YouTube gameplay videos. As the best-selling video game of all time, Minecraft's open-ended gameplay, exploratory nature, and creative possibilities have attracted a vast player community. With its rich built-in commands and support for player-made mods, almost any concept imaginable can be recreated within Minecraft. Within the game, the "redstone circuit system" is a mechanism that simulates real-world circuits, allowing players to design and implement various complex logic devices such as computers, pianos, elevators, etc. This feature makes it an ideal choice for STEM (Science, Technology, Engineering, Mathematics) game-based learning.

Thus, this research aims to design a series of game levels and scaffolding using Minecraft's redstone circuit system to study the correlation between players' behaviors in using scaffolding during game-based learning in the STEM domain and their learning styles. The findings aim to assist game developers and educators in more effectively utilizing game-based learning as an educational resource.

2 Literature Review

2.1 Game-Based Learning

Gros et al. [6] proposed a connection between games and learning theories, categorizing game-based learning games into three generations. The first generation is behaviorism, which focuses on player behaviors. Players learn relevant knowledge by practicing a certain skill repeatedly. However, games in this generation often fail due to their simplicity, repetition, and lack of thoughtful design. The second generation emphasizes cognitivism and constructivism, centering on the learner and providing tailored information to assist different learners in their learning process. The third generation goes beyond specific games and emphasizes the learning process through game-based learning. Teachers create a social context to encourage students to ask the right questions and go to the right places, making games in this generation adaptable to the school learning environment.

In the study by Al Azawi et al. [7], game-based learning is used to motivate students to engage in learning through games, making the learning process more enjoyable and interesting. The research shows that this approach has a positive impact on cognitive development and helps learners build foundational skills and knowledge in specific areas. Students generally perceive game-based learning as a means to learn faster, increase interest, and focus on the learning topics.

2.2 Scaffolding

The concept of scaffolding in learning was first introduced by Wood et al. [8]. When solving problems or acquiring skills, scaffolding can assist children or beginners in reaching higher goals that they wouldn't be able to achieve without support.

Quintana et al. [2] proposed a design framework in 2004, referencing several science inquiry software to design scaffolding in that context. Although it was not specifically targeted at game-based learning, it provided guidelines and strategies that are still valuable for designing scaffolding as software tools in game-based learning.

In the research by Ersani et al. [9], the key points in designing scaffolding include contingency, fading, and transfer of responsibility. These aspects are applicable to both synchronous and asynchronous learning and can be achieved through static or dynamic interactions with peer, teacher, or technology support. The research classified scaffolding into four types based on their purpose: procedural, conceptual, metacognitive, and strategic. In online environments, strategic implementation can involve course structuring, accessing resources and tools, developing critical thinking, guiding problem-solving, providing abundant examples and exploratory questions, and offering constructive feedback. These findings highlight the importance of incorporating scaffolding strategies in online learning environments.

2.3 Learning Style and Felder

Richardson's research [5] explores how different personality types affect interpersonal relationships, motivation, and learning outcomes. Based on educational concepts such as brain-based learning, learning styles, and multiple intelligences, the research emphasizes that teachers need to consider students' preferences and potentials and provide diverse teaching strategies. Two commonly used personality assessment tools, the Myers-Briggs Type Indicator (MBTI) and Keirsey Temperament Sorter (KTS), were employed in the study. These tools represent two similar learning style models but with different classification methods. Additionally, Richardson points out that people often tend to socialize with those who are similar to themselves and may not prefer interactions with those who are different. Therefore, compatibility between teachers and students is crucial in creating a positive interaction and atmosphere. The research further indicates that positive teacher-student relationships can enhance students' attitudes, confidence, interest, participation, and academic performance, while negative relationships may lead to learning obstacles and failure.

2.4 Felder and Silverman Learning Style Model (FSLSM)

The FSLSM model was proposed by Felder [10] and primarily investigates the issue of mismatch between students' learning styles and instructors' teaching styles in engineering education. It introduces a four-dimensional model of learning and teaching styles, including Perception, Input, Processing, and Understanding. Felder's research revealed that most engineering students tend to have sensing, visual, inductive, active, and global learning styles, while most engineering instructors lean towards intuitive, verbal, deductive, passive, and sequential teaching styles.

Such mismatches can lead to poor student performance, frustration among instructors, and the loss of potential outstanding engineers in society. Therefore, Felder suggests that engineering education should strive to strike a balance between concrete and abstract information, visual and verbal presentation, inductive and deductive reasoning, active engagement and reflective learning, as well as sequential and global approaches in information processing. By progressively experimenting and retaining effective teaching techniques, instructors can develop a teaching style that suits the students' needs while also making them comfortable, ultimately enhancing the quality of learning.

3 Methodology

This study will employ both questionnaire surveys and game behavior observation methods for data collection. The questionnaire survey will be used to gather relevant information about players' styles, including learning style, gaming style, and gaming experience. On the other hand, the game behavior observation method involves observing and recording players' use of scaffolding in the game through programming. This data will then be cross-analyzed with the player style information collected from the questionnaires to address the research questions. The research platform used is Minecraft Java Edition. Specifically, the researchers will design game tasks on custom maps and host them on a self-set Minecraft multiplayer server. Participants will connect to the server using the Minecraft client and play through the game to complete the tasks.

Before the experiment, participants will complete questionnaires to assess their gaming and learning styles. During the experiment, a plugin installed on the server will record various player behaviors in detail, such as movement paths, item interactions, entity inter-actions, achievement unlocks, etc. This data will be saved in a MySQL database. Finally, the data will undergo preprocessing to identify key decision points in player choices. The statistical analysis will be performed by comparing this data with the initial questionnaire survey results to reveal the correlations related to the research questions.

3.1 Research Process

In this study, Minecraft Java Edition was chosen as the experimental platform, and 39 students from National Yang Ming Chiao Tung University and Tsing Hua University were invited to participate as subjects. Before starting the experiment, the participants will be briefed on the experimental procedures and asked to fill out basic information and learning style questionnaires. Once the briefing and questionnaire are completed, the participants will be allowed to enter the Minecraft map designed for this research and engage in the game. After the game session ends, a brief interview will be conducted, and the participants' game records will be analyzed (see Fig. 1).

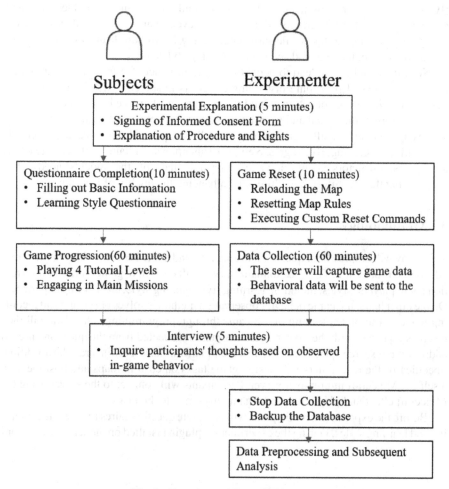

Fig. 1. Experimental Process Diagram.

3.2 Research Tools

3.2.1 Questionnaire Data Collection

The questionnaire survey was conducted using an online platform (Google Forms) and was divided into two parts:

1. Participant Recruitment Questionnaire: This part aimed to gather information about the participants' contact details, age range, gender, and their prior experience with Minecraft game. The consideration of prior knowledge (Prior Knowledge) of players and its potential impact on the study led to the inclusion of questions related to their Minecraft gaming experience.
2. Learning Style Questionnaire: The Index of Learning Styles Questionnaire provided by North Carolina State University was adopted for this part. The questionnaire consists of 40 items, with each item corresponding to one dimension in the learning style model. Respondents were asked to choose one of two options for each item, which would classify them into one of the learning style categories based on the model.

3.2.2 Creation of the Minecraft Map

In this study, the version of Minecraft used was JAVA 1.19.2. The main task assigned to the participants was to repair six automated harvesting machines, which consisted of three automated sugarcane harvesters and three automated bamboo collectors (see Fig. 2).

Before the formal task began, the participants went through four tutorial levels: Planting Tutorial, Transportation Tutorial, Component Tutorial, and Projection Tutorial. The purpose of designing these tutorial levels was not only to teach players how to use the tools designed for this experiment but also to minimize the knowledge gap among different players. The aim was to ensure that all participants had a relatively equal understanding of various aspects in the game, such as redstone mechanics and crop cultivation.

During the formal task, players earned points by completing the machines, which would be converted from harvested sugarcane and bamboo. These points could be used to exchange for projection tools or traded with in-game villagers for materials, tools, component items, material pack items, instructional books, and other items.

3.2.3 Design of the Scaffolding

In this study, four different types of scaffolding were designed, including Projection Tools, Module Scaffolding, Instructional Books, and Material Packs. The Instructional Books scaffolding provided textual instructions to guide players in creating the mechanisms, allowing them to read relevant tutorials and grasp the concepts needed to complete the machine's construction. The Material Packs scaffolding bundled all the materials required to build the mechanisms, reducing players' cognitive load when purchasing materials and serving as a reminder of the necessary items for machine construction.

The Projection Tools were available for purchase and appeared in the player's inventory, with unlimited uses. When used, a floating projection appeared around the player,

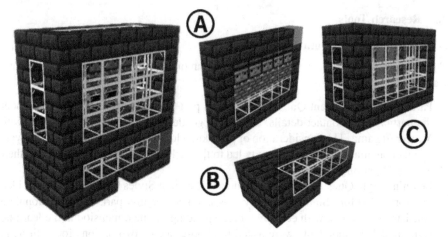

Fig. 2. Automated Collectors (A: Destruct Module, B: Collect Module, C: Plant Module).

displaying the machine's components in layers. Each time the tool was used, a new layer was added to the projection, gradually revealing the complete structure of the machine. Players could browse through each layer of the projection to understand their construction progress.

The Module Scaffolding divided the automated harvesting machines into three parts: Collection, Planting, and Destruction. Players could interact with the corresponding colored blocks using the module tool held in their hands to directly complete specific sections of the machine's structure.

3.2.4 Game Process Recording

In this experiment, a custom plugin was used to capture various events triggered by players in the game on the server-side. Each event data was timestamped, and when the server received these events triggered by the players, they were written into a data table. After the completion of the experiment for each participant, a Python script was run to copy the data from the data table into individual tables specific to each participant. These collected event data were preprocessed to convert them into an analyzable format and were subsequently subjected to further analysis.

4 Results and Analysis

In this chapter, we will conduct correlation analysis between the four dimensions of learning style and game style of the participants and their timing and frequency of purchasing and using scaffolds in the game. As each dimension in the learning style has two complementary tendencies in the scores, the correlation analysis will only differ in the sign of the correlation coefficient. Therefore, we will focus on the "active", "sensing", "visual", and "sequential" tendencies as positive aspects for each learning style in the analysis. As for the purchase of the instruction book scaffold, even though it is a binary value, it should have been analyzed using the point-biserial correlation coefficient. However, since it mathematically equals the Pearson correlation coefficient, we will use the Pearson correlation coefficient for analysis.

4.1 Learning Style and Module Scaffold: Correlation Analysis

After conducting cross-analysis between each variable and the purchase and usage behavior of the module scaffold, the correlation coefficients and p-values of each variable are summarized in Tables 1, 2, 3, 4, 5 and 6. It can be observed that, except for two groups, "Input - Module Usage Frequency" and "Understanding - Module Usage Frequency", which have a significant moderate negative correlation, all other variables show no significant results.

From the results, it can be inferred that in the "Input" dimension of the player's learning style, as the tendency approaches "visual" (i.e., further away from "verbal" tendency), when using scaffolds of the same type as the module, there will be a smaller standard deviation, meaning that the interval between continuous use of the scaffold will be shorter, and the frequency will be higher. This suggests that in digital games, modularity means that after players use a module, the machine will have a significant visual difference from the previous state, stimulating players with "visual" learning style to have a higher willingness to use the scaffold again.

As for the "Understanding" dimension of the player's learning style, as the tendency approaches "sequential" (i.e., further away from "global" tendency), the participants also show a smaller standard deviation in the frequency of using scaffolds. Observing the behavior of participants with different tendencies in this dimension when repairing the machines, it can be noted that participants with a "global" style tend to repair multiple machines simultaneously, rather than completing one before moving on to the next. Their repair approach also alternates between "using modules" and "manual repair", resulting in more dispersed time intervals for using modules.

Table 1. Learning Style and Correlation Analysis of "Module Purchase Quantity".

Dimension	Correlation Coefficient	p-value
Processing (Active–Reflective)	−0.117	0.503
Perception (Sensory–Intuitive)	−0.108	0.539
Input (Visual–Verbal)	−0.163	0.350
Understanding (Sequential–Global)	−0.034	0.846

4.2 Learning Style and Projection Scaffold: Correlation Analysis

After conducting cross-analysis between each variable and the purchase and usage behavior of the projection scaffold, the correlation coefficients and p-values are summarized in Tables 7, 8, 9, 10 and 11. It can be observed that all p-values between the variables are

Table 2. Learning Style and Correlation Analysis of "Module Purchase Average Time Point".

Dimension	Correlation Coefficient	p-value
Processing (Active–Reflective)	0.291	0.125
Perception (Sensory–Intuitive)	0.056	0.773
Input (Visual–Verbal)	−0.122	0.528
Understanding (Sequential–Global)	0.320	0.090

Table 3. Learning Style and Correlation Analysis of "Module Purchase Frequency".

Dimension	Correlation Coefficient	p-value
Processing (Active–Reflective)	−0.007	0.973
Perception (Sensory–Intuitive)	0.056	0.773
Input (Visual–Verbal)	−0.211	0.272
Understanding (Sequential–Global)	−0.151	0.436

Table 4. Learning Style and Correlation Analysis of "Module Usage Count".

Dimension	Correlation Coefficient	p-value
Processing (Active–Reflective)	−0.165	0.343
Perception (Sensory–Intuitive)	−0.184	0.290
Input (Visual–Verbal)	−0.203	0.241
Understanding (Sequential–Global)	−0.207	0.233

Table 5. Learning Style and Correlation Analysis of "Module Usage Average Time Point".

Dimension	Correlation Coefficient	p-value
Processing (Active–Reflective)	0.288	0.145
Perception (Sensory–Intuitive)	−0.101	0.617
Input (Visual–Verbal)	−0.186	0.353
Understanding (Sequential–Global)	0.150	0.455

greater than 0.05. Therefore, it can be concluded that there is no significant correlation between players' learning style, game style, and the purchase and usage of the projection scaffold.

In other words, the results suggest that players' learning and game styles do not have a significant influence on their decision to purchase or use the projection scaffold

Table 6. Learning Style and Correlation Analysis of "Module Usage Frequency".

Dimension	Correlation Coefficient	p-value
Processing (Active–Reflective)	−0.243	0.223
Perception (Sensory–Intuitive)	−0.102	0.613
Input (Visual–Verbal)	−0.437	0.023 **
Understanding (Sequential–Global)	−0.438	0.022 **

** $p < 0.05$ indicates a statistically significant correlation

in the game. The data show that these factors do not play a significant role in determining whether players choose to utilize the projection scaffold during their gameplay experiences.

Table 7. Learning Style and Correlation Analysis of "Projection Scaffold Purchase Quantity".

Dimension	Correlation Coefficient	p-value
Processing (Active–Reflective)	−0.202	0.245
Perception (Sensory–Intuitive)	0.105	0.548
Input (Visual–Verbal)	0.003	0.986
Understanding (Sequential–Global)	0.106	0.544

Table 8. Learning Style and Correlation Analysis of "Projection Scaffold Purchase Frequency".

Dimension	Correlation Coefficient	p-value
Processing (Active–Reflective)	0.105	0.659
Perception (Sensory–Intuitive)	0.292	0.212
Input (Visual–Verbal)	0.131	0.582
Understanding (Sequential–Global)	0.426	0.061

4.3 Learning Style and Material Pack Scaffold: Correlation Analysis

By conducting cross-analysis between each variable and the purchase behavior of the material pack scaffold, the correlation coefficients and p-values are summarized in Tables 12, 13 and 14. It can be observed that except for the "Input - Material Pack Scaffold Purchase Count" group, which shows a significant moderate negative correlation, there are no significant results in the other variables.

From the results, it can be inferred that in the "Input" dimension of players' learning style, the closer they are to the "Visual" tendency (i.e., further away from the "Verbal"

Table 9. Learning Style and Correlation Analysis of "Projection Scaffold Usage Count".

Dimension	Correlation Coefficient	p-value
Processing (Active–Reflective)	−0.181	0.299
Perception (Sensory–Intuitive)	0.229	0.186
Input (Visual–Verbal)	0.048	0.783
Understanding (Sequential–Global)	−0.034	0.845

Table 10. Learning Style and Correlation Analysis of "Projection Scaffold Usage Average Time Point".

Dimension	Correlation Coefficient	p-value
Processing (Active–Reflective)	−0.272	0.307
Perception (Sensory–Intuitive)	−0.113	0.676
Input (Visual–Verbal)	−0.168	0.534
Understanding (Sequential–Global)	0.022	0.935

Table 11. Learning Style and Correlation Analysis of "Projection Scaffold Usage Frequency".

Dimension	Correlation Coefficient	p-value
Processing (Active–Reflective)	−0.079	0.772
Perception (Sensory–Intuitive)	−0.076	0.780
Input (Visual–Verbal)	−0.117	0.666
Understanding (Sequential–Global)	0.025	0.925

tendency), the fewer material pack scaffolds they tend to purchase. The possible reason for this is as follows: Although most of the input in the game comes from visual cues, compared to other visual-based tools such as modules or projections, the material pack does not provide immediate visual feedback. Instead, players need to understand the positions and uses of the various materials provided, which is analogous to processing text by decoding characters into meaningful words. As a result, learners with a visual learning style are less inclined to use this type of scaffold.

4.4 Learning Style and Instructional Book Scaffold: Correlation Analysis

By conducting cross-analysis between each variable and the purchase and usage behavior of the instructional book scaffold, the correlation coefficients and p-values are summarized in Tables 15, 16, 17, 18 and 19. It can be observed that all p-values are greater than 0.05, indicating that there is no significant correlation between players' learning style and game style with the purchase and usage of the instructional book scaffold.

Table 12. Correlation Analysis of Learning Styles with "Material Pack Scaffold Purchase Quantity".

Dimension	Correlation Coefficient	p-value
Processing (Active–Reflective)	−0.138	0.428
Perception (Sensory–Intuitive)	−0.239	0.167
Input (Visual–Verbal)	−0.386	0.022 **
Understanding (Sequential–Global)	−0.122	0.486

** $p < 0.05$ indicates a statistically significant correlation

Table 13. Correlation Analysis of Learning Styles with "Material Pack Scaffold Average Purchase Time Point"

Dimension	Correlation Coefficient	p-value
Processing (Active–Reflective)	−0.073	0.753
Perception (Sensory–Intuitive)	0.149	0.519
Input (Visual–Verbal)	0.241	0.293
Understanding (Sequential–Global)	0.095	0.683

Table 14. Correlation Analysis of Learning Styles with "Material Pack Scaffold Purchase Frequency"

Dimension	Correlation Coefficient	p-value
Processing (Active–Reflective)	−0.081	0.728
Perception (Sensory–Intuitive)	−0.174	0.450
Input (Visual–Verbal)	−0.123	0.594
Understanding (Sequential–Global)	−0.174	0.450

From the results, it can be concluded that players' learning style and game style do not have a significant association with their purchase and usage behavior of the instructional book scaffold. This suggests that players' preferences for learning and game interaction styles do not significantly influence their inclination to use instructional books as scaffolding tools in the game.

Table 15. Correlation Analysis of Learning Styles with "Purchase of Instruction Manual Scaffold".

Dimension	Correlation Coefficient	p-value
Processing (Active–Reflective)	0.143	0.412
Perception (Sensory–Intuitive)	−0.144	0.409
Input (Visual–Verbal)	−0.160	0.359
Understanding (Sequential–Global)	0.012	0.945

Table 16. Correlation Analysis of Learning Styles with "Instruction Manual Scaffold Purchase Time Point"

Dimension	Correlation Coefficient	p-value
Processing (Active–Reflective)	0.225	0.560
Perception (Sensory–Intuitive)	0.641	0.063
Input (Visual–Verbal)	−0.155	0.691
Understanding (Sequential–Global)	0.093	0.811

Table 17. Correlation Analysis of Learning Styles with "Instruction Manual Scaffold Usage Count".

Dimension	Correlation Coefficient	p-value
Processing (Active–Reflective)	0.132	0.449
Perception (Sensory–Intuitive)	−0.208	0.230
Input (Visual–Verbal)	−0.202	0.245
Understanding (Sequential–Global)	−0.078	0.655

Table 18. Correlation Analysis of Learning Styles with "Instruction Manual Scaffold Average Usage Time Point".

Dimension	Correlation Coefficient	p-value
Processing (Active–Reflective)	0.061	0.875
Perception (Sensory–Intuitive)	0.084	0.830
Input (Visual–Verbal)	−0.175	0.652
Understanding (Sequential–Global)	−0.436	0.240

Table 19. Correlation Analysis of Learning Styles with "Instruction Manual Scaffold Usage Frequency".

Dimension	Correlation Coefficient	p-value
Processing (Active–Reflective)	−0.051	0.897
Perception (Sensory–Intuitive)	−0.610	0.081
Input (Visual–Verbal)	0.480	0.191
Understanding (Sequential–Global)	−0.142	0.716

5 Conclusion

In game-based learning, scaffolding serves as a temporary tool to assist learners in overcoming their cognitive limits and advancing their knowledge to higher levels or achieving various tasks and challenges in the game while acquiring game rules and skills. However, designing effective learning scaffolds requires consideration of various factors. Therefore, this study aimed to explore the influence of different learning styles on players' behaviors in purchasing and using scaffolds in the game. Additionally, the study provided a comprehensive process for recording, analyzing, and examining the relationships between various types of scaffolds and players' learning styles in game-based learning.

The study designed four types of scaffolds in the game, including modules, projection models, instructional books, and material packs, and collected players' behavioral data on scaffold purchase and usage, including the frequency and standard deviation of purchases and usage. The FSLSM learning style questionnaire was employed to categorize players' learning styles, and the correlation analysis was used to examine the relationships among variables.

The results of the correlation analysis showed that except for three pairs, "Visual - Module Usage Frequency", "Sequential - Module Usage Frequency", and "Visual - Material Pack Purchase Quantity", which exhibited moderate negative correlations, there were no significant correlations between the "Sensing", "Processing" learning style dimensions, and game style dimensions and players' behaviors in purchasing and using scaffolds in the game.

Since redstone circuits involve logical and engineering principles, this study not only verified Felder's [10] findings that most engineering students tend to have sensing, visual, inductive, active, and global learning styles but also suggested that in game-based learning of this type, providing more visualized and detailed scaffolds, rather than textual descriptions, could motivate learners to actively use scaffolds and enhance learning quality.

References

1. Qian, M., Clark, K.R.: Game-based learning and 21st century skills: a review of recent research. Comput. Hum. Behav. **63**, 50–58 (2016)

2. Spadafora, N., Downes, T.: Scaffolding in learning. In: Shackelford, T.K., Weekes-Shackelford, V.A. (eds.) Encyclopedia of Evolutionary Psychological Science, pp. 1–4. Springer International Publishing, Cham (2019). https://doi.org/10.1007/978-3-319-16999-6_1350-1

3. Quintana, C., et al.: A scaffolding design framework for software to support science inquiry. J. Learn. Sci. 13(3), 337–386 (2004). https://doi.org/10.1207/s15327809jls1303_4

4. Faisal, R.: Influence of personality and learning styles in English language achievement. Open J. Soc. Sci. 7, 304–324 (2019)

5. Richardson, R.C., Arker, E.: Personalities in the classroom: making the most of them. Kappa Delta Pi Record 46(2), 76–81 (2010)

6. Gros, B.: Digital games in education. J. Res. Technol. Educ. 40, 23–38 (2007)

7. Al-Azawi, R., Al-Faliti, F., Al-Blushi, M.: Educational gamification vs. game based learning: comparative study. Int. J. Innov. Manage. Technol. 7, 131–136 (2016). https://doi.org/10.18178/ijimt.2016.7.4.659

8. Wood, D., Bruner, J.S., Ross, G.: The role of tutoring in problem solving. Child Psychol. Psychiatry Allied Disciplines 17, 89–100 (1976)

9. Ersani, N.P.D., et al.: Schemes of scaffolding in online education. RETORIKA: Jurnal Ilmu Bahasa 7(1), 10–18 (2021)

10. Felder, R.M.: Learning and teaching styles in engineering education. Eng. Educ. 78(7), 674–681 (1988)

Real-Time Simulation for Controlling the Mobility of an Unmanned Ground Vehicle Based on Robot Operating System

Abubaker Ahmed H. Badi⬤, Salinda Buyamin$^{(\boxtimes)}$⬤,
Mohamad Shukri Zainal Abidin⬤, and Fazilah Hassan⬤

Department of Control and Mechatronics, Faculty of Electrical Engineering,
Universiti Teknologi Malaysia, Johor Bahru 81310, Johor, Malaysia
salinda@utm.my

Abstract. Robot Operating System (ROS) is a framework platform that runs on UNIX operating systems and enables more efficient operation and administration of different kinds of robots. The Jaguar 4×4 Wheel Mobile Robot by Dr. Robot Inc. is one example of an unmanned ground vehicle (UGV) that can be controlled using ROS. Dr. Robot Inc. has released a Windows-based software development kit (SDK) and open-source code to allow developers to develop this kind of robot. However, to manipulate this robot's control and monitoring by ROS, a set of Linux command lines should be executed to launch the robot without the ability to use Dr. Robot's SDK. Therefore, in order to benefit from the provided open-source code with ROS packages with Python and to avoid the need for running multiple Linux command lines, this paper presents a Python-based SDK and a real-time simulation for controlling and monitoring the movement of the Jaguar robot. The methodology technique that is considered in this paper is to analyze the open-source code to access all the robot's sensor data and nodes using the proposed SDK, then simulate the hardware (the Jaguar Mobile Robot) with ROS Visualization (RViz) 3D software in real-time. The practical experiment is carried out by using a keyboard and a gamepad as controllers to achieve mobility control. The results show the robot's position and orientation visualized by RViz, as well as its turn angle (theta), linear and angular velocities in a 2D space x-y plot. Using the RViz visualization tool and with the aid of the proposed SDK, it is straightforward to access and monitor all topics and nodes in real-time. Consequently, algorithms such as obstacle avoidance and object tracking could be readily tested and implemented in future works.

Keywords: Robot Operating System (ROS) · ROS Topics · ROS Visualization (RViz) · Unmanned Ground Vehicles (UGVs)

1 Introduction

Robot Operating System (ROS) [1] is a framework platform that runs on UNIX operating systems and enables more efficient operation and administration of

F. Hassan et al. (Eds.): AsiaSim 2023, CCIS 1912, pp. 377–395, 2024.
https://doi.org/10.1007/978-981-99-7243-2_32

robots. It provides various tools and libraries for programmers, including access to standardized hardware devices across multiple computers. Additionally, ROS helps organize code into packages so it can be easily shared between different applications or machines [2].

Willow Garage [3] and Stanford University have developed the ROS [4] as a free and open-source robotic middleware for the extensive creation of complex robotic systems as part of the STAIR project. The main advantage of ROS is that it eliminates the need for dealing with hardware drivers and enables manipulation of robot sensor data as a labeled abstract data stream, known as a "topic". As a result, software developers do not have to deal with hardware drivers and interfaces when programming robots [1,3]. Because of its flexibility and processing reliability, the ROS platform is used to control the majority of UGVs, unmanned aerial vehicles (UAVs), unmanned underwater vehicles (UUVs) and unmanned surface vehicles (USVs) [5].

The Jaguar 4×4 Wheel Mobile Robot [6] is one example of UGVs based on the ROS platform. Dr. Robot Inc. was founded in 2001 by a group of scientists and engineers who had prior experience in designing robotic systems for various space agencies. Their main goal is to offer economical and efficient robotic systems that cater to the needs of industrial, commercial and academic consumers. Dr. Robot has been at the forefront of conducting research in the field of dynamic bipedal robotics and autonomous navigation since its inception. Robots have been developed by their team to cater to diverse applications and a wide spectrum of clientele [6]. To give developers a chance to control this kind of robot, [6] has released open-source code that can be implemented by different programming packages on UNIX or Windows operating systems.

In order to gain benefits from the ROS-based packages and the open-source code provided by Dr. Robot, this paper presented a Python-based SDK and a real-time simulation for controlling and visualizing the movement of the Jaguar robot. The main objective of this paper is to examine the open-source code provided by [6] to communicate with and control the Jaguar robot using Python and simulate the movement using ROS Visualization (RViz) 3D software in real-time operating mode. Python is a popular interpreted language with thousands of libraries, such as those for artificial intelligence, image processing, and optimization. Because of this, the approach that has been taken is to develop the primary open-source code so that Python can access all sensor data and nodes through the suggested SDK. This was done by writing publisher and subscriber nodes in Python that connect to the main open-source code. The robot's movement is then simulated in real-time by linking the proposed SDK to RViz software. RViz is a software tool that enables the visualization of robots in 3D space. It is a very flexible tool with a wide variety of visualizations and plugins [7].

The rest of the paper is organized as follows: Related works are presented in Sect. 2. The Jaguar 4×4 wheel robot's hardware architecture is described in Sect. 3. Section 4 illustrates the network configuration, ROS packages required to visualize and control the robot and the proposed Jaguar Robot SDK. The experimental findings of our work are discussed in Sect. 5. Section 6 concludes with conclusions and recommendations for future research.

2   Related Works

Dr. Robot Inc. has created a SDK for the purpose of controlling and supervising the Jaguar robot [6], but it is exclusively compatible with Windows operating system. Therefore, the only way to interact with the Jaguar Robot using ROS is by executing a series of Linux command lines, which are very difficult to handle by the operator or normal user. Another SDK was presented by [8] which is also based on Windows. On the other hand, [9] has used the same robot based on ROS, but the controlling operation was done by applying Linux command lines. Therefore, as mentioned in Sect. 1, in order to interface the Jaguar robot using Python and to simulate the movement using RViz software in real-time, this work will look in depth at the open-source code offered by [6] to attain the objective. Many input methods have been employed for the robot's movement, either through manual intervention such as a keypad, game-pad controller (joystick), and touch screen or autonomously by utilizing various algorithms in this field [8–12].

3   Jaguar 4 × 4 Wheel Robot Hardware Architecture

This section presents the primary and supplementary hardware components of Jaguar mobile robot.

3.1   Hardware Architecture

The Jaguar 4×4 platform is designed to be used for indoor and outdoor operations requiring fast maneuverability and high ground guard. In addition, it is designed to cope with atmospheric changes being water resistant. Moreover, it is ideal for operations on rugged terrain being able to climb stairs up to 110 mm high. The control of the robot is made through a wireless connection, for space orientation is used a global positioning system (GPS) for the outside, an inertial measurement unit (IMU) sensor with 9°C of freedom [6]. The main characteristics are:

- Practical mobile platform for both indoor and outdoor applications
- High maneuverability
- Resistant to water and atmospheric changes
- Can go over stairs with a maximum height of 110 mm
- Wireless connection

The hardware structure of the robot connects sensors (GPS, IMU), actuators and network modules to enable communication capabilities. An ARM®
Cortex®-M4F Based MCU TM4C123G Launchpad Evaluation Kit is the main
controller [13]. Figure 1 illustrates a graphical representation of the connections
that are made between the primary controller and the modules. This diagram
also includes supplementary components such as a laser scanner device. Table 1
contains a listing of the Jaguar's core components, while Table 2 lists the optional
components of the Jaguar robot [14].

Table 1. Jaguar's core components [14].

Device	Description	No
JAGUAR4 × 4W-ME	Jaguar 4 × 4 Wheel Chassis (including motors and encoders)	1
PMS5005-J4W	Motion and Sensing Controller (Jaguar 4 × 4 Wheel Version)	1
WFS802G	WiFi 802.11b/g Wireless Module	1
DMD1200	12A (peak 25A) Dual-channel DC Motor Driver Module	2
PMCHR12	DC-DC Power Board	1
AXCAM-A	640 × 480 Networked Color Camera (max. 30fps) with Two-Way Audio	1
OGPS501	Outdoor GPS Receiver with 5 Hz Update Rate and WAAS	1
IMU9000	9 degrees of freedom (DOF) IMU (Gyro/ Accelerometer/ Compass)	1
WRT802G	802.11b/g wireless AP/router	1
BPN-LP-10	22.2 V 10 AH LiPo Battery Pack	1
LPBC5000	2A LiPo Battery Charger	1
GPC0010	Gamepad Controller	1

Table 2. Jaguar's optional components [14].

Description	Device
Laser Scanner (Range 4 m) for Indoor Application	LAS04M
Laser Scanner (Range 5.6 m) for Indoor Application	LAS04M
Laser Scanner (Range 30 m) for Outdoor Application	LAS30M
22.2V 20 AH Li-Polymer Battery Pack Upgrade	BPN-LP-20
Head Mounted Display (800 × 600)	HMD8H6H
802.11N Wireless AP/Router	WRT802N
Host Controller PC	HCPC1008

3.2   The PMS5006 Controller

The PMS5006 controller is the main component of the Jaguar 4 × 4 robot platform. It communicates via a serial UART port with the host computer. The

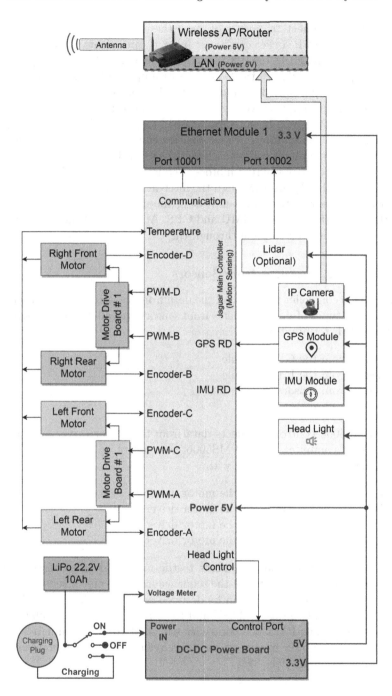

Fig. 1. Inter-connection between the main controller and modules.

serial port will connect to port 1 of the Ethernet network module. From the host computer, a TCP socket can be set up to communicate with this controller using the IP address 192.168.0.XX and port 10001. (XX-varies depending on the robotic platform. The PMS5006 can control up to 4 SDC2130 motor drivers via UART serial ports 2, 3, 4 and 5 [15]. The controller can read the motor current, power consumption, voltage, motors temperature, encoder data and the state of the motor driver. It can support open loop control, closed loop encoder speed control and closed loop encoder position control. The default mode for the motor driver is open loop control, where driver 1 will control the front motors and driver 2 will control the rear motors. The PMS5006 controller reads all data from the 9-DOF IMU sensor through the inter-integrated circuit (I2C) port. The orientation of the robot is estimated using the direction cosine matrix (DCM) algorithm using data from the IMU and GPS. At the same time, it will read the messages from the GPS module through the UART port.

3.3   Decoding Messages from Sensors

The PMS5006 controller will send all the data from the sensors to the user. There are four main types of data messages from sensors and all of these messages end with the CRLF termination.

- Messages from the GPS Module
- Messages from the IMU Module
- Messages from sensors
- Messages from motor drivers

The PMS5006 controller collects data from the motor driver and transmits it to the host computer. Since PMS5006 can control up to 4 SDC2130 motor drivers, these messages will start with:

- MM0 - data is received from the motor driver 1
- MM1 - data is received from the motor driver 2
- MM2 - data is received from the motor driver 3
- MM3 - data is received from the motor driver 4

The controller will send commands to the motor driver to access the data from their sensors. An example of the sequence from the motor sensors: "MM1 V = 105,228,4830". The message can be decoded as follows: the voltage for driver 2 is 10.5V, the total voltage is 22.8V and the output of the 5V regulator on the board is 4.830V. The PMS5006 controller will send 10 loop commands for each motor driver at a 50 Hz frequency, so the refresh rate for each command from the sensor is approximately 5 Hz.

3.4   Control Commands

All system or motor controls are delimited by the CRLF end. The following lines will illustrate the controls for the Jaguar 4X4 mobile platform [6]:

- Ping Command **Format:ping**
- GPIO control **Format:SYS GPO n**
- Power Control Command **Format: SYS MMC n**
- Control of the IMU sensor **Format: SYS CAL**
- Control of GPS **Format: EGPS n**
- Set the initial value of DCM **Format: DCM n**
- Motor Control Commands

Because PMS5006 can control up to 4 SDC2130 motor drivers, there are 5 types of motor control commands, examples are listed in Table 3.

Table 3. Examples of 5 commands for Jaguar robot's DC motors [14].

Command	Effect
"MM0 !G 0 220"	Control only the left front motor
"MM0 !G 1 -240"	Control only the right front motor
"MM1 !G 0 170"	Control only the left rear motor
"MM1 !G 1 -156"	Control only the right rear motor
"MMW !M 205 -205"	Control all four motors at the same time and cause the robot to move forward
"MMW !EX"	Stop all motors on the Jaguar 4 × 4 platform
"MMW !MG"	Give access to control all motors ("MMW !M 0 0")

3.5 Motors Control Modes

For each motor, the controller supports several modes of operation. In the initial configuration of the controller, the open loop control mode is set for each motor. In open loop control mode, the controller drives the motor using a power proportional to the information in the control, the motor speed is not measured. This way, the motor will slow down if there is a change in load such as encountering an obstacle or changing the angle of a ramp. This mode is suitable for most applications where the operation visual contact with the robot. In this mode an optical encoder is used to measure the current speed of the motor. If the speed changes due to a change in load, the controller will automatically offset the output power. This mode is preferred in precision control mode or in autonomous robot applications. On the other hand, the Closed loop control mode with encoder speed control and closed loop mode with encoder position control.

3.6 Optical Encoder

Another important hardware component in the Jaguar Robot is the optical encoder that is integrated into the motor shafts. These modules help measure the distance traveled by a motor and its speed of rotation. The rotational speed can be calculated by counting the number of pulses delivered over a given time period. Because incremental encoders are digital devices, they will measure speed and distance with high accuracy. Since the motors can move both backward and forward, it is necessary to differentiate how pulses are counted so that they decrease or increase a number depending on two channels, A and B, with a 90-degree phase shift. [15].

3.7 Power Supply

Jaguar's robotic platform has a lot of different hardware parts that can be used to make many different kinds of apps. The sensors (GPS and IMU), actuators and network modules are all connected by the robot's hardware structure. The way that feeds all the components is the DC-DC power board that receives 22.2 V at the input from the battery when the switch is in the ON position and provides at the output a voltage of 5 V from which the controller, headlights, video cameras, router and IMU and GPS sensors are powered. The power board also provides an output and 3.3 V of voltage for the power of the Ethernet module. The two motor drivers have a dual character, resulting in independent handling of each motor. The first driver controls the left- and right-back motors and the second controls the left- and right-front motors.

4   Network Configuration, ROS Package and the Proposed SDK

4.1 Network Configuration

A serial-to-ethernet module [16] is attached serially to the launchpad in order to transmit and receive data from and to the MCU board. So, this Ethernet module is a vital element for connecting the hardware (robot main board) to the software on the host PC. On the other hand, the Ethernet module (port 10001) is attached to a LAN switch along with the IP camera and the access point device. The IP camera is located at the base of the robot with resolution of 640X480 pixels, captures 30 frames per second and is able to record audio [17]. Table 4 lists the network settings for Jaguar Robot.

4.2 ROS Control and Visualization Packages

The ROS Noetic framework has been installed on Ubuntu 20.04 Linux operating systems [18]. After installation, a new "catkin" workspace must be created to build the open-source codes created by [6]. All publishers and subscribers nodes

Table 4. Jaguar robot's network settings [14].

Ethernet Module 1	192.168.0.60
Port 1	Port Number 10001, UDP 115200. 8, N, 1, no flow control
Port 2	Not Used
Camera	192.168.0.75 Port 8081
User ID	root
Password	drrobot

can be displayed by using **rqt_graph** Graphical User Interface (GUI) which is a plugin for displaying the ROS computation graph. Its components are made generic so that other packages requiring graph representation may rely on this package [19]. Moreover, sensor and pose signals can be viewed by **rqt_plot** GUI [20]. In addition, the joy package consists of (**joy_node**), which serves as a node that facilitates the interaction between a Linux joystick to ROS. This node publishes a message labeled as "Joy" that include the recent status of every button and axis of the joystick [21]. Another important package is RViz **rviz_node**, which is used for robot simulation in 3D space. To startup all nodes, the Linux command line **roscore** [22] must be first initiated, then each node must be executed one by one using the command line **rosrun** [23] followed by package name and the desired node name. The disadvantage of using **rosrun** command line is that the user will spend long time to run all the nodes manually. **rosrun <package> <node>**

Therefore, **roslaunch** command line [24] is considered as an alternative method to run **roscore** and overall nodes that are declared in the launch file respectively. **roslaunch <package> <launchfile>.launch**

In our work, the launch file contains the following nodes is created to simulate and control the Jaguar robot's movement [4].

- jaguar4 × 4_2014_node
- drrobot_keyboard_teleop_node
- drrobot_ros_sensor_node
- rqt_graph_node
- rqt_plot_node
- joy_node
- rviz_node

4.3 The Proposed SDK and RViz Simulator Tool

The proposed Python-based SDK environment shown in Fig. 2 was created to simplify robot control by the user. One of the primary aspects of this SDK is that it includes publisher and subscriber topics that are used to connect to the robot core topics provided by [6]. Moreover, it collects all the robot's sensor data and messages to visualize the robot in the RViz software tool. The user can control

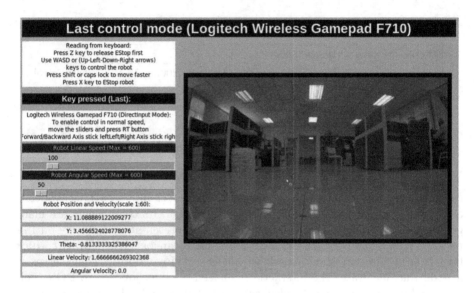

Fig. 2. The proposed Jaguar robot SDK.

the robot's movement using a keyboard or game-pad controller. The user can also adjust the speed for both linear and angular directions by using the two sliders in the SDK. Moreover, the user can track the robot's movements via the SDK by monitoring the IP camera installed on the robot. In addition, the proposed SDK displays the robot's status, including its position, current linear and angular velocities. On the other hand, the proposed SDK is designed to interface with Dr. Robot's open-source code, allowing it to both send and receive commands and data through the creation of new publisher and subscriber topics. Hence, all necessary topics and nodes can be accessed using this SDK to be ready for simulation on RViz. Figure 3 is a flowchart depicting the steps required to execute all ROS packages, including the SDK. The user must first turn on the robot and then establish a wireless connection between the robot and the host PC. If the connection cannot be established, the user must retry until the connection is verified. In the second step, two terminals are required to be opened, one of which corresponds to launching the ros packages and the other to execute the Python script for the proposed SDK. If neither the ROS packages nor the SDK is terminated by the user, the system will continue to operate and await the operator's commands to move the robot; otherwise, the system is terminated.

The controlling techniques include utilizing the keyboard's WASD or arrow buttons and a game-pad controller to manipulate the movement operations. This work's methodology was derived from the open-source code released by [6]. The main open-source code has been modified so that data may be read from and written to the proposed SDK. Furthermore, to be able to communicate with and receive joystick data from and to the SDK environment, another rewritten piece of code has been modified further. For visualization, the data which was received

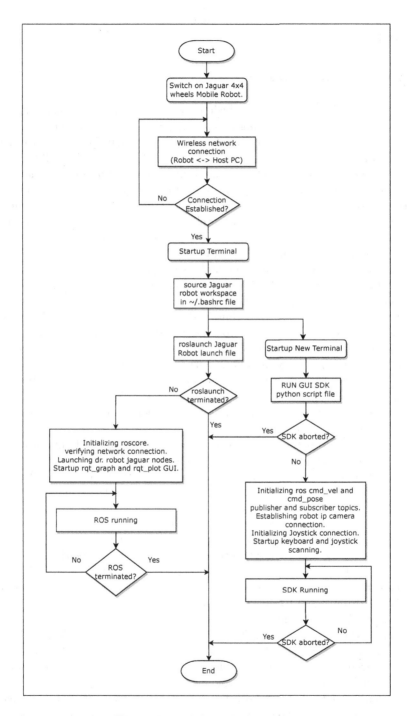

Fig. 3. Operating mode flowchart.

from the robot is processed and then broadcast to the RViz using **tf Transform** package [25].

5   Experimental Results

To startup the system, the steps which are mentioned in the flowchart in Fig. 3 must be followed. **roslaunch jaguar4x4 all_nodes.launch** command line is executed in the terminal window to startup roscore, jaguar main controller and sensor controller nodes. The SDK publisher and subscriber nodes will be started up after **roscore** and all nodes in the launch file are initialized. The second step is done by running Jaguar SDK node publishes by typing a commanad line **python jaguar_robot_sdk.py** in a new terminal window, after running the SDK, five topics (/sdk_kepad_cmd_vel, /sdk_joy_cmd_vel, /Jaguar_Pose, /Jaguar_point_stamped and /Jaguar_vis_marker) are published. All nodes can be viewed by rqt_graph as shown in Fig. 4.

Both of /sdk_kepad_cmd_vel and /sdk_joy_cmd_vel topics will subscribed to the /Jaguar4x4_2014_node node which is represent the main controller with four publishers topics. The four topics (/drrobot_gps, /drrobot_motorboard, /drrobot_imu and /drrobot_motor) are subscribed to /drrobot_ros_sensor_node node to send the data and messages back to the host PC in consequence to the SDK as callback logging data or messages that can be processed by the SDK. On the other hand, Jaguar_Pose is subscribed to rqt_plot GUI to display the position and turn angle, in addition the linear and angular velocities. The robot can be controlled either by a keyboard or a game-pad controller as shown in Fig. 5, the movement controlling by keypad and joystick nodes can be run simultaneously.

The linear and angular speeds can be adjusted in the SDK by moving the two sliders in the range from 0 to 600 directed to the Twist geometry message. Table 5 lists the robot operation modes, WASD keyboard arrows or WASD keys in both capital or small letters. Using game pad controller is more flixeble for controlling because of the two axis analog sticks, forward/backward by the left stick (up/down) and turning angle left/right by the right stick (left/right).

Figure 6 depicts the robot position (x and y) and turn angle (Theta) in real time by rqt_plot graph GUI, while the linear and angular velocities in real time are shown in Fig. 7 The data posed by Jaguar_Pose topic is callback logged messages that are sent from the robot through IMU sensor, so it is processed by the SDK and published to rqt_plot graph.

Figure 8 illustrates the RViz simulation results for the position and orientation of the Jaguar robot. Figure 9 and Fig. 10 provide additional information, by using geometry_msgs/PointStamped or nav_msgs/Odometry, the robot's movement could be represented graphically in RViz. Point stamped (/Jaguar_point_stamped) topic plots points (dots) on the grid as the robot moves from point to point at both linear and angular velocities. In contrast to the Odometry (/odom) topic, arrows or axes (in this case, the arrows option) are displayed on the grid; the arrows indicate precisely where the robot is located

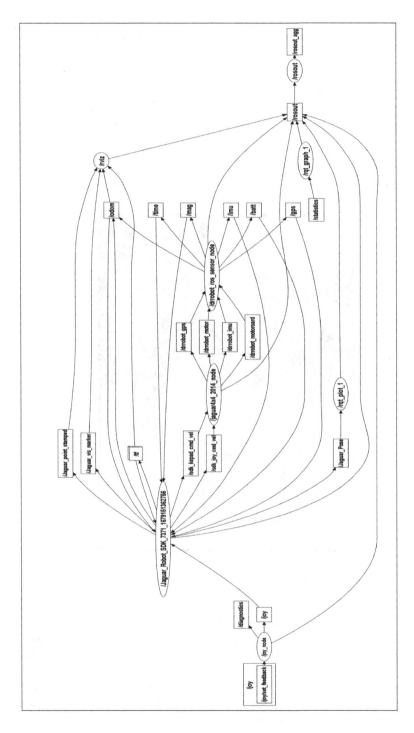

Fig. 4. Jaguar main controller & SDK nodes and topics viewed by rqt_graph GUI.

Table 5. Robot operating modes.

Direction	Keyboard	Game-pad Controller (RB − > Normal Speed) (RT − > High Speed)
Forward	\<w\>, \<W\> or \<Up arrow\>	Up Axis (stick left)
Backward	\<s\>, \<S\> or \<Down arrow\>	Down Axis (stick left)
Turning Left	\<a\>, \<A\> or \<Left arrow\>	Left Axis (stick right)
Turning Right	\<d\>, \<D\> or \<Right arrow\>	Right Axis (stick right)
Robot Stop	\<x\> or \<X\>	LB button
Robot Release	\<z\> or \<Z\>	LT button

Fig. 5. Movement control using a joystick.

(its position and turn angle). As shown in Fig. 10, a Marker visualization message is used to visualize the robot in various shapes and colors. In our project, a directional arrow is also used as a marker. If the robot is in stop or ready mode, a green arrow will appear. The blue arrow indicates that the robot is moving at a normal rate of speed, whereas the red arrow indicates that it is moving at a rapid rate. Both blue and red colors are displayed for linear and angular velocities (forward, reverse, left, or right).

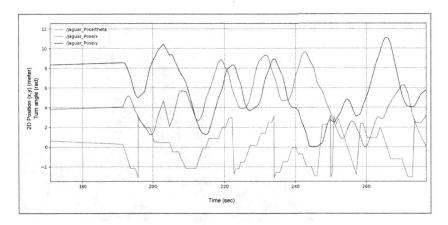

Fig. 6. Jaguar robot pose viewed by rqt_plot GUI.

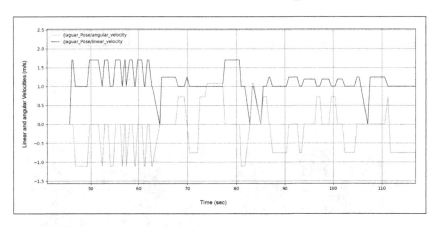

Fig. 7. Linear and angular velocities for the Jaguar robot viewed by rqt_plot GUI.

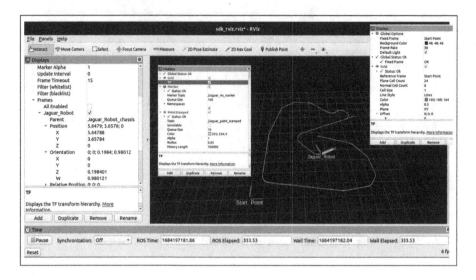

Fig. 8. RViz simulation results (position and orientation of the Jaguar robot).

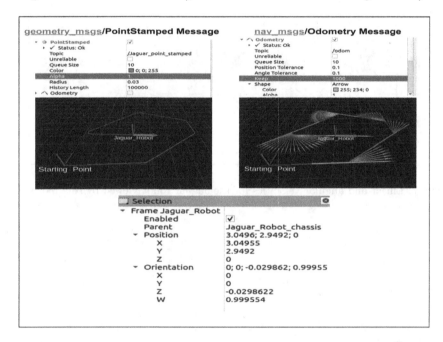

Fig. 9. RViz point stamped and odometry visualization of the Jaguar robot.

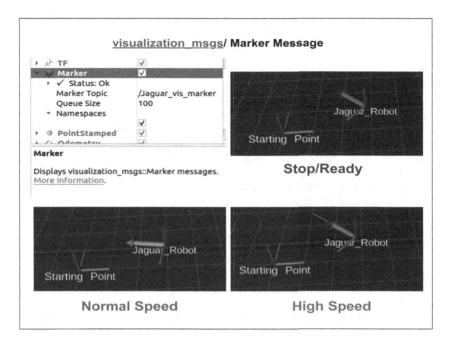

Fig. 10. RViz marker visualization of the Jaguar robot.

6   Conclusion and Future Work

In this paper, simulation processes are implemented to visualize and control the Jaguar 4×4-wheel mobile robot. The simulation is done by the RViz software tool, and the controlling is performed by interfacing Jaguar Robot to the proposed SDK. All the robot's sensor data is utilized by the SDK, and then the simulated results are transmitted to the visualization tool. With this work, the user can use a keyboard or game-pad controller to direct the movement of the robot. Using the two sliders in the SDK, the user may additionally modify the speed for both linear and angular directions. Moreover, the user can use the SDK to track the robot's movements by viewing the IP camera that is attached to it. The Jaguar Robot SDK additionally displays the current position of the robot and its linear and angular velocities. The current position and orientation could also be monitored by the RViz visualization tool in 3D space. The results show the robot's turn angle (Theta), linear and angular velocities in real time by plotting an x and y graph in two-dimensional space. In conclusion, by interfacing the Jaguar SDK with the RViz tool, there is an opportunity to access all nodes in the main open-source code and then simulate them in real-time 3D space. For future works, with the aid of ROS packages and the huge Python libraries, several algorithms such as obstacle avoidance and object tracking, could be easily implemented.

Acknowledgements. The authors are grateful to the Universiti Teknologi Malaysia and the Ministry of Higher Education Malaysia for their support and facilites.

References

1. ROS: Home ros.org. http://www.ros.org, (Accessed Feb 2023)
2. Alajlan, M., Koubâa, A.: Writing global path planners plugins in ros: a tutorial, vol. 625 (2016)
3. Wyrobek, K.A., Berger, E.H., Van der Loos, H.M., Salisbury, J.K.: Towards a personal robotics development platform: rationale and design of an intrinsically safe personal robot. In: IEEE International Conference on Robotics and Automation, pp. 2165–2170. IEEE (2008)
4. Quigley, M., et al.: ROS: an open-source Robot Operating System, vol. 3, p. 5 (2009)
5. Balestrieri, E., Daponte, P., De Vito, L., Lamonaca, F.: Sensors and measurements for unmanned systems: an overview. Sensors **21**(4), 1518 (2021)
6. Dr Robot Inc: WiFi 802.11 robot, Network-based Robot, robotic, robot kit, humanoid robot, OEM solution drrobot.com. http://www.drrobot.com, (Accessed Feb 2023)
7. RViz - ROS Wiki wiki.ros.org. http://wiki.ros.org/rviz, (Accessed Feb 2023)
8. Vaz, D.J., João, F.A., Serralheiro, A.J., Gerald, J.A.: Autopilot-An autonomous navigation system, pp. 1–5 (2016)
9. Wunderlich, S., Schmolz, J., Kühnlenz, K.: Follow me: a simple approach for person identification and tracking, pp. 1609–1614 (2017)
10. Ortega-Garcia, J.L., Gordillo, J.L., Soto, R.: A new method to follow a path on indoor environments applied for mobile robotics. In: 11th IEEE International Conference on Control & Automation (ICCA), pp. 631–636 (2014)
11. Niroui, F., et al.: A graphical user interface for multi-robot control in urban search and rescue applications, pp. 217–222 (2016)
12. Hönig, W., Ayanian, N.: Flying multiple uavs using ros. In: Robot Operating System (ROS) The Complete Reference, vol. 2, pp. 83–118 (2017)
13. EK-TM4C123GXL   Evaluation   board.   https://www.ti.com/tool/EK-TM4C123GXL, (Accessed Feb 2023)
14. Dr Robot Inc: WiFi 802.11 robot, Network-based Robot, robotic, robot kit, humanoid robot, OEM solution jaguar. http://jaguar.drrobot.com/specification_4x4w.asp, (Accessed Feb 2023)
15. Centiva, Documentation Download roboteq.com. https://www.roboteq.com/support/documentation-download, (Accessed Feb 2023)
16. WiPort NR Wireless Serial to Ethernet Device Server Lantronix lantronix.com, https://www.lantronix.com/products/wiport-nr/, (Accessed Feb 2023)
17. AXIS M1065-LW Network Camera Axis Communications axis.com. https://www.axis.com/products/axis-m1065-lw, (Accessed Feb 2023)
18. noetic - ROS Wiki wiki.ros.org. http://wiki.ros.org/noetic, (Accessed Feb 2023)
19. rqt_graph - ROS Wiki wiki.ros.org. http://wiki.ros.org/rqt_graph, (Accessed Feb 2023)
20. rqt_plot - ROS Wiki wiki.ros.org. http://wiki.ros.org/rqt_plot, (Accessed Feb 2023)
21. joy_node - ROS Wiki wiki.ros.org. http://wiki.ros.org/joy, (Accessed Feb 2023)
22. roscore - ROS Wiki wiki.ros.org, http://wiki.ros.org/roscore, (Accessed Feb 2023)

23. rosrun - ROS Wiki wiki.ros.org. http://wiki.ros.org/rosrun, (Accessed Feb 2023)
24. roslaunch - ROS Wiki wiki.ros.org. http://wiki.ros.org/roslaunch, (Accessed Feb 2023)
25. tf - ROS Wiki wiki.ros.org. http://wiki.ros.org/tf, (Accessed Feb 2023)

Scattering Source Determination Using Deep Learning for 2-D Scalar Wave Propagation

Takahiro Saitoh[1]([✉])[iD], Shinji Sasaoka[1], and Sohichi Hirose[2][iD]

[1] Gunma University, 1-5-1, Tenjin, Kiryu, Gunma, Japan
`t-saitoh@gunma-u.ac.jp`
[2] Tokyo Institute of Technology, 2-12-1, Ookayama, Meguro, Tokyo, Japan
`https://civil.ees.st.gunma-u.ac.jp/~applmech/index_e.html`

Abstract. Simulations focusing on wave propagation and inverse problems involving the estimation of a scattering source have been conducted for a long time. As a method for estimating a scattering source of wave propagation, there are known techniques such as time-reversal methods that utilize simulation results and inverse scattering analysis methods based on the Born approximation. However, these methods are mathematically complex and require much computation time. Therefore, in this study, we aim to develop a scattering source estimation method using deep learning, which has been receiving increasing attention in recent years. However, the waveforms used for this scattering source estimation are generated using simulations. In this paper, we first simulate the 2-D scalar waves from the scattering source using the convolution quadrature time-domain boundary element method (CQBEM). The received waveforms at observation points are transformed into image data. These image data are utilized for the deep learning to estimate the actual position of the scattering source. As numerical examples, some unlearned waveforms by a scattering source are given to the created deep learning model and the position and size of the scattering source are estimated to verify the proposed method.

Keywords: Deep learning · Convolution quadrature time-domain boundary element method · Inverse problem · 2-D scalar wave propagation · Ultrasonic non-destructive testing

1 Introduction

In recent years, the importance of non-destructive testing [25] has been increasing, particularly for the purpose of maintaining and managing social infrastructure structures and materials. In particular, the ultrasonic non-destructive testing is one of the most widely used methods in the field of non-destructive testing. Generally, the ultimate goal of the ultrasonic non-destructive testing is to accurately estimate the position, size, and other characteristics of the scattering source, which is typically a defect. To achieve this goal, simulations of wave

F. Hassan et al. (Eds.): AsiaSim 2023, CCIS 1912, pp. 396–407, 2024.
https://doi.org/10.1007/978-981-99-7243-2_33

propagation and scattering source estimation have been conducted for a long time [5,6,13,15]. The time-reversal method [7,8,11,22] and inverse scattering technique [12,28] based on the Born approximation [10] are well known methods for estimating scattering sources. However, time-reversal methods require the measured waveform to be time-reversed and the wave to be re-incident in a simulated region that mimics the actual experiment target. Therefore, estimating scattering sources using the time-reversal method requires much computation time. In addition, the inverse scattering technique based on the Born approximation can accurately reconstruct a scattering source shape and position considering phase information. However, from the perspective of the ultrasonic non-destructive testing site, this method also requires much computation time and demands complex mathematical theories. Therefore, there is a demand for the development of new methods that can easily estimate scattering sources with reduced computational time.

Indeed, in recent years, there has been a growing interest in the machine learning [2] and deep learning [4] in various engineering fields, which forms the foundation of AI(Artificial Intelligence) development. In general, deep learning requires a significant amount of data to be prepared for training purposes [26]. Therefore, using the deep learning to create an AI model requires a substantial amount of time for initial development. However, once the deep learning model is created, the time required for estimating scattering sources can be significantly reduced. Furthermore, in the field of non-destructive testing, there is a growing concern about the future shortage of inspection technicians at job sites. Therefore, there is an increasing momentum to introduce AI into non-destructive testing as a means to address the potential shortage of inspection technicians. Indeed, in recent times, there has been a growing focus on utilizing AI for ultrasonic non-destructive testing [3,9,17,23].

Based on the aforementioned factors, this study proposes a new method that utilizes deep learning to estimate a scattering source in 2-D scalar wave fields. In the following sections, we first provide a brief explanation of the problem we need to solve. Then, we describe the numerical simulation method used in this study, which is the convolution quadrature time-domain boundary element method (CQBEM). Afterwards, we provide a brief explanation of the deep learning and image data used in this study. Then, we demonstrate some numerical examples to examine the effectiveness of our proposed method. Finally, we briefly mention the future research plans.

2   Problem Statement

In this paper, for simplicity, we consider a 2-D scalar wave propagation in a domain D as shown in Fig. 1. In this study, the superposition of ultrasonic waves excited simultaneously by multiple ultrasonic transducers is considered as a plane wave. Therefore, as shown in Fig. 1, we consider the scattering problem caused by a defect (scattering source) with the surface S when an incident plane wave is present. The spacing between each ultrasonic element is assumed to be $0.5a$

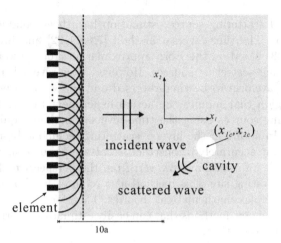

Fig. 1. Forward and inverse scattering analysis model.

and the circular scattering source has a radius r. Furthermore, each ultrasonic element is assumed to be placed on the $x_1 = -10a$ and the center coordinates of a scattering source, which is a defect in the ultrasonic non-destructive testing, are denoted as (x_{1c}, x_{2c}) as shown in Fig. 1.

In this case, the incident plane wave is scattered by a defect, and the scattered waves are received by each ultrasonic element. The displacement of the scalar wave u satisfies the following wave equation and boundary conditions in the time domain:

$$\nabla^2 u = \frac{1}{c^2}\frac{\partial^2 u}{\partial t^2} \quad \text{in } D \tag{1}$$

$$q = \frac{\partial u}{\partial n} = 0 \quad \text{on } S \tag{2}$$

where t represents time, c denotes the scalar wave speed, $\partial/\partial n$ represents the outward normal derivative taken from the region D to the boundary S of the scattering source. The boundary condition in Eq. (2) represents the rigid body condition on the surface S of a scattering source. To determine the displacement $u(\boldsymbol{x}, t)$ in the infinite domain D, the convolution quadrature time-domain boundary element method (CQBEM) [1, 19, 21] discussed in the next section is utilized in this study.

3   Convolution Quadrature Time-Domain Boundary Element Method (CQBEM)

The CQBEM is a relatively new time-domain boundary element method that allows for stable numerical solutions compared to the conventional time-domain boundary element method (BEM) [29]. A significant difference in the formulation compared to the classical time-domain BEM is that the CQBEM uses the convolution quadrature method (CQM) [16] for the time discretization of the time-domain boundary integral equation. In this paper, we briefly describe the basic equations. For more detail formulations, see the papers [1,21,24].

3.1   CQBEM Formulation for 2-D Scalar Wave Propagation

Assuming the zero-initial condition, the scalar wave displacement $u(\boldsymbol{x},t)$ in the domain D can be obtained using the time-domain boundary integral equation as follows:

$$C(\boldsymbol{x})u(\boldsymbol{x},t) = u^{\text{in}}(\boldsymbol{x},t) - \int_S H(\boldsymbol{x},\boldsymbol{y},t) * u(\boldsymbol{y},t)dS_y \tag{3}$$

Here, $C(\boldsymbol{x})$ represents the source term which depends on the point $\boldsymbol{x}$, and $*$ denotes the convolution integral. $u^{\text{in}}(\boldsymbol{x},t)$ shows the incident plane wave as mentioned in the previous section. Additionally, $H(\boldsymbol{x},\boldsymbol{y},t)$ represents the time-domain double-layer kernel for 2-D scalar wave propagation, which can be computed using the corresponding fundamental solution $G(\boldsymbol{x},\boldsymbol{y},t)$. For 2-D scalar wave propagation, they are given by the following closed forms:

$$G(\boldsymbol{x},\boldsymbol{y},t) = \frac{1}{2\pi\sqrt{t^2 - r^2/c^2}} H\left(t - \frac{r}{c}\right) \tag{4}$$

$$H(\boldsymbol{x},\boldsymbol{y},t) = \frac{\partial}{\partial n} G(\boldsymbol{x},\boldsymbol{y},t) \tag{5}$$

where r is given by $r = |\boldsymbol{x} - \boldsymbol{y}|$ and $H(t)$ is the step function. The time-domain boundary integral Eq. (3) may lead to unstable numerical solutions when discretizing the time variable using the conventional scheme [29]. Furthermore, depending on problems, the numerical solution obtained by the classical time-domain BEM may diverge. Therefore, in this study, the CQM is applied to the convolution of the time-domain boundary integral Eq. (3) to improve the numerical stability. For the spatial discretization, the following expression for the displacement using the interpolation function $\phi(\boldsymbol{x})$ as follows:

$$u(\boldsymbol{x},t) = \sum_i \phi_i(\boldsymbol{x})u_i(t) \tag{6}$$

Taking the limit for observation point $\boldsymbol{x}$ as $\boldsymbol{x} \in S$, Eq. (3) can be rewritten as follows:

$$\frac{1}{2}\sum_i \phi(\boldsymbol{x})u_i(n\triangle t) = u^{\text{in}}(\boldsymbol{x},n\triangle t) - \sum_i \sum_{k=1}^{n} \left[B_i^{n-k}(\boldsymbol{x})u_i(k\triangle t)\right] \tag{7}$$

where $B_i^m(\boldsymbol{x})$ represents the influence function, which is given by:

$$B_i^m(\boldsymbol{x}) = \frac{\mathcal{R}^{-m}}{L} \sum_{l=0}^{L-1} \int_S \hat{H}(\boldsymbol{x},\boldsymbol{y},s_l)\phi_i(\boldsymbol{y})e^{\frac{-2\pi iml}{L}} dS_y \tag{8}$$

where i represents the imaginary unit, s_l is defined as $s_l = \gamma(z_l)/\Delta t$, $\gamma(z_l)$, $\mathcal{R}$, and L are parameters of the CQM [21]. Moreover, Δt is the time increment. In Eq. (8), the double layer kernel $\hat{H}(\boldsymbol{x}, \boldsymbol{y}, s_l)$ in the Laplace-domain can be obtained from that in time-domain using the Laplace-transform as follows:

$$F(s) = \int_0^\infty f(t)e^{-st}dt \qquad (9)$$

The Laplace-transformed fundamental solution $\hat{G}(\boldsymbol{x}, \boldsymbol{y}, s)$ and the double-layer kernel $\hat{H}(\boldsymbol{x}, \boldsymbol{y}, s)$ can be expressed using the Laplace parameter s as follows:

$$\hat{G}(\boldsymbol{x}, \boldsymbol{y}, s) = \frac{1}{2\pi} K_0(sr) \qquad (10)$$

$$\hat{H}(\boldsymbol{x}, \boldsymbol{y}, s) = -\frac{s}{2\pi} \frac{\partial r}{\partial n_y} K_1(sr) \qquad (11)$$

where K_n represents the modified Bessel function of the second kind of order n. Equation(8) is the form of the Fourier transform. Therefore, Eq. (8) can be quickly calculated using the fast Fourier transform if we set the CQM parameter L as $L = N$ where N is the total time step number. If we solve Eq. (7) and obtain the displacement $u_i(n\Delta t)$ for all time-steps on the boundary S, we can calculate Eq. (3) to determine the displacement $u_i(n\Delta t)$ in the infinite domain D.

3.2   Scattering of an Incident Plane Wave by a Scattering Source

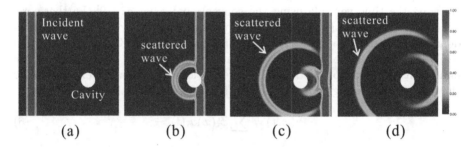

(a)   (b)   (c)   (d)

Fig. 2. Time variations of 2-D scalar wave fields around a scatterer obtained by the CQBEM.

Simulation of 2-D scalar wave scattering by a defect is demonstrated in this section. Figure 2 shows the total displacement fields around a defect in region of $-8.0 \leq x_1/a, x_2/a \leq 8.0$, as shown in Fig. 1. The center coordinates of the scattering source (x_{1c}, x_{2c}) are set as $(x_{1c}/a, x_{2c}/a) = (3.1, -2.5)$. The time increment $c\Delta t/a$ is set to $c\Delta t/a = 0.05$, and the period cT_0/a is set to $cT_0/a = 2.0$. The total time step number N is 1024.

From Fig. 2(a), we can observe the incident plane wave being transmitted. In Fig. 2(b), we can see the generation of scattered waves. The backward and forward scattered waves propagate in the domain D, as shown in Figs. 2(c) and (d). In the following, we will describe a method for estimating the scattering source using deep learning based on the simulation results as shown in Fig. 2.

4 Deep Learning for Scattering Source Determination

In the previous section, the simulation of 2-D scalar wave propagation was demonstrated. In this section, the deep learning and learning data used in this study are discussed.

4.1 Image Data for Deep Learning

As shown in Figs. 2(b)-(d), the scattered waves were successfully simulated. These scattered waves further propagate to far away and are received at observation points. As shown in Fig. 1, the observation points are placed on $x_1 = -10a$. Therefore, it is possible to create a B-scan image of the scattered waveforms obtained at the observation points $o_i^{obs} = (-10a, y_i)$ (where $i = 1, \ldots, N_f$), as shown in Fig. 3(a). In Fig. 3(a), the vertical and horizontal axes denote y_i of the observation point o_i^{obs} and the nondimensional time ct/a, respectively. As evident from Fig. 3(a), it can be observed that the peaks of the scattered waves appear earlier when the distance between the observation points and the scattering source is shorter. By utilizing this characteristic, it is possible to estimate the approximate position of the scattering source or other related information from waveforms as shown in Fig. 3(a). Therefore, employing deep learning with a large dataset of B-scan images, as depicted in Fig. 3(a), can be an effective approach for estimating the position of a scattering source. In that case, for the B-scan

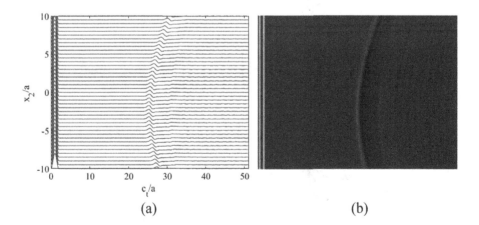

(a)   (b)

Fig. 3. An example of the simulated B-scan data obtained by the CQBEM and its transformation.

images as shown in Fig. 3(a), a significant portion of each image is white and does not carry meaningful information. In general B-scan images in the ultrasonic non-destructive testing, there are commercial devices available that can produce color-coded images if we use an array transducer with many elements. Therefore, in this study, we transform the B-scan images as shown in Fig. 3(a) into color-coded images as shown in Fig. 3(b) and using such color-coded images as training data for deep learning. In this process, any captions or labels on the vertical and horizontal axes, as shown in Fig. 3(b), are omitted to serve as training data.

4.2   Image Data Preparation Using the CQBEM

In the CQBEM calculations, the time increment is set to $c\Delta t/a = 0.05$, and the total number of time steps is set to $N = 1024$. Therefore, Eq. (8) is calculated using the fast Fourier transform with $L = N$. For spatial discretization, a piecewise constant element with $\phi_i = 1$ in Eq. (6) is used. The coordinates of the observation points are set as $o_i^{obs} = (-10a, -10a + (0.5a)*(i-1))(i = 1, \ldots, 41)$ as shown in Fig. 4. Therefore, the total 41 observation points are considered ($N_f = 41$). Furthermore, the radius of the scattering source, r/a, is varied from 0.5 to 1.1 with an increment of $\Delta(r/a) = 0.1$. This resulted in seven variations of the scattering source radius. Additionally, the center position of the scattering source (x_{1c}, x_{2c}) is varied in the range of $-7.9 \leq x_{1c}/a, x_{2c}/a \leq 7.9$, which is denoted by the dashed-line in Fig. 4, with increments of 0.2 in both x_1/a and x_2/a directions. This generated a total of $81 \times 81 = 6561$ variations for each scatterer radius, resulting in a total of $7 \times 6480 = 45360$ CQBEM calculations. The obtained images as shown in Fig. 3(b), are used for this deep learning. In

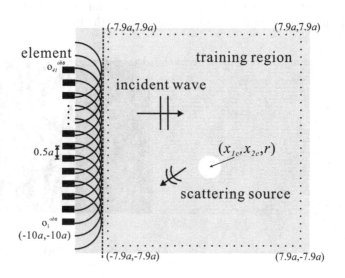

Fig. 4. Observation points and training region in this study.

conventional deep learning, images as shown in Fig. 3(b) are used as training data. However, in this study, the positions (x_{1c}, x_{2c}) and the radius r of the scattering sources used in each image creation process are also included as part of the training data during sequential analysis.

The CNN architecture used in this study is shown in Fig. 5.

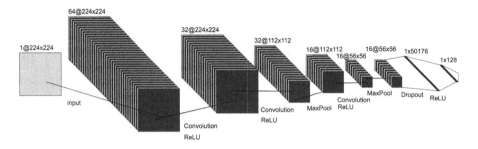

Fig. 5. CNN architecture used in this study.

5   Numerical Examples

In practical non-destructive testing, it is important not only to determine the location of defects but also to estimate their size. Therefore, in this section, we present the results of estimating both the position and size of a defect.

5.1   CNN Parameters and CNN Conditions

The input image size is not specifically constrained, but following the paper [14], we set it to 224 × 224 pixels. The Adam optimization algorithm is used, and weight decay regularization with a weight decay rate of 1.0e-4 and a learning rate of 1.0e-3 are applied. The ReLU activation function is used. A linear function is applied to the final output of the CNN as shown in Fig. 5. The mean squared error is used as the loss function, and mini-batch learning is employed. Dropout is applied with a dropout rate of 40 percent. In mini-batch learning, the entire training image dataset is randomly divided into multiple batches. One round of learning (updating the weights, etc., of the CNN) is performed for each batch, and this process is repeated for all the divided batches, which constituted one epoch. The image dataset used for training is divided into training and validation sets using the hold out method. The ratio of the number of images between the training and validation sets is set to 8:2. The deep learning calculations are performed using a GPU. The specific GPU used in this study is the GeForce RTX 3090, which has NVIDIA CUDA Cores of 10496 and 24 GB GDDR6X memory. For training, the images, such as those shown in Fig. 3(b), and the corresponding scattering source positions (x_{1c}, x_{2c}) and radius r used in generating those images with CQBEM are provided together as the training data. It should be noted that the linear function, mini-batch learning, and other parameters described here are commonly used in deep learning. For more detailed information on these parameters, please refer to the literature [4] or other relevant sources.

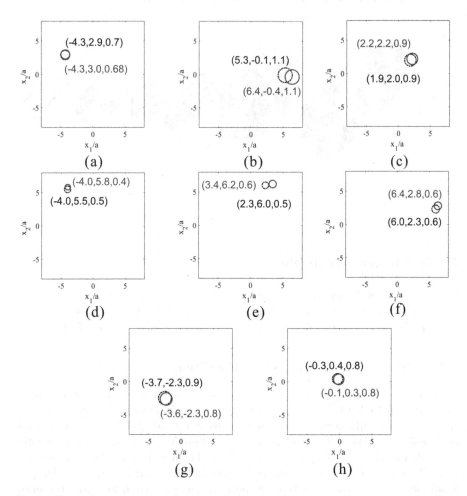

Fig. 6. Scattering source location and size estimation results obtained by the deep learning.

5.2 CNN Results

Figure 6 shows eight examples of scattering source estimation results for unlearned waveform data. In this case, the random scattering source positions $(x_{1c}/a, x_{2c}/a)$ and the radius r/a are given, and the deep learning-based inverse scattering analysis results, along with the corresponding actual solutions, are presented in Fig. 6 in the order of $(x_{1c}/a, x_{2c}/a, r/a)$. The estimated results obtained by the deep learning are highlighted in the red circles in Fig. 6.

From the forward analysis model as shown in Fig. 1, it can be observed that the incident plane wave reaches the left side of the scatterer. In the case of an isotropic solid, the scattered waves propagate in concentric circles, and their energy decays on the order of $\log \bar{r}$ where $\bar{r}$ represents the distance between the incident point and the observation point. Therefore, as a scattering source moves farther away from the observation points, the amplitude of the received waveform decreases, making it more challenging to reconstruct the scattering source accurately. Indeed, from Fig. 6, it can be seen that the accuracy of the scattering source reconstruction is higher for $x_{1c} \leq 0.0$, indicating that the scattering source located closer to the observation points are relatively well reconstructed.

On the other hand, for $x_{1c} \geq 0.0$, it can be observed that the size of the scattering source is approximately estimated, but the positions of the scattering sources are slightly shifted. This could be attributed to the lower amplitude of the received waveform, which may affect the accuracy of the scattering source estimation.

6   Conclusions

In this study, we attempted to develop a novel deep learning-based inverse scattering technique. Considering that this is the first step of our this kind of research, we used simulated received waveforms obtained using CQBEM. Generally, the deep learning improves in accuracy as it is trained sufficiently. Therefore, the accuracy of the scattering source estimation results tends to be higher as the scattering source position approaches the training region as shown in Fig. 4. In this study, a relatively simple deep learning model, as shown in Fig. 5, was used. Therefore, depending on the required accuracy, if sufficient training is conducted for the scattering source reconstruction, this method could be utilized as one of the defect shape reconstruction methods in the ultrasonic non-destructive testing.

In the future, we plan to expand this proposed technique for 2-D to 3-D using the CQBEM [20], and consider the case using waveform data obtained from actual array ultrasonic measurements. Moreover, even if actual experimental waveforms are used as the training data, implementing numerous experiments and creating a sufficient amount of training data may not be appropriate in terms of time and cost. Therefore, we are also considering the utilization of the transfer learning [18] to supplement the insufficient images. Additionally, we are considering the application of deep learning to nonlinear inverse scattering formulation based on the Born approximation.

With the advancement of sensing technology [27], big data has become manageable in various fields. Consequently, we are going to actively utilize this research for evaluating the integrity of social infrastructure structures.

Acknowledgements. This work was supported by "Joint Usage/Research Center for Interdisciplinary Large-scale Information Infrastructures", and "High Performance Computing Infrastructure" in Japan (Project ID: jh220033 and jh230036). Additionally, funding from JSPS KAKENHI (21K0423100) and the SECOM Science and Technology Foundation supported this work.

References

1. Abreu, A.I., Carrer, J.A.M., Mansur W.J.: Scalar wave propagation in 2D: a BEM formulation based on the operational quadrature method. Eng. Anal. Boundary Elem. **27**(2), 101–105 (2003). https://doi.org/10.1016/S0955-7997(02)00087-5
2. Müller A.C., Guido, S.: Introduction to Machine Learning with Python: A Guide for Data Scientists (English Edition). O'Reilly Media (2016)
3. Chen, X., Wei, Z., Li, M., Rocca, P.: A review of deep learning approaches for inverse scattering problems. Prog. Electromag. Res. **167**, 67–81 (2020)
4. Chollet, F.: Deep learning with Python. Manning Publications (2017)
5. Colton, D., Coyle, J., Monk, P.: Recent developments in inverse acoustic scattering theory. SIAM Rev. **42**(3), 369–414 (2000)
6. Doctor, S.R., Hall, T.E., Reid, L.D.: Saft - the evolution of a signal processing technology for ultrasonic testing. NDT Int. **19**(3), 163–167 (1986)
7. Dominguez, N., Gibiat, V., Esquerre, Y.: Time domain topological gradient and time reversal analogy: an inverse method for ultrasonic target detection. Wave Motion **42**(1), 31–52 (2005). https://doi.org/10.1016/j.wavemoti.2004.09.005
8. Fink, M.: Time reversal of ultrasonic fields - part I: Basic principles. IEEE Trans. Ultrason. Ferroelectr. Freq. Control **39**(5), 555–566 (1992)
9. Han, X., Yang, Y., Liu, Y.: Determining the defect locations and sizes in elastic plates by using the artificial neural network and boundary element method. Eng. Anal. Boundary Elem. **139**, 232–245 (2022). https://doi.org/10.1016/j.enganabound.2022.03.030
10. Hudson, J.A., Heritage, J.R.: The use of the Born approximation in seismic scattering problems. Geophys. J. Int. **66**(1), 221–240 (1981). https://doi.org/10.1111/j.1365-246X.1981.tb05954.x
11. Kimoto, K., Nakahata, K., Saitoh, T.: An elastodynamic computational time-reversal method for shape reconstruction of traction-free scatterers. Wave Motion **72**, 23–40 (2017)
12. Kitahara, M., Achenbach, J., Guo, Q., Peterson, M.L., Notake, M., Takadoya, M.: Neural network for crack-depth determination from ultrasonic backscattering data. In: Review of Progress in Quantitative Nondestructive Evaluation, vol. 11, pp. 701–708 (1992)
13. Kitahara, M., Achenbach, J., Guo, Q., Peterson, M.L., Ogi, T., Notake, M.: Depth determination of surface-breaking cracks by a neural network. In: Review of Progress in Quantitative Nondestructive Evaluation, vol. 10(A), pp. 686–696 (1991)
14. Krizhevsky, A., Sutskever, I., Hinton, G.E.: Imagenet classification with deep convolutional neural networks. Commun. ACM **60**(6), 84–90 (2017). https://doi.org/10.1145/3065386
15. Langenberg, K.J., Berger, M., Kreutter, T., Mayer, K., Schmitz, V.: Synthetic aperture focusing technique signal processing. NDT Int. **19**(3), 177–189 (1986)

16. Lubich, C.: Convolution quadrature and discretized operational calculus. I. Numerische Mathematik **52**(2), 129–145 (1988). https://doi.org/10.1007/BF01398686
17. Meng, M., Chua, Y.J., Wouterson, E., Ong, C.P.K.: Ultrasonic signal classification and imaging system for composite materials via deep convolutional neural networks. Neurocomputing **257**, 128–135 (2017). https://doi.org/10.1016/j.neucom.2016.11.066
18. Pan, S.J., Yang, Q.Y.: A survey on transfer learning. IEEE Trans. Knowl. Data Eng. **22**, 1345–1359 (2009)
19. Saitoh, T., Chikazawa, F., Hirose, S.: Convolution quadrature time-domain boundary element method for 2-D fluid-saturated porous media. Appl. Math. Model. **38**(15), 3724–3740 (2014). https://doi.org/10.1016/j.apm.2014.02.009
20. Saitoh, T., Hirose, S.: Parallelized fast multipole bem based on the convolution quadrature method for 3-D wave propagation problems in time-domain. IOP Conf. Ser. Mater. Sci. Eng. **10**, 012242 (2010)
21. Saitoh, T., Hirose, S., Fukui, T., Ishida, T.: Development of a time-domain fast multipole BEM based on the operational quadrature method in a wave propagation problem. In: Advances in Boundary Element Techniques VIII, pp. 355–360 (2007)
22. Saitoh, T., Ishiguro, A.: Surface crack detection in a thin plate using time reversal analysis of SH guided waves. Inter. J. Struct. Eng. Mech. **80**(3), 243–251 (2021)
23. Saitoh, T., Kato, T., Hirose, S.: Deep learning for scattered waves obtained by time-domain boundary element method and an attempt to classify defect types. J. JSNDI **70**(7), 272–279 (2021). https://doi.org/10.11396/jjsndi.70.272
24. Schanz, M., Antes, H.: Application of 'Operational Quadrature Methods' in time domain boundary element methods. Meccanica **32**, 179–186 (1997). https://doi.org/10.1023/A:1004258205435
25. Schmerr, L.W.: Fundamentals of Ultrasonic Nondestructive Evaluation. SSMST, Springer, Cham (2016). https://doi.org/10.1007/978-3-319-30463-2
26. Simonyan, K., Zisserman, A.: Very deep convolutional networks for large-scale image recognition. CoRR abs/ arXiv: 1409.1556 (2014)
27. Testa, A., Cinque, M., Coronato, A., De Pietro, G., Augusto, J.C.: Heuristic strategies for assessing wireless sensor network resiliency: an event-based formal approach. J. Heurist. **21**, 145–175 (2015). https://doi.org/10.1007/s10732-014-9258-x
28. Wang, B., Da, Y., Qian, Z.: Forward and inverse studies on scattering of rayleigh wave at surface flaws. Appli. Sci. **8**(3) (2018). https://doi.org/10.3390/app8030427
29. Mansur, W.J., Brebbia, C.A.: Transient elastodynamics using a time-stepping technique. In: Brebbia, C.A., Futagami, T., Tanaka, M. (eds.) Boundary Elements, pp. 677–698 (1983)

Optimal Control of Double Pendulum Crane Using FOPID and Genetic Algorithm

Mohamed O. Elhabib[1]📷, Herman Wahid[1(✉)]📷, Zaharuddin Mohamed[1]📷, and H. I. Jaafar[2]📷

[1] Control and Mechatronics Engineering Division, Faculty of Electrical Engineering, Universiti Teknologi Malaysia, 81310 Johor Bahru, Malaysia
herman@utm.my
[2] Faculty of Electrical Engineering, Universiti Teknikal Malaysia Melaka, Melaka, Malaysia

Abstract. Gantry cranes are commonly utilized to transfer huge loads in construction projects as well as essential sectors such as petrochemical and nuclear power. The objective of their operation is to achieve high levels of precision in trolley positioning while simultaneously reducing the amplitudes of sway oscillations. The goal of control approaches is to achieve optimum operational efficiency by maintaining accurate trolley placement while simultaneously meeting safety requirements by reducing sway-induced oscillatory oscillations. To satisfy control objective this paper considered the design of fractional order PID (FOPID) with help of genetic algorithm to compute controller parameters using MATLAB Software. Simulation results showed the controller performance is better than classic PID and also it was capable of providing good response for different payload masses.

Keywords: Fractional order PID · Double pendulum crane · Genetic algorithm · Optimal control

1 Introduction

Cranes are widely recognized as prominent tools utilized in various industrial sectors, including construction, marine industries, and port locations, owing to their exceptional capability to transport substantial loads or hazardous materials from one area to another [1]. A crane is made up of two basic parts: a hoisting mechanism (usually a hoisting line and a hook) and a support mechanism (such as a trolley-girder, trolley-jib, or a boom). The cable-hook-payload assembly is suspended from the support mechanism at a predetermined location. The suspension point traverses the crane's working zone through the support mechanism, while the hoisting mechanism raises and lowers the payload to pass obstacles along the way and securely deposit the payload at the designated location [2].

Various methodologies can be employed to control the operations of cranes, typically involving a series of actions such as gripping, lifting, transporting the

F. Hassan et al. (Eds.): AsiaSim 2023, CCIS 1912, pp. 408–420, 2024.
https://doi.org/10.1007/978-981-99-7243-2_34

load, and subsequently lowering and releasing the load [3]. Regrettably, cranes are responsible for a significant portion of fatalities and long-term injuries in the construction and maintenance sectors. The gantry crane system represents a fully non-linear and under-actuated system, featuring a solitary input (the force exerted on the trolley) and three outputs, namely the positions of the trolley, hook, and payload angles. Furthermore, the system exhibits notable instability due to oscillations in the payload when the trolley is in motion. Therefore, it be-comes imperative to mitigate the swing angles of the payload, enhance the accuracy of trolley positioning, and facilitate swift movement towards the intended destination, thereby minimizing or eliminating sway and expediting the transfer of the load.

In literature many different control approaches applied to gantry crane system, William [4] examined different types of input shaping for residual vibration reduction in a planar gantry crane with load hoisting and discovered that input shaping offered considerable decrease in both residual and transient oscillations. JieHuang [5] introduced command smoothing scheme to suppress the payload oscillations in double pendulum bridge crane, the results showed the robustness of the controller in suppression the vibrations. To improve the robustness authors [6] used cascaded SMC with input shaping to control gantry crane, author found that this method can assure accurate positioning of the trolley with notable reduction in payload sway angle, in addition it could successfully eliminates the chattering effect of SMC and improve the robustness. In [7] adaptive control applied to an under-actuated overhead crane in the presence of parametric uncertainty. Results demonstrated that controller was capable of establishing a precise cart position and eliminating residual swing. Due to its immediate feedback technique for control system resilience, the approach has provided an effective control strategy. Mahdieh Adeli [8] proposed a hybrid controller that integrates both position regulation and anti-swing control and the experimental results demonstrated the effectiveness of the proposed control algorithm.

In paper [9], PID and PD applied to gantry crane to control trolley position and damping the sway oscillations, respectively. The authors used genetic algorithm to find optimal gains and tested objective functions, the main observation was the selection of appropriate fitness function is the key of GA, which will lead to better performance as simulation results showed. Jaafar [10], developed an improved PSO based on vertical distance oscillations and crane potential energy, and compared it to the classic PSO. The proposed objective function gave a quick convergence solution, and it was able to decrease fitness value as well. The simulation results indicated that the proposed controller outperforms the PSO-based horizontal distance controller in terms of trolley position responsiveness and hook and payload oscillations. In [11], Mahmoud introduces two schemes to control double pendulum gantry crane. At the beginning he used three PIDs and then he used one PID for trolley and two PDs for hook and payload sway angles. To find optimal gains the author used MOGA (NSGA-II) with five different fitness functions and one weighted function based on ITAE of each controller. The author discovered that Scheme 1 outperformed Scheme

2 in terms of performance (lower hook and payload oscillations, reduced over-shoot percentage, and settling time of the trolley position response). Despite the advent of many controllers approaches, PID controller is still one of the most widely used in industries due to its simplicity, low cost and effective. Thus, this paper aims to investigate the performance of fractional order PID since it has similar structure and controller design process same as PID, additionally the extended parameters give more flexibility in its design than classic PID.

2   Mathematical Modeling of Double Pendulum Crane System

The system illustration depicted in Fig. 1 represents a double pendulum crane system. The movement of the crane is facilitated by an externally applied force, denoted as F, which acts on the trolley. The system can be divided into three subsystems: the trolley, the hook, and the payload. Consequently, there are three variables required to describe the crane system. Each subsystem is associated with a specific variable: the position of the trolley $x(meters)$, which is relative to the origin, the hook angle $\theta_1(radians)$, and the payload angle $\theta_2(radians)$ in relation to the vertical axis.

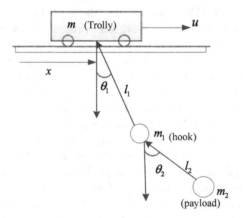

Fig. 1. Diagram of Double Pendulum Crane System

In Fig. 1 let $M(kg)$, $m_1(kg)$, and $m_2(kg)$, are the mass of trolley, hook, and payload, respectively. $L_1(m)$ is the distance between the trolley and hook, while $L_2(m)$ is the distance between hook and payload. During modeling some assumptions are made such as,

– Nonlinear frictions are neglected.
– Hook and payload are mass points.

The Lagrangian approach is utilized to create the dynamic model of this crane system.

The Lagrange equation for this model may be constructed with regard to the three generalized coordinates q_i, which are x, θ_1, and θ_2, respectively:

$$\frac{d}{dt}\left(\frac{\partial L_{ad}}{\partial \dot{q}_i}\right) = T_i \tag{1}$$

where, $L_{ad} = K_d - P_d$ (K_d is the kinetic energy of the system and P_d is the potential energy.), and T_i is the external force. According to the assumptions the total kinetic energy of gantry crane system is:

$$K_d = \frac{1}{2}M\dot{x}^2 + \frac{1}{2}m_1v_1^2 + \frac{1}{2}m_2v_2^2 \tag{2}$$

The trolley subsystem's potential energy remains constant, since it is moving in horizontal plane only. As a result, the system potential energy is simply represented by the potential energies of the hook and payload subsystems, which are specified as

$$P_d = m_1gL_1\left(1 - \cos\theta_1\right) + m_2g\left[L_1\left(1 - \cos\theta_1\right) + L_2\left(1 - \cos\theta_2\right)\right] \tag{3}$$

Then, L_{ad} has the form:

$$\begin{aligned}\frac{1}{2}M\dot{x}^2 + \frac{1}{2}m_1v_1^2 + \frac{1}{2}m_2v_2^2 - m_1gL_1\left(1 - \cos\theta_1\right) \\ -m_2g\left[L_1\left(1 - \cos\theta_1\right) - L_2\left(1 - \cos\theta_2\right)\right]\end{aligned} \tag{4}$$

The Lagrangian equation with respective to x is:

$$\begin{aligned}\left(M + m_1 + m_2\right)\ddot{x} + \left(m_1 + m_2\right)L_1\ddot{\theta}_1\cos\theta_1 - \left(m_1 + m_2\right)L_1\dot{\theta}_1^2\sin\theta_1 \\ +m_2L_2\ddot{\theta}_2\cos\theta_2 - m_2L_2\dot{\theta}_2^2\sin\theta_2 = F\end{aligned} \tag{5}$$

The Lagrangian equation with respective to θ_1 is:

$$\begin{aligned}\left(m_1 + m_2\right)L_1\ddot{x}\cos\theta_1 + \left(m_1 + m_2\right)L_1^2\ddot{\theta}_1 + m_2L_1L_2\ddot{\theta}_2\cos\left(\theta_1 - \theta_2\right) \\ +m_2L_1L_2\dot{\theta}_2^2\sin\left(\theta_1 - \theta_2\right) + \left(m_1 + m_2\right)gL_1\sin\theta_1 = 0\end{aligned} \tag{6}$$

The Lagrangian equation with respective to θ_2 is:

$$\begin{aligned}m_2L_2\ddot{x}\cos\theta_2 + m_2L_2^2\ddot{\theta}_2 + m_2L_1L_2\ddot{\theta}_1\cos\left(\theta_1 - \theta_2\right) - m_2L_1L_2\dot{\theta}_1^2 \\ \sin\left(\theta_1 - \theta_2\right) + m_2gL_2\sin\theta_2 = 0\end{aligned} \tag{7}$$

Equations (5), (6) and (7) represent the nonlinear dynamics equations of double pendulum crane system.

3   Fractional Order PID Design

The proportional integral derivative which is known as PID controller is widely recognized as the predominant form of feedback control. In the realm of process control, PID controllers account for over 95% of control loops and are utilized across various domains where control is employed. The enduring popularity of PID controllers can be attributed to their uncomplicated structure, as well as the simplicity of the design procedures, which result in effective system performance [12]. These factors have allowed PID controllers to withstand numerous technological advancements.

In control engineering, the main goal is to improve the behavior of a system. One way to do this is to generalize classical PID controllers to non-integer orders of integration and differentiation. This was first proposed in [13]. Fractional Order Proportional-Integral-Derivative (FOPID) control is an extension of the traditional PID as in Fig. 2, where the derivative and integral actions are generalized to non-integer orders. It is based on the concept of fractional calculus as in Eq. 8, which enables the incorporation of non-integer differentiation and integration orders in control systems.

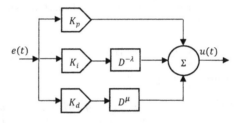

Fig. 2. Fractional Order PID Structure

$$G_c(s) = K_p + \frac{K_i}{S^\lambda} + K_d S^\mu \tag{8}$$

This papers takes into account a two-FOPID control scheme, where two FOPID controllers used for position and hook oscillation, as illustrated in Fig. 3. In actual use, this method tries to regulate the oscillations of the payload, hook, and position without requiring a sensor to monitor the oscillation of the payload [10].

4   Genetic Algorithm Optimization

Genetic algorithms (GAs) are optimization techniques inspired by the process of natural selection and genetics [14]. They are widely used in various fields, including computer science, engineering, economics, and biology, to solve complex optimization problems. The fundamental principle behind genetic algorithms is to mimic the process of natural evolution to find optimal solutions. The algorithm begins with a population of potential solutions, referred to as individuals or chromosomes, which represent different candidate solutions to the problem at hand.

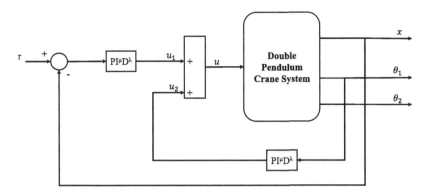

Fig. 3. Control Scheme of FOPID and Crane System

Each individual is encoded as a string of genes, typically binary digits, but other encoding schemes such as real-valued or permutation-based representations are also used [15].

The GA operates through a series of iterative steps to find optimal solutions, in previous section controllers designed but to find optimum response we have to tune ten gains concurrently $(K_p, K_i, \lambda, K_d, \mu, K_{p1}, K_{i1}, \lambda_1, K_{d1}, \mu_1)$. These steps include evaluation, selection, crossover, and mutation. During the evaluation stage, each individual's fitness is assessed using a fitness function as in Eq. 9 that quantifies how well it solves the problem. Individuals with higher fitness scores are more likely to be selected for further breeding.

$$FitnessFunction = \int_0^{T_s} t|e_x|dt + \int_0^{T_s} t|e_\theta|dt \qquad (9)$$

where T_s is the simulation time, e_x represents the error of the trolley position and e_θ is the error of hook angel. The optimization ran for 100 generations using MATLAB GA solver with the following parameters (Population size = 200, selection: binary tournament method, mutation fraction = 0.7, crossover fraction = 0.8). Table 1 shows the parameters of FOPID after optimization.

5   Simulation Results and Discussion

To control the double pendulum crane system two FOPIDs are designed using MATLAB Simulink as in Fig. 3. First one for trolley position control, which takes the difference between actual system position and the desired one as input and generates u_1. While the second FOPID is for sway oscillations reduction, and it is connected as positive feedback and produces u_2 and the total control signal obtained by summing FOPID outputs.

Table 1. The values of FOPID

	Parameter	Value
FOPID (Trolly)	K_p	5.67
	K_i	0.34
	λ	1.87
	K_d	10.75
	μ	0.98
FOPID (Oscillations)	K_p	24.680
	K_i	8
	λ	0.02
	K_d	35
	μ	1.07

The controller is tested using nominal parameters as in Table 2, the step response is applied with amplitude equal to 0.6 m. Figure 4 and 5 and illustrates the response of both FOPID and PID [10], where PID tuned using PSO. For FOPID the trolley could reach steady state within 1.9 s with no overshoot and hook angle took around 2.1 s with peak value 0.11 rad, while payload oscillation peak was 0.15 rad. On the other hand, PID trolley settling time was 2.5 s, which is greater than FOPID and hook maximum sway angle was 0.11 rad and also took longer time to reach steady state. Hook oscillations were also higher than FOPID.

Table 2. Double pendulum gantry crane parameters [10]

Parameter	Value
M (Trolley Mass)	6.5 kg
m_1 (Hook Mass)	2 Kg
m_2 (Payload Mass)	0.6 kg
L_1	0.53 m
L_2	0.4 m
g	9.8 m/s^2

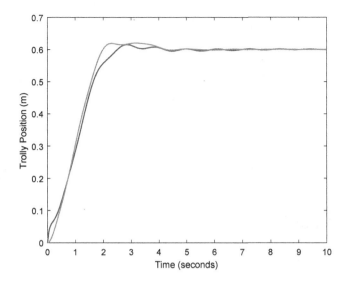

Fig. 4. Step response of double pendulum crane: Trolley position

Table 3. Responses of the system for various masses

	Hook Angle		Payload Angle	
	Maximum Sway (rad)	Settling Time (s)	Maximum Sway (rad)	Sway Angle (s)
FOPID (m = 2 kg)	−0.081	3	−0.12	3.2
PID	0.085	4.5	−0.15	4.8
FOPID (m = 4 kg)	−0.081	2.6	−0.12	3
PID	−0.09 rad	4.5	0.11	5

To evaluate the capability of the controller for oscillations, various masses were simulated. Usually in construction industries the mass of the hook mass m_1 is constant, while payload mass m_2 is varying. In the simulation payload mass increased to 2 kg and 4 kg. Table 3 summarizes the results of responses for various masses (Fig. 6 and 7).

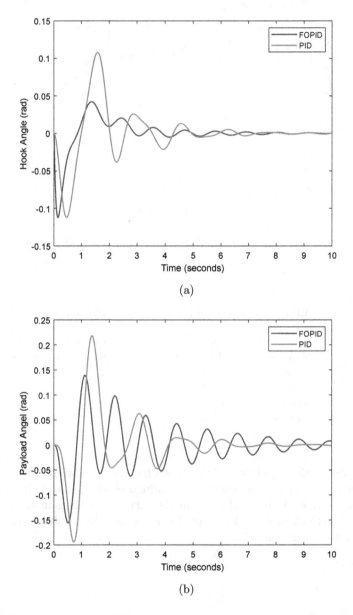

Fig. 5. Step response of double pendulum crane: hook and payload oscillations

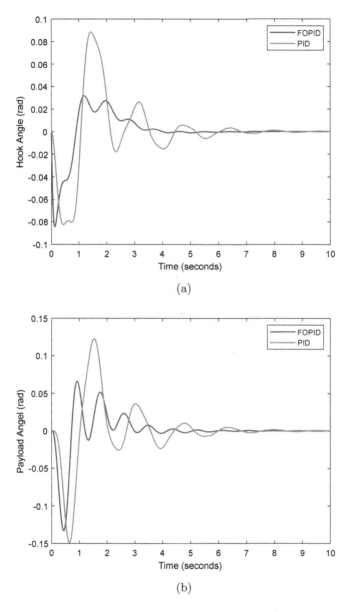

Fig. 6. Response of the system when payload mass = 2 Kg: (a) hook oscillations (b) payload oscillations

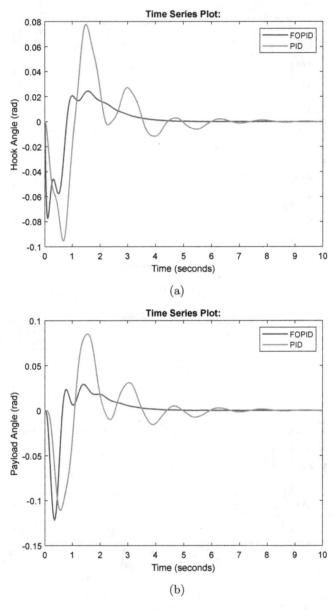

Fig. 7. Response of the system when payload mass = 4 Kg: (a) hook oscillations (b) payload oscillations

6 Conclusion

This paper represents the design of fractional order PID for double pendulum gantry crane system to control trolley position and suppress sway angles for hook and payload. The paper started by deriving the mathematical model of the system using Lagrange approach followed by controller design. The system was built and tested in MATLAB SIMULINK environment. The simulation results showed that the controller successfully achieved control objectives and also it provided better results than classic PID.

Acknowledgment. This work is funded by the Ministry of Higher Education under FRGS, Registration Proposal No. FRGS/1/2020/ICT02/UTM/02/5 & UTM.

References

1. Hong, K.S., Shah, U.H.: Dynamics and Control of Industrial Cranes. Springer, Singapore (2019). https://doi.org/10.1007/978-981-13-5770-1
2. Abdel-Rahman, E.M., Nayfeh, A.H., Masoud, Z.N.: Dynamics and control of cranes: a review. J. Vib. Control **9**(7), 863–908 (2003)
3. Ramli, L., Mohamed, Z., Abdullahi, A.M., Jaafar, H.I., Lazim, I.M.: Control strategies for crane systems: a comprehensive review. Mech. Syst. Signal Process. **95**, 1–23 (2017)
4. Singhose, W., Porter, L., Kenison, M., Kriikku, E.: Effects of hoisting on the input shaping control of gantry cranes. Control. Eng. Pract. **8**(10), 1159–1165 (2000)
5. Huang, J., Liang, Z., Zang, Q.: Dynamics and swing control of double-pendulum bridge cranes with distributed-mass beams. Mech. Syst. Signal Process. **54**, 357–366 (2015)
6. Chen, Z.M., Meng, W.J., Zhao, M.H., Zhang, J.G.: Hybrid robust control for gantry crane system. Appl. Mech. Mater. **29**, 2082–2088 (2010)
7. Sun, N., Fang, Y., Chen, H.: Adaptive control of underactuated crane systems subject to bridge length limitation and parametric uncertainties. In: Proceedings of the 33rd Chinese Control Conference, pp. 3568–3573. IEEE (2014)
8. Adeli, M., Zarabadipour, H., Zarabadi, S.H., Shoorehdeli, M.A.: Anti-swing control for a double-pendulum-type overhead crane via parallel distributed fuzzy LQR controller combined with genetic fuzzy rule set selection. In: 2011 IEEE International Conference on Control System, Computing and Engineering, pp. 306–311. IEEE (2011)
9. Solihin, M.I., Kamal, M., Legowo, A., et al.: Objective function selection of GA-based PID control optimization for automatic gantry crane. In: 2008 International Conference on Computer and Communication Engineering, pp. 883–887. IEEE (2008)
10. Jaafar, H., et al.: Efficient control of a nonlinear double-pendulum overhead crane with sensorless payload motion using an improved pso-tuned pid controller. J. Vib. Control **25**(4), 907–921 (2019)
11. Abdel-razak, M.H., Ata, A.A., Mohamed, K.T., Haraz, E.H.: Proportional-integral-derivative controller with inlet derivative filter fine-tuning of a double-pendulum gantry crane system by a multi-objective genetic algorithm. Eng. Optim. (2019)

420   M. O. Elhabib et al.

12. Tejado, I., Vinagre, B.M., Traver, J.E., Prieto-Arranz, J., Nuevo-Gallardo, C.:
    Back to basics: meaning of the parameters of fractional order PID controllers.
    Mathematics **7**(6), 530 (2019)
13. Podlubny, I.: Fractional-order systems and pi/sup/spl lambda//d/sup/spl mu//-
    controllers. IEEE Trans. Autom. Control **44**(1), 208–214 (1999)
14. Mitchell, M.: An Introduction to Genetic Algorithms. MIT Press, Cambridge
    (1998)
15. Katoch, S., Chauhan, S.S., Kumar, V.: A review on genetic algorithm: past,
    present, and future. Multimedia Tools Appl. **80**, 8091–8126 (2021)

SHADOW: Silent-based Hybrid Approach for Dynamic Pseudonymization and Privacy Preservation in Vehicular Networks

Zahra Kadhum Farhood[1][(✉)], Ali A. Abed[1], and Sarah Al-Shareeda[2]

[1] Department of Computer Engineering, College of Engineering, University of Basrah, Basrah, Iraq
zahra.farhood@uobasrah.edu.iq
[2] Department of Electrical and Electronics Engineering, University of Bahrain, Isa Town, Bahrain

Abstract. Preserving location privacy in Vehicular Ad hoc NETwork (VANET) is crucial for gaining public acceptance of this emerging technology. Many privacy schemes focus on periodically changing pseudonyms to prevent message linking. However, the spatiotemporal information contained in beacons allows vehicles to be traced, compromising driver privacy. Therefore, pseudonym changes should be performed in a mix-context, which disrupts the spatial and temporal correlation of subsequent beacons. This mix-context is commonly achieved through methods such as a silence period or predetermined locations (e.g., Mix-Zone). In this paper propose a new privacy scheme named SHADOW (Silent-based Hybrid Approach for Dynamic Pseudonymization and Privacy Preservation in Vehicular Networks), which is a location privacy scheme that allows vehicles to determine when to change their pseudonyms based on the number of neighboring vehicles and a silent period. Evaluated this scheme against a global multitarget tracking adversary using simulated and realistic vehicle traces and compare it with twelve previous privacy schemes in the Enhanced Privacy Extension Model (EPREXT) by using privacy metrics. The simulation results demonstrate that SHADOW achieves the highest level of security and privacy compared to the other twelve schemes.

Keywords: VANET · Privacy schemes · Pseudonym · Silent period

1 Introduction

In recent years, there has been significant interest and progress in the field of VANETs. These networks offer various applications that can enhance safety, efficiency, and convenience in transportation through Vehicle-to-Vehicle (V2V) and Vehicle-to-Infrastructure (V2I) communications. Many of these applications rely on sharing vehicular status and time information, which requires each vehicle to regularly broadcast beacon messages to its neighboring vehicles. These beacon messages include the vehicle's position, velocity, direction, and a timestamp indicating when the message was generated, and they are openly shared with all nearby vehicles [1]. All of this information is broadcasted

F. Hassan et al. (Eds.): AsiaSim 2023, CCIS 1912, pp. 421–440, 2024.
https://doi.org/10.1007/978-981-99-7243-2_35

in plaintext format [2]. The broadcasting of these beacon messages in plaintext format poses a threat to the privacy of the driver. Eavesdroppers have the potential to collect and analyze these broadcasted messages, allowing them to track the whereabouts of individual drivers by linking subsequent beacon messages. Therefore, it is imperative to ensure robust protection of the driver's location privacy before deploying any VANET applications [3]. Preserving location privacy has emerged as a significant research trend, capturing the attention of researchers. Existing location privacy schemes such as Mix-Zone and synchronized schemes have proven ineffective in achieving a high level of privacy due to the precise location information contained in Beacon Safety Messages (BSMs) and their resource-intensive nature. Silent period schemes, which temporarily cease broadcasting BSMs until reaching a new location with a new pseudo-identifier, are considered a more viable option. However, a major drawback of this approach is the tradeoff between safety and privacy. Since safety is a crucial requirement in Vehicle-to-Everything (V2X) communication, the research community has expressed reservations about silent period schemes. The motivation is to find a solution that allows nearby vehicles to maintain awareness for safety purposes while minimizing the opportunities for eavesdropping attacks by adversaries. Protecting user location privacy in an Internet of Vehicles (IoV) context is essential to mitigate the risk of private information disclosure. Location privacy breaches can reveal sensitive details such as home and work addresses, visits to sensitive locations, travel patterns, and times of absence from home. Correlating this spatiotemporal information with other data enables adversaries to conclude health habits, social contacts, religious beliefs, and more [4]. Preserving user location privacy yields several benefits for both users and the system. Improved privacy protection enhances the performance of the IoV system and alleviates users' concerns about security and privacy. This, in turn, encourages more users to utilize IoV functions and applications, especially those related to safety, thereby fostering further innovation and development in the automobile industry [5].

In this paper, SHADOW, a novel location privacy scheme that allows vehicles to make informed decisions about when to change their pseudonyms, taking into account the number of neighboring vehicles and a silent period, is presented. To assess the effectiveness of SHADOW, it is subjected to rigorous evaluation against a global multi-target tracking adversary, and simulated vehicle traces are utilized to test its performance. Additionally, SHADOW is compared with twelve existing privacy schemes in EPREXT using privacy metrics. The simulation results unequivocally demonstrate that SHADOW achieves the highest level of security and privacy when compared to the other schemes. Furthermore, this scheme effectively strikes a balance between security, privacy, and safety considerations, making it a robust and efficient solution for vehicular networks.

Hence, the paper makes the following key contributions:

1. Proposes a novel location privacy scheme that leverages silent periods and considers the number of neighboring vehicles to determine when to change pseudonyms.
2. Conducts a comprehensive evaluation of the proposed scheme using simulated vehicle traces.
3. Analyzes the effects of the proposed privacy scheme on the security, privacy, and safety aspects of the traffic.

4. Finally, compares the proposed privacy scheme with twelve existing privacy schemes in the EPREXT framework, utilizing various privacy and safety metrics.

The remaining sections of this article are organized as follows: Sect. 2 explores the most recent methods for safeguarding privacy in the safety applications of VANETs. Section 3 describes the system model, encompassing the network and EPREXT models. Section 4 explains the details of the proposed scheme (SHADOW). Sections 5 and 6 explain the simulation steps and results and conclude the analysis of SHADOW's performance and compare different privacy schemes. Section 7 provides a discussion. Lastly, Sect. 8 explains the conclusion.

2 Related Work

In recent years, there has been a significant discussion surrounding location privacy, with numerous studies focusing on pseudonym changes as a means to enhance syntactic unlinkability. One approach suggested in these studies involves periodic certificate changes, considering factors such as the vehicle's driving behavior and Dedicated Short-Range Communication (DSRC) properties like speed, transmission range, and transmission rate [6]. Inspired by the mix-network concept, Beresford and Stajano introduced the concept of mix-groups, where vehicles are effectively blended to ensure untraceable communication [7]. However, the inclusion of precise location information in BSM messages does enable user tracking. Expanding on this concept, Freudiger et al. proposed the CMIX protocol [8]. CMIX utilizes Mix-Zones and cryptography to preserve drivers' location privacy. By encrypting their BSMs while within these zones, the protocol guarantees the confidentiality of drivers' locations. The effectiveness of the protocol relies on infrastructures that provide the necessary shared keys, which presents a challenging task. Sampigethaya et al. introduced the CARAVAN scheme, which involves grouping all vehicles to protect their anonymity [9]. Within the group, vehicles utilize silent periods between pseudonym changes, with the group leader acting as a representative for the other members. The concept of silent periods was initially proposed by Huang et al. in the context of wireless LAN systems [10]. The principle involves keeping a node silent, meaning it does not communicate for a short period, making it difficult for adversaries to link its new and old pseudonyms. This approach is particularly effective against correlation attacks [11]. In their work [12], Ullah et al. proposed a group based strategy for pseudonym changes specifically in the context of Location-Based Service (LBS) access. However, this strategy may not be applicable when vehicles are broadcasting Basic BSMs. In [13], Buttyan et al. introduced the SLOW protocol, which enables vehicles to independently determine the appropriate moment to change their pseudonyms based on their speeds. When the speed drops below a certain threshold, indicating a lower risk of accidents, the vehicle can utilize silent periods. The main challenge with this approach lies in the contradiction with standardization requirements, which mandate a minimum beaconing frequency of once per second. Eckhoff et al. proposed Slotswap [14], a privacy preserving scheme that utilizes a time-slotted pool of pseudonyms for each vehicle. Each slot is dedicated to a specific period of time, with the ability to reuse pseudonyms when reaching the last time slot. The exchange of pseudonyms per timeslot between vehicles is also suggested. Lu et al. explored the concept of social spots, which are locations with high vehicle densities such as intersections and parking lots [15]. Changing pseudonyms in such places confuses adversaries. Similarly, Babaghayou et al. addressed the identification problem of a person leaving a specific district and proposed a scheme called Extreme Points Privacy (EPP) [16]. In the EPP scheme, vehicles cease beaconing while in the district due to the low probability of crashes. Simulation results demonstrate that the more vehicles adhere to the EPP scheme, the longer it takes for the adversary to identify the quitting event of their target. Tomandl et al. investigated the effects of Mix-Zones and silent periods, a concept further implemented by Emmara et al. in their Privacy Extension called PREXT [17]. This approach is referred to as Coordinated Silent Period (CSP) [18]. Emmara et al. also proposed a privacy scheme called Context-Aware Privacy Scheme (CAPS) [19], which enables vehicles to choose the appropriate context for entering the silent period and changing their pseudonyms. Zidane et al. presented the ENeP-AB strategy, which triggers pseudonym changes based on the number of neighbors and their predicted positions. This strategy is compared to other approaches such as mix-context enhanced and CAPs [20]. In [2], Ruqayah Al-ani et al. presented the (SRPS) Safety Related Privacy Scheme with the primary goal of mitigating the negative impact of silence periods in current pseudonym changing systems on the safety applications of VANETs. One of the factors that facilitate tracking of targets by adversaries is the eavesdropping coverage. To address this issue, Babaghayou et al. introduced a Transmission Range Adjustment (TRA) mechanism in two well-known privacy schemes, CAPS and SLOW [21]. TRA dynamically reduces the transmission range when vehicles are moving at low speeds, thus reducing the likelihood of intercepted packets compared to the conventional 300 m safety message range. After implementing TRA in CAPS and SLOW, the authors observed a decrease in traceability, indicating the potential for incorporating such methods in future privacy preserving schemes. Another scheme presented by Babaghayou et al. is WHISPER (WSP) [5]. In addition to maintaining or improving location privacy during driving, WHISPER utilizes the shift in transmission power based on the neighborhood and recommended speeds of vehicles. The assumption behind this approach is that adversaries strategically and efficiently deploy eavesdropping stations, assuming that vehicles transmit within a range of 300 m. In another work [22], Babaghayou presented the Overseers (OVR) scheme to enhance location privacy. In this scheme, regular vehicles can depend on role vehicles and public vehicles, which serve as overseers. However, in certain scenarios, the involvement of regular vehicles may be required to attain the desired level of location privacy if role vehicles and public vehicles alone are inadequate. Vehicles continuously adjust their broadcasting status based on the surrounding environment and current conditions.

WHISPER-N (WSPN), proposed by H. Aidjouli et al. [23], provides vehicle privacy by utilizing a specific number of neighboring vehicles and speed within a defined range.

3 System Model

3.1 Network Model

In Fig. 1, present an illustration of a Vehicular Social Network (VSN) deployed in an urban area. The VSN comprises several elements, including vehicles, roadside infrastructures, and an Intelligent Transportation System (ITS) data center. Provides a brief description of each component below.

1. Every vehicle has an On-Board Unit (OBU) capable of storing, processing, and communicating with other VANET entities. Based on the safety application requirements [24], the OBU will periodically broadcast Beacon Messages (BMs) to nearby entities within a communication range of 300 m. These BMs will contain the vehicle's current position, speed, and heading. Communication will be facilitated through Dedicated Short-Range Communication (DSRC) [25], and the broadcasting frequency will range from 1 to 10 Hz.

2. Roadside Units (RSUs) fixed in physical locations along the roadside or highway serve various functions. They are responsible for routing messages, extending the communication range, providing internet access to vehicles through the road, acting as intermediaries between proxy vehicles and trusted authorities, and more [26]. A roadside infrastructure consists of two primary components: an RSU serving as a wireless communication interface, and a Front-Computing Unit (FCU) responsible for local data processing. These roadside infrastructures are strategically placed at sparse intervals along the road due to economic considerations. As a result, vehicles only have intermittent access to coverage on the road. All roadside infrastructures are connected to the ITS data center through wired backhauls. The data center serves as the central hub for all ITS-related data aggregation. Within the data center, you can find the Trusted Registration Authority (RA), the location server, and the pseudonym database. It is the responsibility of the data center to facilitate global decision-making processes, including the generation and revocation of pseudonyms [27].

3. To ensure secure communications in VANET, approach the use of the PKI protocol or public key infrastructure. According to this protocol, every vehicle and RSU must receive certified public keys and register with a certain authority. These keys enable secure message exchange [28] and allow participation in various applications of VANET.

4. To preserve the anonymity of the vehicle and the driver, verified publicly accessible keys are anonymized by excluding any identifying information and utilized as identities [29]. Following that, these pseudonyms are kept in the Trusted Pseudonym Database (TPD). Role isolation among authorities has also been advocated and is generally used as a way to protect vehicle privacy in the case of an authority breach. The existence of at least three authorities is usually required for this: one for providing Long Term Pseudonyms (LTPs), one for providing Short Term Pseudonyms (STPs), with the third for overseeing a resolution center and a key cancellation procedure. The granting authorities keep a database that contains links between real identities and LTPs, as well as links between LTPs and STPs, allowing for accountability in cases of misbehavior.

5. The next subsection will further elaborate on the pseudonym management system. RSUs typically communicate with each other and with authorities through wired connections. OBUs (or vehicles) instead establish wireless contact with RSUs and other OBUs

using Vehicle-to-Vehicle (V2V) and Vehicles-with-Roadside (V2R) communication, respectively.

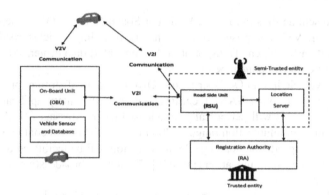

Fig. 1. The structure of a vehicular social network.

3.2 Enhanced Privacy Extension (EPREXT)

EPREXT is a highly flexible framework specifically developed for simulating privacy schemes in VANETs. It features a modular design that facilitates the easy integration of new privacy schemes. Furthermore, it offers comprehensive support for twelve widely recognized privacy schemes, in addition to the newly proposed scheme, each employing distinct approaches. To enhance its tracking capabilities, EPREXT incorporates an efficient adversary module that utilizes the nearest neighbor probabilistic data association algorithm. Additionally, the framework empowers users to accurately measure and evaluate various privacy metrics (More details about EPREXT can be found in the paper titled "EPREXT: Enhanced Privacy Extension for Veins of Ad hoc Network Simulator"). The architecture of EPREXT encompasses the following key components:

3.2.1 Global Passive Adversary (GPA)

A global passive adversary (GPA) deploys inexpensive receivers across a significant portion of the road network to intercept all message exchanges. While it may initially seem challenging for an adversary to cover the entire network, this scenario becomes achievable with the involvement of an untrusted service provider and its deployed roadside units (see Fig. 2). The main objective of this adversary is to obtain a comprehensive sequence of all vehicle movements. Therefore, the level of protection against this attack determines the driver's location privacy, irrespective of the adversary's ability to re-identify the driver from each trace. To achieve its goal, the adversary correlates the beacons emitted by a vehicle through pseudonym matching. Subsequently, it employs a multi-target tracker algorithm such as NNPDA to correlate beacons associated with different pseudonyms. If the adversary only covers a small portion of the road network, it can still track vehicles within that limited area.

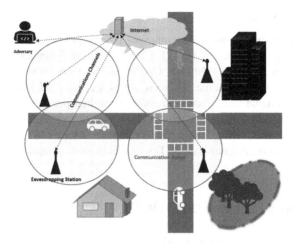

Fig. 2. Model illustration of Adversary.

3.2.2 Vehicle Tracker

To examine the framework of any message, referring to the status of vehicles, it is assumed that each vehicle is equipped with a multi-target vehicle tracker. This tracker has the responsibility of monitoring nearby vehicles, even during periods of silence. This capability significantly contributes to safety enhancement and enables vehicles to make informed decisions regarding changing their status. Karim et al. [33, 31] introduced and put into practice Vehicle Tracking (VTr), which comprises the four key stages of state estimating, gating, data association, and track maintenance. The summarized description of these phases is as follows:

1. A Kalman filter is used to estimate the vehicle's state, which includes its location, speed, and direction. Both the imperfect measurements collected from the vehicle's sensor at every step and estimated measurements generated from a predetermined kinematic model are used in this estimating procedure. The most precise assessment of the vehicle's status is obtained using these inputs.
2. By constructing an assignment probability matrix, messages from the same vehicle albeit with multiple pseudonyms are attempted to be connected to their original vehicle throughout the data association process. The nearest-neighbor Probability Data Association (NNPDA) approach, first used in [33], is used to simplify real-time computations and enable effective management of many vehicles. As an alternative, messages can only be linked together if they have the same pseudonym.
3. A gating mechanism is used before association to maximize the efficacy of the data association process. This entails eliminating improbable linkages and improving the process's overall accuracy.
4. The final phase involves eliminating vehicles that are beyond the communication range, focusing solely on tracking neighboring vehicles, even if they are in a silent period. This step helps streamline the tracking process by disregarding vehicles that are not within proximity.

3.2.3 Privacy Schemes

The EPREXT privacy extension encompasses several privacy schemes, such as Co-operative Pseudonym Neighbors (CPN), CSP, Periodical Pseudonym Change (PPC), Random Silent Periods (RSP), SLOW, CAPS, SRPS, TRA, WSP, OVR, and WSPN in addition to SHADOW scheme. These schemes are designed based on the number of neighbors and enable local decision-making regarding pseudonym changes and the duration of vehicle silence, as per the specified parameters. Additionally, these systems utilize Mix-Zones, where RSUs collaborate with vehicles to facilitate pseudonym switching and encrypt communication. The details of these schemes are given in [2, 5, 6, 13, 18, 19, 21, 22, 30].

4 The Proposed Scheme (SHADOW)

Privacy schemes play a crucial role in vehicle networks, aiming to ensure protection and privacy for vehicles, drivers, and safety applications to enhance traffic safety. However, these schemes have a significant impact on safety applications and often fail to provide the necessary level of privacy. In the SHADOW scheme, through previous analysis of privacy schemes, it is observed that schemes incorporating a period of silence, particularly a fixed period, offer a higher level of privacy compared to schemes without such a feature. Nonetheless, this silence period also affects traffic safety, as vehicles cease to transmit messages during this period. Based on these findings, the task of developing a new scheme that addresses the privacy requirements without compromising safety has been undertaken. The SHADOW scheme involves a fixed period of silence and several neighboring vehicles (see Fig. 3). Extensive evaluation against various privacy standards has confirmed that this new scheme significantly enhances privacy compared to previous approaches. The details of this comparison will be delved into in a subsequent section. Moreover, the scheme successfully strikes a balance between privacy and safety. Let us now outline the scenario in which this scheme operates.

In the scenario implemented by this scheme, the vehicle establishes an internal flag known as the Ready_Flag and continuously monitors the neighboring vehicles within a radius of R. It remains in a waiting state until the number of neighbors either reaches a specified threshold or becomes equal to K. Once this condition is met, the following steps are executed:

When the vehicle wants to send a beacon to check the message type, either Exit_Silence or Chang_Psynm, the vehicle follows specific steps depending on the message type, as outlined in Algorithm 1 (lines 1–11) for the Exit_Silence case:

1. Firstly, the vehicle confirms the receipt of the OwnReady_Flag.
   by neighboring vehicles, indicated by the condition (OwnReady_Flag == 1).
2. It then checks if the silence period has expired, denoted by (bSilent = 1).
3. Once these conditions are met, the vehicle proceeds to change its pseudonym using the function get_new_pseudonym. It exits the silent period and resumes message transmission by a new name.

The OwnReady_flag and bSilent are reset to zero. See more details in algorithm1 Alternatively, if the vehicle has already received a flag from neighboring vehicles, as determined by (ReceivedNeighborReady_Flag = 1), it changes its pseudonym, exits the silent period, and resumes transmitting.

For beacons of type Chang_Psynm, the vehicle follows the steps described in lines (12–21):

1. It checks if the number of neighboring vehicles meets or exceeds a threshold (K) within the specified range (R), indicated by (SIZE(neighbors) ≥ K). Additionally, it verifies that the vehicle is currently not in the silent period, i.e., actively sending messages (Silent_Period == 0).
2. If these conditions are satisfied, the vehicle enters a fixed duration silence period, such as 5 s. It initiates a countdown while refraining from message transmission. Furthermore, it resets the OwnReady_Flag to 0 and sets both K and ReceivedNeighborReady_Flag(neighbors) to 0.

These steps ensure that the vehicle's behavior aligns with the specific requirements and actions associated with each message type in the given scheme.

When the vehicle receives a message, it follows these steps in lines (22–25):

1. It checks the beacon from the neighboring vehicle to determine the position of the neighbor within range, denoted as R, and verifies that the NeighborReady_Flag is set to 1.
2. It sets the ReceviedNeighborReady_Flag to 1 and increments the count of neighbors.

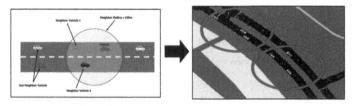

Fig. 3. Neighbor and not neighbor vehicles.

Algorithm: SHADOW

Inputs: Beacon, OwnReady_Flag, R, K, Silent_Period, silent, RecceviedNeighborReady_Flag, PseudonymLifeTime, Status

1: if status = Sending_Beacon then

2:　if Beacon = Exit_Silent then

3:　　if ((OwnReady_Flag = 1 or ReceivedNeighborReady_Flag = 1) and bSilent = 1) then

4:　　　bSilent = 0;

5:　　　Silent _Start _Time = 0;

6:　　　psynm := get_new_pseudonym();

7:　　　PseudonymLifeTime = 0;

8:　　　send_beacon (psynm, Ready_Flag);

9:　　　OwnReady_Flag = 0;

10:　　end if

11:　end if

12:　else if Beacon = Chang_Psynm then

13:　　　if ((Number of neighbor >= K) and bSilent = 0) then

14:　　　　bSilent = 1;

15:　　　　Silent_Period = 0;

16:　　　　OwnReady_Flag = 0;

17:　　　　ReceviedNeighborReady_Flag = 0;

18:　　　　Number _neighbor = 0;

19:　　　end if

20:　　end if

21:　end if

22: if status = Received_beacon then

23:　if ((Neighbor position \in　R)and NeighborReady_Flag = 1) then

24:　　ReceivedNeighborReady_Flag = 1;

25:　　Number of neighbors ++;

27:　end if

28: end if

---

5 Simulation and Performance Analysis

5.1 Simulation Setup

To implement the SHADOW algorithm, the following components are utilized.

1. A discrete event simulator used for building wireless communication networks between vehicles is the network simulation program OMNeT + + version 5.0, also referred to as Object-oriented Modular NETwork.
2. The Mobility Simulator, also known as Simulations of Urban Mobility (SUMO) version 0.25.0, is a time-dependent discrete simulator used to create large-scale road traffic networks.
3. A common protocol called TraCI, or the Communications Protocols for Traffic Controls Interface, is used to link OMNeT + + with SUMO in both directions.
4. The vehicular network is simulated using Veins, version 4.4 of the Vehicular Simulators framework that combines OMNeT + + and SUMO.
5. The proposed privacy solution is evaluated using the Privacy Simulator EPREXT (Enhanced PRivacy EXTension for Veins), which is furnished with a variety of privacy metrics and methods.

Downloading the map area of roads in Iraq, specifically Basrah, Al-Ashar, with dimensions of 2.7 km by 2.4 km, is done using the Open Street Map (OSM) database. OSM is a freely editable map covering the entire world. The conversion of the OSM data into the SUMO network format is carried out using two command-line applications: "net convert" and "poly-convert". Figure 4 displays the downloaded map with its corresponding readable file format for the SUMO road network. Following this, vehicles are generated with randomly selected trips within the provided network. The source and destination of each vehicle are determined using Python scripts (randomTrips.py). By default, the arrival rate of vehicles is one per second (v/s). However, the rate is also increased to one vehicle per 6, 3, 2, 1.5, and 1 s. Refer to Table 1 to examine the performance of SHADOW scheme in various traffic scenarios, particularly when the vehicle density increases. The statistics obtained from OMNeT + + and EPREXT are elaborated upon below. Regarding privacy metrics, several well-known privacy metrics commonly employed to evaluate the effectiveness of privacy schemes are introduced.

Table 1. Different numbers of vehicles, arrival times, and map sizes.

Inter-arrival time (s/v)	6	3	2	1.5	1
Vehicle generated by SUMO (s/v)	50	100	150	200	300
Al-Ashar map size	2.7 km * 2.4 km				

- Max-Anonymity Set Size: The anonymity set of a target vehicle v refers to the group of vehicles where vehicle v cannot be identified or differentiated in terms of its location. Therefore, it signifies the number of nodes that are indistinguishable from each other.

For example, if n neighboring vehicles simultaneously change their pseudonyms at a specific time t, can determine that the size of the anonymity set is n. In such a scenario, the tracker's task is to correctly identify the position of the vehicle among the locations of n members within the same anonymity set. The higher the value of the Max-Anonymity set size the better privacy level is achieved.

- Max-Entropy: Unlike the size of the anonymity set, this metric considers the likelihood of certain vehicles resembling the monitored vehicle more than others. Entropy is a measure of uncertainty in the value of a random variable. In this context, entropy quantifies the level of anonymity of a vehicle within a specific group of vehicles, taking into account that not all of these vehicles are similar to the monitored vehicle v_i. Equation (1) presents the entropy equation as follows:

$$Hp = - \sum p_i \log p_i \tag{1}$$

Sample Head As the likelihood of vehicles increases. The maximum value of entropy (Hmax) is attained when all vehicles have an equal probability of being the monitored vehicle v_i, as illustrated in Eq. (2):

$$\forall i : p_i = \frac{1}{|As|} pma = - \sum p_i \log_2 p_{i = \log_2 |As|} \tag{2}$$

The higher value of Max-Entropy the better privacy level is achieved.

- Traceability: This metric is referred to as the private location metric, which quantifies the adversary's ability to reconstruct a vehicle's trajectory based on the transmitted beacons. Equation (3) presents the traceability equation as follows:

$$\Pi = \frac{1}{N} \lambda_v * 100 \tag{3}$$

where $\lambda_v = \begin{cases} 1, \frac{T_v}{L_v} \geq 0.90 \\ 0, otherwise \end{cases}$

- Normalized Traceability: The traceability metric focuses exclusively on vehicles that have changed their pseudonyms, excluding those that have not. This exclusion enhances the significance of the measure, as depicted in Eq. (4):

$$D_n = \sum \alpha_v^{norm} * 100, and \tag{4}$$

where $\alpha_v^{norm} = \begin{cases} 1, \alpha_v = 1 \wedge psd_v(q) \neq psd_v(n) \\ 0, otherwise \end{cases}$

The lower the values of traceability and normalized traceability the better privacy level is achieved.

5.2   Simulation Results and Comparison of SHADOW'S Scenarios

The simulations are performed on an Intel-based computer equipped with a Core i7-1065GG7 CPU running at a speed of 1.3GHz and a memory size of 12GB. The simulation

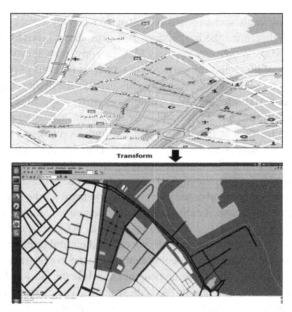

Fig. 4. The conversion process of the OSM map into a readable file format for the SUMO road network.

Table 2. The simulation parameters specific to the SHADOW scheme.

Number of Scenarios	Number of Vehicles	Silent Period	Number of Neighbors	Pseudonym Lifetime
S1	50,100,150,200,300	5	2	60
S2		10	2	60
S3		3	2	60
S4		5	6	60
S5		10	6	60
S6		5	2	30
S7		10	2	30
S8		5	6	30
S9		10	6	30
S10		5	2	10
S11		5	2	20

parameters for the SHADOW scheme, examined in this paper, are outlined in Table 2. The subsequent sections provide comprehensive results and discussions of this scheme. Extensive testing has been conducted under various scenarios to ensure a thorough assessment of their performance based on the aforementioned metrics.

After analyzing each scenario and its impact on privacy levels, a comparison is now conducted to determine the best option. This assessment is based on the following privacy metrics: Max-Anonymity set size, Max-Entropy, traceability, and normalized traceability. Figure 4 illustrates the variation in scenarios in terms of all metrics, particularly in low densities. The second scenario demonstrates the highest degree of entropy, surpassing the first, sixth, tenth, and eleventh scenarios, respectively. As the number of vehicles increases, there is a noticeable rise in entropy across all densities, ranging from 100 to 300 vehicles. The first, sixth, tenth, eleventh, seventh, and second sequences exhibit progressively increasing degrees of entropy. The Max-Anonymity set size metric is compared for different traffic densities, ranging from 50 to 300 vehicles. The second and seventh scenarios achieve the highest scores, followed by the first, tenth, eleventh, and third scenarios in sequential order. The comparison involves the tracking percentage based on vehicle behavior, specifically whether they change or maintain their pseudonym during the simulation period. In low densities, such as 50 vehicles, the sixth scenario exhibits the lowest tracking percentage. However, as the number of vehicles increases, the first, tenth, and eleventh scenarios consistently have the lowest tracking percentages. These scenarios share the same tracking percentage, while the third and seventh scenarios follow closely behind, and finally, the sixth scenario. The comparison is focused on the tracking percentage of vehicles that only changed their pseudonym. The first, tenth, eleventh, third, seventh, sixth, and finally the second scenarios yielded the best results in terms of reduced tracking percentage.

After comparing different scenarios, the first, tenth, and eleventh scenarios stand out as the best options, as they yield identical results. In these scenarios, the required number of neighboring vehicles within a radius of R is 2, and the silence period is 5 s. The only difference lies in the lifetime of the pseudonym, as indicated in Table 2. Based on this analysis, it is concluded that the lifetime of the pseudonym does not significantly affect the level of privacy. Instead, the most influential parameter is the number of neighboring vehicles. The impact becomes evident when considering scenarios where the number of neighboring vehicles increases to 6, specifically the fourth, fifth, eighth, and ninth scenarios (see Fig. 5). The various privacy measures employed show that the reduction in pseudonym changes in these scenarios, where 6 nearby vehicles are required, negatively affects privacy due to the increased time required and potential unavailability of such vehicles. It's important to note that any difference in the silence period has a minimal or negligible effect on privacy levels. Consequently, the first scenario proves to be the best choice for comparison and evaluation. The compatibility of this scheme with previous schemes will be discussed in the next section.

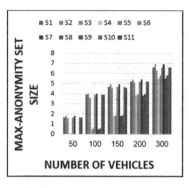

(a) Max-Anonymity set size metric.

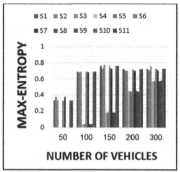

(b) Max-Entropy metric.

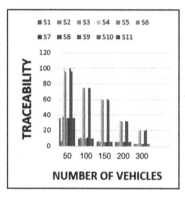

(c) Traceability metric.

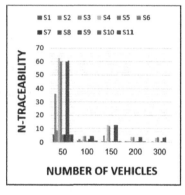

(d) N-Traceability metric.

Fig. 5. Result for all scenarios.

6   Comparison of SHADOW Scheme with Previous Privacy Schemes

In a previous study, twelve different schemes have been examined and a comprehensive comparison has been conducted among them. However, the level of privacy provided by these schemes has been unsatisfactory, and numerous issues have remained unresolved. Consequently, it has become imperative to develop a new scheme that would address these concerns and deliver the required level of privacy. Building upon the insights gained from the previous schemes, the SHADOW scheme has been devised. The scenario of this scheme has been extensively described in the preceding sections, and through careful analysis and selection of the best scenario, it has been ensured that it outperforms all other schemes in terms of privacy.

Figure 6 (a) illustrates this improvement, as depicted by the Max-Anonymity set size metric. Previously, the CPN scheme was considered the best, but now the SHADOW scheme has claimed the top position, followed by the CPN scheme. Figure 6 (b) presents the disparity based on the Max-Entropy metric. While the SHADOW scheme has achieved remarkable results, it has fallen second to the PPC chart in this particular metric. Moving on to Fig. 7(a) and (b), they illustrate the distinction in terms of reducing the tracking percentage. Notably, the SHADOW scheme has emerged as the superior choice when compared to the other twelve schemes. It has been closely followed by the CSP, WSP, and WSPN schemes.

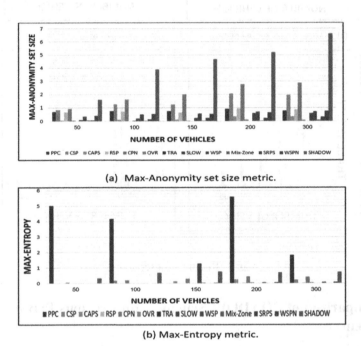

(a) Max-Anonymity set size metric.

(b) Max-Entropy metric.

Fig. 6. Comparison between the SHADOW scheme and 12 schemes by Max-Entropy metric, Max-Anonymity set size.

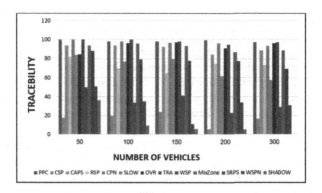

(a) Traceability.

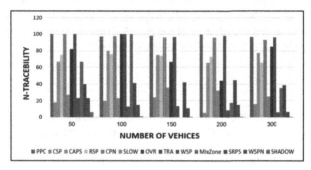

(b) N-Traceability.

Fig. 7. Comparison between the SHADOW scheme and 12 schemes by Traceability, N-Traceability.

7 Discussion

Depending on the results in Sect. 5.2 and the comparison in Sect. 6, we find that the SHADOW scheme achieves the best level of privacy without affecting safety. After testing the SHADOW with several scenarios, noticed that the best time for the fixed silence period is 5 s, and it is the scenario used in the SADOW that greatly reduced the effect of the silence period on The level of safety, and therefore this scheme has achieved good results in all measures of privacy, and this is not found in other schemes, where find that one achieved a good level in one metric, but in the other it is very weak, and these are the most important weaknesses of privacy schemes, but in The SADOW matter is different, and this is what made it the best scheme to preserve privacy. Also, increasing the number of neighboring vehicles negatively affects privacy, because the vehicle has to wait until there are many required to change the pseudonym, but the number 2 is the best for the neighbors' vehicles, and within a certain area, this contributed to reducing tracking increased by 85%, thus increasing the level of privacy.

8 Conclusion

This paper introduced a novel privacy scheme called SHADOW, which achieves a higher level of privacy compared to previous schemes. The effectiveness of the SHADOW scheme is demonstrated through extensive comparisons Furthermore, this scheme addressed the issues that are prevalent in earlier approaches. Following a comprehensive study presented in paper titled " EPREXT: Enhanced Privacy Extension for Veins of Ad-hoc Network Simulator", and after conducting thorough testing with various scenarios, it is observed that the fixed period of silence does not impact the level of privacy significantly due to the specific scenario employed in the SHADOW scheme. However, an increase in the number of neighboring vehicles affects privacy. It is advisable to maintain the number of neighboring vehicles at a maximum of 2. This limitation ensures that the vehicle does not have to wait excessively until the required number of neighbors is reached, thus reducing the frequency of pseudonyms changes and minimizing the risk of tracking. Additionally, the lifetime of the pseudonyms has no role in the pseudonyms change process and there-fore does not have any impact.

References

1. Pan, Y., Li, J., Feng, L., Xu, B.: An analytical model for random pseudonym change scheme in VANETs. Cluster Comput. **17**(2), 413–421 (2014). https://doi.org/10.1007/s10586-012-0242-7
2. Al-ani, R., Baker, T., Zhou, B., Shi, Q.: Privacy and safety improvement of VANET data via a safety-related privacy scheme. Int. J. Inf. Secur. **22**, 763–783 (2023). https://doi.org/10.1007/s10207-023-00662-6
3. Mundhe, P., Verma, S., Venkatesan, S.: A comprehensive survey on authentication and privacy-preserving schemes in VANETs. Comput. Sci. Rev. **41**, 100411 (2021). https://doi.org/10.1016/j.cosrev.2021.100411
4. Eckhoff, D., Sommer, C.: Readjusting the privacy goals in vehicular ad-hoc networks: a safety-preserving solution using non-overlapping time-slotted pseudonym pools. Comput. Commun. **122**, 118–128 (2018). https://doi.org/10.1016/j.comcom.2018.03.006
5. Babaghayou, M., Labraoui, N., Ari, A.A.A., Ferrag, M.A., Maglaras, L., Janicke, H.: Whisper: a location privacy-preserving scheme using transmission range changing for the internet of vehicles. Sensors **21**(7), 1–21 (2021). https://doi.org/10.3390/s21072443
6. Buttyán, L., Holczer, T., Weimerskirch, A., Whyte, W.: SLOW: a practical pseudonym changing scheme for location privacy in VANETs. In: 2009 IEEE Vehicular Networking Conference (VNC), pp. 1–8. IEEE (2009). https://doi.org/10.1109/VNC.2009.5416380
7. Görlach, A., Heinemann, W., Terpstra, W., Mühlhäuser, M.: Location privacy. Handb. Algorithms Wirel. Netw. Mob. Comput. 17–393–17–411 (2005). https://doi.org/10.1201/9781420035094
8. J. Freudiger, R. Shokri, and J. P. Hubaux, "On the optimal placement of mix zones," Lect. Notes Comput. Sci. (including Subser. Lect. Notes Artif. Intell. Lect. Notes Bioinformatics), vol. 5672 LNCS, pp. 216–234, 2009, doi: https://doi.org/10.1007/978-3-642-03168-7_13
9. Sampigethaya, K., Huang, L., Li, M., Poovendran, K., Matsuura, R., Sezaki, K.: CARAVAN: providing location privacy for VANET. Embed. Secure. Cars 1–15 (2005)
10. Huang, L., Matsuura, K., Yamanet, H., Sezaki, K.: Enhancing wireless location privacy using silent period. IEEE Wirel. Commun. Netw. Conf. WCNC **2**(February), 1187–1192 (2005). https://doi.org/10.1109/WCNC.2005.1424677

11. Babaghayou, M., Labraoui, N., Adamou, A., Ari, A., Lagraa, N.: Journal of information security and applications pseudonym change-based privacy-preserving schemes in vehicular ad-hoc networks : a survey. J. Inf. Secure. Appl. **55**(October), 1–17 (2020)
12. Ullah, A., Wahid, M., Shah, A., Waheed, A.: VBPC: Velocity based pseudonym changing strategy to protect location privacy of vehicles in VANET. In: 2017 International Conference on Communication Technologies (ComTech) (2017)
13. Eckhoff, D., German, R., Sommer, C., Dressler, F., Gansen, T.: SlotSwap: strong and affordable location privacy in intelligent transportation systems. IEEE Commun. Mag. **49**(11), 126–133 (2011). https://doi.org/10.1109/MCOM.2011.6069719
14. Rongxing, L., Lin, X., Luan, T.H., Liang, X., Shen, X.: Pseudonym changing at social spots: an effective strategy for location privacy in VANETs. IEEE Trans. Veh. Technol. **61**(1), 86–96 (2012). https://doi.org/10.1109/TVT.2011.2162864
15. Babaghayou, M., Labraoui, N.: Transmission range adjustment influence on location privacy-preserving schemes in VANETs. 2019 International Conference on Networking and Advanced Systems (ICNAS), pp. 1–6 (2019)
16. Babaghayou, M., Labraoui, N., Abba Ari, A.A.: EPP: extreme points privacy for trips and home identification in vehicular social networks. In: CEUR Workshop Proceeding, vol. 2351, pp. 1–10 (2019)
17. Emara: Poster: PREXT: Privacy extension for Veins VANET simulator. In: 2016 IEEE Vehicular Networking Conference (VNC), pp. 1–2 (2016). https://doi.org/10.1109/VNC.2016.783 5979
18. Tomandl, A., Scheuer, F., Federrath, H.: Simulation-based evaluation of techniques for privacy protection in VANETs. In: 2012 IEEE 8th International Conference on Wireless and Mobile Computing, Networking and Communications (WiMob), pp. 165–172 (2012). https://doi.org/10.1109/WiMOB.2012.6379070
19. Emara, K., Woerndl, W., Schlichter, J.: CAPS: context-aware privacy scheme for VANET safety applications. In: Proceedings of the 8th ACM Conference on Security and Privacy in Wireless and Mobile Networks, pp. 1–12 (2015). https://doi.org/10.1145/2766498.2766500
20. Zidani, F., Semchedine, F., Ayaida, M.: Estimation of neighbors position privacy scheme with an adaptive beaconing approach for location privacy in VANETs. Comput. Electr. Eng. **71**(June), 359–371 (2018). https://doi.org/10.1016/j.compeleceng.2018.07.040
21. Babaghayou, M., Labraoui, N., Ari, A.A.A., Gueroui, A.M.: Transmission range changing effects on location privacy-preserving schemes in the internet of vehicles. Int. J. Strateg. Inf. Technol. Appl. **10**(4), 33–54 (2019). https://doi.org/10.4018/IJSITA.2019100103
22. Babaghayou, M., Chaib, N., Lagraa, N., Ferrag, M.A., Maglaras, L.: A safety-aware location privacy-preserving IoV scheme with road congestion-estimation in mobile edge computing. Sensors **23**(1), 531 (2023). https://doi.org/10.3390/s23010531
23. Mathematics, D., Sciences, C., Sciences, C., Networks, O.: Domain Mathematics and Computer Sciences Defended Publicly in Front of the Jury in 4th July 2022 Composed of (2022)
24. Moubayed, A., Shami, A., Heidari, P., Larabi, A., Brunner, R.: Edge-enabled V2X service placement for intelligent transportation systems. IEEE Trans. Mobile Comput. **20**, 1–13 (2021)
25. Jiang, D., Taliwal, V., Meier, A., Holfelder, W., Herrtwich, R.: Design of 5.9 GHz DSRC-based vehicular safety communication. IEEE Wirel. Commun. **13**, 36–43 (2006)
26. Aljabry, I., Al-Suhail, G.: A simulation of AODV and GPSR routing protocols in VANET based on multimetrices. Iraqi J. Electr. Electron. Eng. **17**(2), 66–72 (2021). https://doi.org/10.37917/ijeee.17.2.9
27. Li, M., Sampigethaya, K., Huang, L., Poovendran, R.: Swing & swap: user-centric approaches towards maximizing location privacy. In: Proceedings of the 5th ACM Workshop on Privacy in Electronic Society, pp. 19–28 (2006). https://doi.org/10.1145/1179601.1179605

28. Fischer, L.: Secure revocable anonymous authenticated inter-vehicle communication (SRAAC). In: 4th Conference on Embedded Security in Cars (ESCAR 2006), Berlin, Germany (2014)

29. Bißmeyer, N., Petit, J.: CoPRA : conditional pseudonym resolution algorithm in VANETs. In: 2013 10th Annual Conference on Wireless On-demand Network Systems and Services (WONS), pp. 9–16 (2013)

30. Al-Shareeda, S., Özgüner, F.: Preserving location privacy using an anonymous authentication dynamic mixing crowd. In: 2016 IEEE 19th International Conference on Intelligent Transportation Systems (ITSC), pp. 545–550 (2016). https://doi.org/10.1109/ITSC.2016.779 5607

Control of Electrical and Calcium Alternans in a One-Dimensional Cardiac Cable

Jin Keong[1] , Boon Leong Lan[1] , Einly Lim[2] , Duy-Manh Le[3,4] , and Shiuan-Ni Liang[1(✉)]

[1] School of Engineering, Monash University Malaysia, Bandar Sunway 47500, Selangor, Malaysia
liang.shiuan-ni@monash.edu
[2] Faculty of Engineering, University of Malaya, 50603 Kuala Lumpur, Malaysia
[3] Institute of Theoretical and Applied Research, Duy Tan University, Hanoi 100000, Vietnam
[4] Faculty of Natural Sciences, Duy Tan University, Da Nang 550000, Vietnam

Abstract. Cardiac alternans is a beat-to-beat alternation in the electrical properties of the heart, such as membrane potential and intracellular calcium (Ca) cycling in myocytes. Due to the bi-directional coupling between voltage and calcium, it is not easy to decide which mechanism drives the alternans and thus, affect the effectiveness of controlling alternans. In this study, the control of cardiac alternans in a one-dimensional short cable was investigated numerically using the $T \pm \epsilon$ feedback control, where T is the basic cycle length and ϵ is a pre-set control parameter. The effectiveness of controlling alternans between action potential duration (APD) and peak value of intracellular calcium concentration (peak-[Ca]) being used as the feedback control variables were compared. Results showed that the effectiveness of APD-based and Ca-based feedback controls were the same when ϵ was less than the critical value (ϵ_c), however, the APD-based control performed better than the Ca-based feedback control when $\epsilon > \epsilon_c$. This study may improve the understanding of which method is a better feedback control and lead to the development of a better alternans control scheme.

Keywords: Nonlinear dynamics and chaos · Cardiac dynamics · Feedback control · Applications of chaos

1 Introduction

Cardiac alternans [1,2] is a heart arrhythmia, where the action potential duration (APD) can alternate from beat to beat. It can be a precursor to more dangerous arrhythmias [3,4], such as ventricular fibrillation and sudden cardiac death [4, 5]. Alternans can be caused by the unstable membrane potential dynamics or the instability in calcium cycling in the cardiac myocytes [6]. The membrane

© The Author(s), under exclusive license to Springer Nature Singapore Pte Ltd. 2024
F. Hassan et al. (Eds.): AsiaSim 2023, CCIS 1912, pp. 441–451, 2024.
https://doi.org/10.1007/978-981-99-7243-2_36

potential (V) and intracellular calcium (Ca) are bi-directionally coupled [7]. The V-Ca coupling is positive if a large (small) Ca transient corresponds to a long (short) APD in the same beat and the V-Ca coupling is negative if a large (small) Ca transient corresponds to a short (long) APD in the same beat. Due to the bi-directional coupling between V and Ca, it is difficult to determine if alternans is driven by the mechanism due to V or Ca.

In this study, the effectiveness of controlling APD alternans using a feedback control scheme was analyzed. In particular, the effectiveness of alternans control when APD or the peak value of Ca was used as the feedback control variable were compared. Instead of applying the commonly used control scheme, i.e. the delayed feedback control (DFC) [8–10], where two successive APDs have to be recorded at every beat, the $T \pm \epsilon$ feedback control [11,12], where T is the basic cycle length and ϵ is a pre-set control parameter, was used in this study. The advantage of using the $T \pm \epsilon$ feedback control is that the control parameter (ϵ) is a fixed value and is decided before the control is turned on. This makes the control easier to be carried out when the peak value of Ca is used as the feedback control variable because only one control variable (i.e. peak value of Ca) needs to be observed instead of two control variables (i.e. APD and peak value of Ca) have to be observed. The $T \pm \epsilon$ feedback control was previously showed that it could control the APD alternans for a map-model and a single cell [11–13], and the Ca alternans for a single cell [13,14] effectively. However, it is not clear if it can also perform effectively for a multi-cellular system.

Therefore, in this work, the control of alternans in a one-dimensional (1D) short cable using $T \pm \epsilon$ feedback control was studied. The alternans was generated using a human ventricular model and the V-Ca coupling was negative based on the model parameter values used. Since alternans is a precursor to ventricular fibrillation and sudden cardiac death, this work may provide guidance to the development of a better control protocol for cardiac alternans.

2   Action Potential Model

In order to investigate the effect of different feedback control variables used for $T \pm \epsilon$ feedback control, a 1D cardiac cell cable was simulated using

$$C_m \frac{\partial V}{\partial t} = -(I_{\text{ion}} + I_{\text{stim}}) + \nabla D \nabla V, \tag{1}$$

where V (mV) is the cell membrane potential, I_{ion} ($\mu A/cm^2$) is the total current density, I_{stim} ($\mu A/cm^2$) is the stimulus current, C_m ($\mu F/cm^2$) is the transmembrane capacitance, D (cm^2/ms) is the diffusion coefficient and t (ms) is the time. The 2006 ten Tusscher & Panfilov human ventricular model (TP06) [15] was used to calculate I_{ion}, and the model was set to produce the dynamics of an endocardial cell. The following parameters were set according to ten Tusscher and Panfilov [15] to get the maximal APD restitution slope of 1.8 so that alternans cases could be generated with a wider range of basic cycle length, i.e. the maximum conductance of the rapid delayed rectifier current $G_{\text{Kr}} = 0.172$ nS/pF, the maximum conductance of the slow delayed rectifier current $G_{Ks} = 0.441$ nS/pF, the maximum conductance of the sarcolemmal calcium ion pump current $G_{pCa} = 0.8666$ nS/pF, the maximum conductance of the plateau potassium current $G_{pK} = 0.00219$ nS/pF and the time constant of slow voltage inactivation gate $\tau_{f\ inact} = 2 \times$ (the original value of $\tau_{f\ inact}$). The setting of transmembrane capacitance and diffusion coefficient were set to $C_m = 1$ μF/cm^2 and $D = 0.00154$ cm^2/ms respectively [15].

The cable length was chosen to be 1 cm. This is because feedback control method using a single electrode are only effective on cardiac cell cable around 1 cm to 2 cm long [9]. Equation (1) was solved using the explicit Euler method, with time-step $\Delta t = 0.01$ ms and space-step $\Delta x = 0.01$ cm. With this value of space-step and by setting the length of each cell as 0.01 cm [16], there were 100 cells (labelled as cell-1 to cell-100) in the cable. Furthermore, the Neumann no-flux boundary was implemented at both ends of the cable according to Sato [17].

To stimulate the cable, a stimulus current density I_{stim} was applied to five cells at one end of the cable. In order to generate a steadier alternans, instead of applying the stimulus current to the cells immediately at one end of the cable as the conventional way [17, 18], the stimulus delivery location was shifted from the first five cells to five cells centered at cell-20. The stimulus current density I_{stim}, which was a rectangular pulse train of magnitude -90 μA/cm^2, was applied to each of the five cells for a duration of 4 ms. Using the above cardiac cable setup, it was found that the basic cycle length, T, that produced alternans lied in the range of 285 ms to 305 ms.

3   Control Methods

In this study, the $T \pm \epsilon$ feedback control method [11–14] was used to suppress the APD alternans. The control method was carried out by making minor perturbation to the basic cycle length (T) at every beat number n. When using the APD as the feedback control variable, the control was called the APD-based control and the pacing interval at each beat n had the following values

$$T_n = T + \epsilon, \qquad \text{if } a_n > a_{n-1},$$
$$= T - \epsilon, \qquad \text{if } a_n < a_{n-1}, \tag{2}$$

where a_n is the APD of the n-th beat and was measured at 90% repolarization of the action potential, and ϵ is the pre-set control parameter, where $\epsilon \ll T$. When using the peak value of intracellular calcium concentration (peak-[Ca]) as the feedback control variable, the control was called the Ca-based control and was carried out as follows

$$T_n = T + \epsilon, \qquad \text{if peak-[Ca]}_n < \text{peak-[Ca]}_{n-1},$$
$$= T - \epsilon, \qquad \text{if peak-[Ca]}_n > \text{peak-[Ca]}_{n-1}, \tag{3}$$

where peak-$[Ca]_n$ is the peak-$[Ca]$ of the n-th beat. Since the APD and peak-$[Ca]$ were negatively coupled based on the TP06 model parameter values used, both APD-based and Ca-based controls used $T_n = T + \epsilon$ when the APD at the n-th beat was longer than that at the previous beat (that is the peak-$[Ca]$ was smaller at the n-th beat) and used $T_n = T - \epsilon$ when the APD at the n-th beat was shorter than that at the previous beat (that is the peak-$[Ca]$ at the n-th beat was larger).

In this study, both APD-based and Ca-based control schemes were tested on two different basic cycle lengths, which were $T = 290$ ms and $T = 300$ ms. The cell-40 was used as a feedback reference for the control because it was found that the dynamic of peak-$[Ca]$ was not period-2 at the stimulus current delivery location (the five cells centered at cell-20). The tissue was paced at a fixed T until the APD was at steady state for at least 200 beats, that is, the variability (relative percentage difference from the mean) of APD of all cells was < 1%. Specifically, in the simulations, the tissue was paced at a fixed T for 1300 beat, and the control was started at $n = 1301$. The simulations were stopped at $n = 3000$.

To omit the control transient dynamics, the APDs from $n = 1501$ to $n = 3000$ were used to evaluate the effectiveness of control. The control was evaluated by how much it could reduce the amplitude of APD alternans. The amplitude of APD alternans when there was no control was recorded as follows: first, from $n = 1101$ to $n = 1300$, the difference between the maximum and minimum APDs for each cell (cell_diff) was calculated. Then, the largest difference was searched among the 100 differences (cell_diff) calculated from the individual cells and was recorded as the amplitude of alternans when there was no control ((cell_diff)$^{max}_{no\ control}$). Similar calculations were done for the APDs during control from $n = 1501$ to $n = 3000$ and was recorded as ((cell_diff)$^{max}_{with\ control}$). The effectiveness of control was measured by a percentage reduction from the amplitude of alternans when there was no control as follows:

$$\text{Percentage reduction} = \frac{(\text{cell\_diff})^{max}_{no\ control} - (\text{cell\_diff})^{max}_{with\ control}}{(\text{cell\_diff})^{max}_{no\ control}} \times 100\%. \quad (4)$$

4   Results

For the TP06 human ventricular model parameter values used, the V and Ca were negatively coupled for all cells in the 1 cm cable when alternans existed. Examples of negative V-Ca coupling for the alternans when $T = 290$ ms and $T = 300$ ms are given in Fig. 1, where the long APD corresponded to a smaller peak-$[Ca]$ in one beat and the short APD corresponded to a larger peak-$[Ca]$ in the next beat.

When the APD-based and Ca-based feedback controls were turned on, and if the control parameter ϵ was less than the critical value (ϵ_c), both feedback controls gave the same percentage reduction for the APD. However, if $\epsilon > \epsilon_c$, the two feedback controls gave different percentage reductions for the APD. Figure 2a shows the control results for $T = 290$ ms, where 2 ms $< \epsilon_c < 2.2$

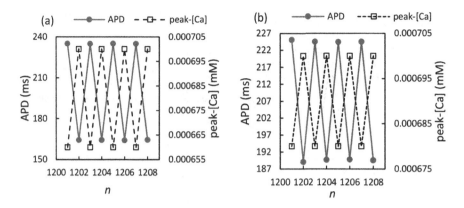

Fig. 1. Negative V-Ca coupling for the alternans when (a) $T = 290$ ms and (b) $T = 300$ ms. No control in these cases.

ms. When $\epsilon \leq 2$ ms, both feedback controls reduced the amplitudes of the APD alternans with the same percentage reduction. When $\epsilon \geq 2.2$ ms, Fig. 2a shows that the APD-based control performed better (percentage reductions were higher) than the Ca-based control (percentage reductions were lower). Similar results were found for $T = 300$ ms where 1.2 ms $< \epsilon_c < 1.4$ ms - see Fig. 2b.

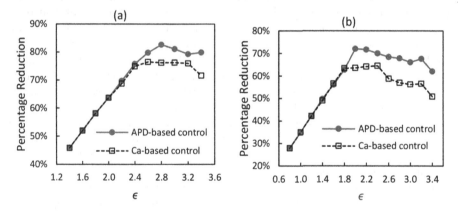

Fig. 2. The percentage reduction for APD when APD-based (red circles) and Ca-based (black squares) feedback controls were applied for (a) $T = 290$ ms with 2 ms $< \epsilon_c <$ 2.2 ms and (b) $T = 300$ ms with 1.2 ms $< \epsilon_c < 1.4$ ms. (Color figure online)

The above results can be understood as follows. When the controls were carried out with $\epsilon < \epsilon_c$, the V-Ca coupling remained negative (see Fig. 3) during the controls and thus, both controls had the same pacing period T_n at every beat during the controls and gave the same percentage reductions for the APDs.

However, when the controls were applied with $\epsilon > \epsilon_c$, the V-Ca coupling was not always negative. In Fig. 4a, compare to the previous beat number $n = 2609$,

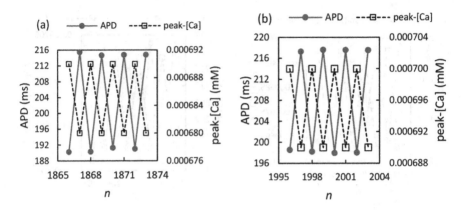

Fig. 3. V-Ca coupling was negative during the controls when $\epsilon < \epsilon_c$. (a) $T = 290$ ms, $\epsilon = 2$ ms where 2 ms $< \epsilon_c < 2.2$ ms and (b) $T = 300$ ms, $\epsilon = 1.2$ ms where 1.2 ms $< \epsilon_c < 1.4$ ms. The controls were applied from $n = 1301$ to 3000.

it shows that the longer APD corresponded to a higher peak-[Ca] at $n = 2610$. As such, at $n = 2610$, the APD-based control decided the pacing interval was $T_n = T + \epsilon$ but the Ca-based control decided the pacing interval was $T_n = T - \epsilon$. Due to using different pacing interval T_n at $n = 2610$, the values of APD and peak-[Ca] for the subsequent beats were different for the two controls as the calcium cycling and other dynamics in the cardiac myocytes started to be different from $n = 2610$. Thus, the percentage reductions for APDs were different for the two control schemes. Similar cases were also found for $T = 300$ ms with $\epsilon = 2.6$ ms. In Fig. 4b, compare to $n = 2702$, a shorter APD corresponded to a smaller peak-[Ca] at $n = 2703$, where the peak-[Ca] changed from 6.92839×10^{-4} mM at $n = 2702$ to 6.928323×10^{-4} mM at $n = 2703$. Similar case was also found at $n = 2708$ where a shorter APD corresponded to a smaller peak-[Ca].

It was checked that the above results were similar for all the cells in the 1 cm cable for both APD-based and Ca-based control schemes. An example for the APD-based control is given in Fig. 5 to show that the APDs of different cells in the cable were very similar before and after the control.

In this study, the effectiveness of controlling alternans using the peak value of intracellular sodium concentration (peak-[Na]) was also examined. Opposite to the V-Ca coupling, the membrane potential (V) and intracellular sodium (Na) concentration coupling was positive. The Na-based feedback control was applied in the similar way as the APD-based control, that was

$$T_n = T + \epsilon, \qquad \text{if peak-[Na]}_n > \text{peak-[Na]}_{n-1},$$
$$= T - \epsilon, \qquad \text{if peak-[Na]}_n < \text{peak-[Na]}_{n-1}, \qquad (5)$$

where peak-[Na]$_n$ is the peak-[Na] of the n-th beat. Since the APD and peak-[Na] were positively coupled, both APD-based and Na-based controls behaved in the same way, i.e. both controls used $T_n = T + \epsilon$ when the APD at the n-th beat was longer than that at the previous beat (meaning the peak-[Na] was larger at

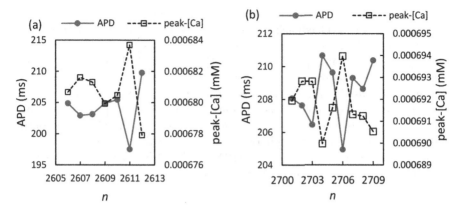

Fig. 4. V-Ca coupling during the controls when $\epsilon > \epsilon_c$. (a) $T = 290$ ms, $\epsilon = 2.6$ ms where 2 ms $< \epsilon_c <$ 2.2 ms. Compare to $n = 2609$, a longer APD corresponded to a larger peak-[Ca] at $n = 2610$. (b) $T = 300$ ms, $\epsilon = 2.6$ ms where 1.2 ms $< \epsilon_c <$ 1.4 ms. Compare to $n = 2702$, a shorter APD corresponded to a smaller peak-[Ca] at $n = 2703$. Similar case was also found at $n = 2708$ where a shorter APD corresponded to a smaller peak-[Ca]. The controls were applied from $n = 1301$ to 3000.

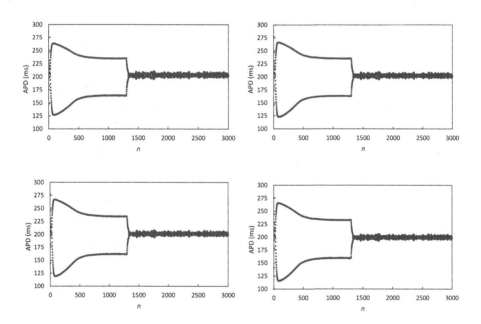

Fig. 5. The APDs of the (Upper left) 40th, (Upper right) 60th, (Lower left) 80th and (Lower right) 100th cell of the cable were very similar before and after the APD-based control with $T = 290$ ms and $\epsilon = 2.6$ ms.

the n-th beat) and used $T_n = T - \epsilon$ when the APD at the n-th beat was shorter (meaning the peak-[Na] at the n-th beat was smaller).

When the Na-based feedback control was turned on, and if the control parameter ϵ was less than the critical value (ϵ_c), similar to the Ca-based feedback control, it gave the same percentage reduction for the APD as the APD-based feedback control. However, if $\epsilon > \epsilon_c$, also similar to the Ca-based feedback control, the APD-based control performed better than the Na-based control. The reason for APD-based and Na-based controls gave different percentage reductions for the APDs when $\epsilon > \epsilon_c$ was similar to that for the Ca-based control. Figure 6 shows that the coupling between APD and peak-[Na] was not positive at a certain beat number n causing the APD-based and Na-based control schemes used different pacing interval T_n from that beat onwards.

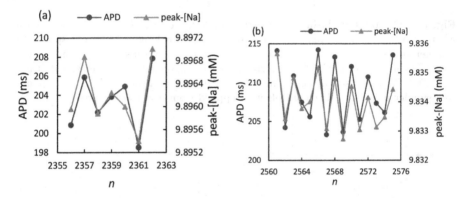

Fig. 6. V-Na coupling during the controls when $\epsilon > \epsilon_c$. (a) $T = 290$ ms, $\epsilon = 2.6$ ms where 2 ms $< \epsilon_c < 2.2$ ms. Compare to $n = 2359$, a longer APD corresponded to a smaller peak-[Na] at $n = 2360$. (b) $T = 300$ ms, $\epsilon = 2.6$ ms where 1.2 ms $< \epsilon_c <$ 1.4 ms. Compare to $n = 2564$, a shorter APD corresponded to a larger peak-[Na] at $n = 2565$. Similar case was also found at $n = 2574$ where a shorter APD corresponded to a larger peak-[Na]. The controls were applied from $n = 1301$ to 3000.

By ranking the effectiveness among the three feedback controls, i.e. APD-based, Ca-based and Na-based controls, Fig. 7 shows that the APD based control was the most effective one as it always gave the highest percentage reductions, followed by the Na-based control. The Ca-based control was the least effective control.

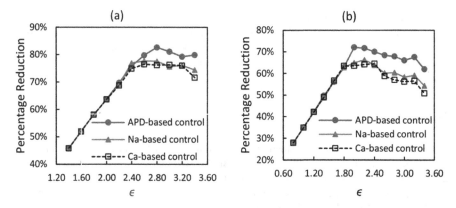

Fig. 7. The percentage reduction for APD when APD-based (red circles), Na-based (blue triangles) and Ca-based (black squares) feedback controls were applied for (a) $T = 290$ ms with 2 ms $< \epsilon_c < 2.2$ ms and (b) $T = 300$ ms with 1.2 ms $< \epsilon_c < 1.4$ ms. (Color figure online)

5   Conclusion

In this study, the control of APD alternans in a 1D short cardiac cable using $T \pm \epsilon$ feedback control was explored. In particular, the effectiveness of controlling alternans using APD or peak-[Ca] as the feedback control variable were compared. The alternans was simulated using the TP06 human ventricular model and the V-Ca coupling was negative. The results showed that the percentage reduction for APD was the same for APD-based and Ca-based feedback control if $\epsilon < \epsilon_c$, and the APD-based control performed better than the Ca-based control if $\epsilon > \epsilon_c$. It was showed that this was due to the V-Ca coupling remained negative when the control was applied with $\epsilon < \epsilon_c$, causing the APD-based and Ca-based controls using the same pacing interval T_n for every beat and thus, had the same percentage reduction for APD. When the control was applied with $\epsilon > \epsilon_c$, the V-Ca coupling was not always negative during the control. This led to the results that the APD-based and Ca-based controls started to use a different pacing interval T_n from a particular beat onwards and thus, having different percentage reductions for APD.

The feedback control using peak-[Na] as the feedback control variable was also examined for the purpose of comparison. It was found that among the three feedback controls, the APD-based control was the most effective one, followed by the Na-based control. The Ca-based control was the least effective one.

Further studies can be done to explore the detailed reasons for why Ca-based control was not performing as good as the APD-based control and the Na-based control. Studying the dynamics of other state variables in the cardiac myocytes may help to understand the reasons and improve the Ca-based control.

Acknowledgements. This work has been supported by the Ministry of Higher Education Malaysia under the Fundamental Research Grant Scheme (FRGS) with project code FRGS/1/2019/STG02/MUSM/03/2. It was also supported in part by the Monash University Malaysia Advanced Computing Platform through the use of the MONTAGE HPC Cluster.

References

1. Shiferaw, Y., Karma, A.: Turing instability mediated by voltage and calcium diffusion in paced cardiac cells. PNAS **103**(15), 5670–5675 (2006)
2. Livshitz, L.M., Rudy, Y.: Regulation of Ca^{2+} and electrical alternans in cardiac myocytes: role of CAMKII and repolarizing currents. Am. J. Physiol. Heart Circ. Physiol. **292**(6), H2854–H2866 (2007)
3. Merchant, F.M., Sayadi, O., Puppala, D., Moazzami, K., Heller, V., Armoundas, A.A.: A translational approach to probe the proarrhythmic potential of cardiac alternans: a reversible overture to arrhythmogenesis? Am. J. Physiol. Heart Circ. Physiol. **306**(4), H465–H474 (2014)
4. Qu, Z., Xie, Y., Garfinkel, A., Weiss, J.N.: T-wave alternans and arrhythmogenesis in cardiac diseases. Front. Physiol. **1**, 1–15 (2010)
5. Walker, M.L., Rosenbaum, D.S.: Repolarization alternans: implications for the mechanism and prevention of sudden cardiac death. Cardiovasc. Res. **57**(3), 599–614 (2003)
6. Kanaporis, G., Martinez-Hernandez, E., Blatter, L.A.: Calcium- and voltage-driven atrial alternans: insight from [Ca]$_i$ and V_m asynchrony. Physiol. Rep. **11**(10), e15703 (2023)
7. Qu, Z., Shiferaw, Y., Weiss, J.N.: Nonlinear dynamics of cardiac excitation-contraction coupling: an iterated map study. Phys. Rev. E **75**(1), 011927 (2007)
8. Hall, G.M., Gauthier, D.J.: Experimental control of cardiac muscle alternans. Phys. Rev. Lett. **88**(19), 198102 (2002)
9. Echebarria, B., Karma, A.: Spatiotemporal control of cardiac alternans. Chaos **12**(3), 923–930 (2002)
10. Jordan, P.N., Christini, D.J.: Adaptive diastolic interval control of cardiac action potential duration alternans. J. Cardiovasc. Electrophysiol. **15**(10), 1177–1185 (2004)
11. Sridhar, S., Le, D.-M., Mi, Y.-C., Sinha, S., Lai, P.-Y., Chan, C.K.: Suppression of cardiac alternans by alternating-period-feedback stimulations. Phys. Rev. E **87**(4), 042712 (2013)
12. Le, D.M., Lin, Y.T., Yang, Y.T., Lai, P.Y., Chan, C.K.: Cardiac alternans reduction by chaotic attractors in $T \pm \epsilon$ feedback control. Europhys. Lett. **117**, 50001 (2017)
13. Liang, S.N., Le, D.-M., Lai, P.-Y.: Cardiac alternans suppression under $T \pm \epsilon$ feedback control: nonlinear dynamics analysis. Int. J. Bifurcat. Chaos **29**(13), 1930036 (2019)
14. Liang, S.N., Le, D.-M., Lai, P.-Y., Chan, C.K.: Ionic characteristics in cardiac alternans suppression using $T \pm \epsilon$ feedback control. Europhys. Lett. **115**, 48001 (2016)
15. ten Tusscher, K.H.W.J., Panfilov, A.V.: Alternans and spiral breakup in a human ventricular tissue model. Am. J. Physiol. Heart Circ. Physiol. **291**(3), H1088–H1100 (2006)

16. Thakare, S., Mathew, J., Zlochiver, S., Zhao, X., Tolkacheva, E.G.: Global vs local control of cardiac alternans in a 1D numerical model of human ventricular tissue. Chaos **30**(8), 083123 (2020)
17. Sato, D.: Computer simulations and nonlinear dynamics of cardiac action potentials. In: Jue, T. (ed.) Modern Tools of Biophysics. HMB, vol. 5, pp. 81–107. Springer, New York (2017). https://doi.org/10.1007/978-1-4939-6713-1_5
18. Zlochiver, S., Johnson, C., Tolkacheva, E.G.: Constant DI pacing suppresses cardiac alternans formation in numerical cable models. Chaos **27**(9), 093903 (2017)

Study the Effect of Acute Stress on Decision Making Using Function Near Infrared Spectroscopy (fNIRS)

Abdualrhman Abdalhadi[1] , Nina Bencheva[2](✉) , Naufal M. Saad[1] ,
Maged S. Al-Quraishi[1] , and Nitin Koundal[1]

[1] Universiti Teknologi PETRONAS, 32610 Seri Iskandar, Perak, Malaysia
[2] University of Ruse "Angel Kanchev", Studentska str. 8, 7017 Ruse, Bulgaria
nina@uni-ruse.bg

Abstract. The prevalence of stress among individuals has become increasingly common, as approximately 40% of the population experiences stress. Given that decision-making under stressful conditions can have catastrophic consequences, it is imperative to devote additional attention to investigating the impact of acute stress on decision-making processes. The present study aims to explore the effects of acute stress, induced in a controlled laboratory environment, on decision-making. The study employed the Balloon Analog Risk Task (BART), a task designed to assess individuals' behaviour and coping strategies during three distinct stages. The participants in this study were individuals aged between 30 and 34 years. During the first stage, participants engaged in the decision-making task without any time constraints, whereas the second stage introduced time limitations. In the third stage, both time constraints and the N-Back memory task were presented simultaneously. Participants were outfitted with the Function Near Infrared Spectroscopy (fNIRS) device throughout the experiment. Multiple repetitions of tasks and measurements were conducted. The findings of this preliminary experiment revealed that participants performed poorly in the second stage and exhibited the lowest scores in the third stage. This diminished performance was attributed to their inability to process all available information due to limited cognitive resources, resulting in increased errors and misconceptions about the main task. Consequently, these cognitive lapses facilitated the occurrence of unsafe decision-making behaviours.

Keywords: Acute stress · Decision making · function near-infrared spectroscopy · Balloon Analog Risk Task

1 Introduction

Stress among people has become a significant issue, impacting individuals' life circumstances and their professions. The World Health Organization (WHO)

F. Hassan et al. (Eds.): AsiaSim 2023, CCIS 1912, pp. 452–463, 2024.
https://doi.org/10.1007/978-981-99-7243-2_37

has reported that during the COVID-19 pandemic in 2019, the prevalence of anxiety and despair increased globally by 25%. The brief also identified who was most affected and summarized the impact of the pandemic on access to mental health services [1]. A study conducted in Malaysia during the COVID-19 lockdown in April 2020 found that out of 716 adults, 70% reported stress, 76% reported anxiety, and 42.3% reported depression, with 29% actively coping, 56% planning, and 40% using humour as an escape [2].   Acute stress is characterized as a short-term and temporary response to a specific stressor. It's conceptualized as the outcome of a cognitive evaluation or interpretation of a psychosocial stressor that exceeds an individual's coping capacity. Examples of acute stressors include traffic congestion, interpersonal conflicts, criticism from a superior etc. Selye was the first researcher to describe stress as "the non-specific response of the body to any demand for change" [3]. The amygdala senses a threat through sensory input, which causes an increase in amygdala activity and activation of the hypothalamic-pituitary Adrenalin (HPA) axis, a key hormonal system implicated in the stress response [4]. On the other hand, decision-making involves the collaboration of several brain systems, including the prefrontal cortex (PFC), basal ganglia, and amygdala. As shown in Fig. 1. Acute stress can be detected using various devices such as Electrocardiogram (ECG) and Gal-

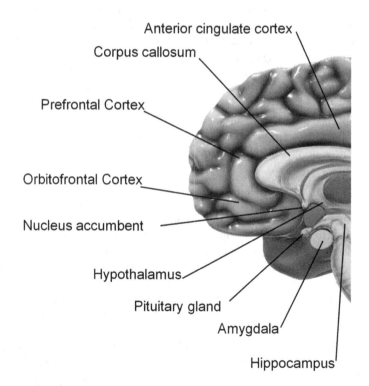

Fig. 1. The Neuroanatomy of The Brain During Decision Making

vanic Skin Response (GSR) [5]. The integration of machine learning techniques and wearable technology in healthcare has the potential to significantly benefit society [6]. Neurophysiology devices, which utilize techniques such as electromyography (EMG), electroencephalography (EEG), and Evoked potential (EP), are utilized to gain a deeper understanding of accurate and unobtrusive information in psychiatry. In this paper, a brief explanation of various emerging experiments, issues, and themes is presented. The proposed work is divided into six sections, the first of which introduces the topic of acute stress, decision-making, and physiological measurements. The second section presents state-of-the-art of recent research on acute stress, risk compensation and risk-taking of decision-making. The third section, the experiment design The fourth section, Results and discussion, and the final section, Conclusion, summarizes the paper.

2   State of the Art

Acute stress impacts various neuronal systems, such as the ventrolateral prefrontal lobe and orbitofrontal cortex. Some of these systems, including the hippocampus and prefrontal cortex, are involved in decision-making processes [7]. In everyday life, we often must weigh the importance of making a quick decision against the need to be cautious, a phenomenon known as the speed-accuracy trade-off. Risk compensation on the other hand known as the Risk Homeostasis theory was introduced by Gerald Wilde in 1982, this theory posits that individuals are inclined to engage in behaviours that involve some degree of risk to achieve potential gains. According to this psychological theory, the amount of danger that individuals perceive as acceptable influences how much risk they take [8]. A related study that used a mixed-reality environment with varying levels of fall risk found that subjects' risk-taking behaviours and perceptions changed while completing the task [9]. These studies suggested that additional safety measures do not always improve safety outcomes.

2.1   Physiological Measurement

EEG is extensively employed in brain-machine interface applications due to its ability to directly capture brain signals without the involvement of the peripheral nervous system [10]. The process involves measuring the electric brain activity generated by the neurons within the brain. By placing electrodes on the scalp, the EEG signal can be detected in a non-invasive manner. Consequently, EEG measurement has become the most widely utilized approach for assessing brain activity due to its cost-effectiveness and excellent temporal resolution, approximately 1 ms as shown in Fig. 2. Nevertheless, it does have limitations in terms of spatial resolution, as it provides a relatively coarse representation of brain activity [11]. Additionally, the EEG signal can be weak and susceptible to various artefacts, further challenging its interpretation and analysis. On the other hand, fNIRS is a relatively new technique that uses near-infrared light, typically with a wavelength of 650–1000 nm, to measure changes in the concentrations of

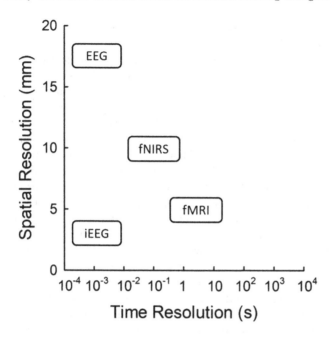

Fig. 2. Physiological Devices in Term of Time and Special Resolution

oxygenated (HbO) and deoxygenated haemoglobin (HbR) [12]. Since more than 20 years ago, fNIRS has been widely used as a measurement technique in cerebral hemodynamics research. It has also been used in various studies to monitor changes in prefrontal brain activity as participants underwent decision-making tasks [13]. For example, the author in [14] used fNIRS to measure changes in the prefrontal cortex during a balloon risk task and successfully detected changes in the brain's connectivity. Similarly, in [15,16] used fNIRS with three wavelengths (780, 805, and 830 nm) to evaluate decision dynamics and record the dyads' concentration changes of oxyhemoglobin (HbO) and deoxyhemoglobin (HbR), a study in [17] used fNIRS to study the effects of a newsvendor problem on the prefrontal cortex. Other studies, such as [18] and [19], have also used fNIRS to measure moral competency and risk-taking mentality, respectively. Another work carried out by [20], has used fNIRS in combination with other methods, such as the Iowa gambling task, to better understand the relationship between personal traits and decision-making. Additionally, [21–23] have used fNIRS in combination with EEG to better understand the brain's processes during decision-making.

2.2 Stressors and Decision-Making Task

The stressor task is an activity that causes stress or anxiety, that can vary greatly in nature and intensity, some common examples might include meeting tight deadlines, giving a presentation, taking a test, or dealing with difficult or demanding people.

Balloon Analogue Risk Task (BART) is a psychological task used to help measure the risk-taking behaviour and decision-making skills [22], recent studies presented a modified version of the BART, The decision-making period was unrestricted, and the distribution of the two keys (pump/cash out) among participants was balanced. [22,24]. The result showed that Participants who were stressed pumped fewer times than the unstressed group, particularly when they had low inclination to take risks. This suggests that they were more inclined to avoid risk-taking. In comparison to the group of stressed individuals with a high inclination to take risks, Salivary cortisol levels were greater and responses to feedback in high-risk scenarios were higher in individuals with a low inclination. Another study used balloon analogue risk tasks for two different groups young and old aged group, The study showed significant differences between old aged and young age participants where the young, aged participant went for a more risky decision. [21]. The N-Back Recall Task is also a cognitive task that is used to measure working memory, which is the capacity to temporarily store and alter information. It is often used as a measure of executive function, which is a set of cognitive processes that are involved in controlling and coordinating various cognitive functions, such as organizing, resolving conflicts, and making decisions. A recent study used N-Back To test, each individual completed a 30-minute breath-count activity, followed immediately by a 21-minute 3-back recall test, the result of this high-engagement memory task resulted in stress, specifically in the prefrontal cortex, which negatively affected performance and increased error percentage [16,25].

3   Experimental Design

The experimental design for this study comprises three distinct components. The first component, BART is conducted for 120 s and adheres to the same methodology as previously established in literature in [14,21,23,24]. The second component introduces a time constraint of 25 s to the BART task. The final component of the experiment combines the BART task with the N-Back task, both of which are conducted under a time constraint of 25 s. The overall experimental design is illustrated in Fig. 3 with a total duration of 260 s for each trial.

3.1   Participants

The present study recruited two healthy patients, both of whom were male. Both participants identified English as their second language. The inclusion criteria for the study stipulated that participants are right-handed and have no history of head injury or neurological disorders. Participants were exposed to stressors, including time pressure and mental demand, during the second and third stages of the experiment. The NASA-TLX was used to assess participants' levels of stress. Physiological measurement was taken at various time points, including

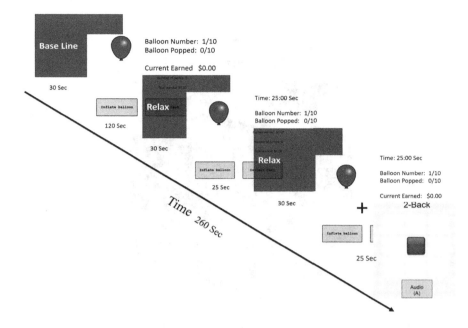

Fig. 3. Experiment Design and Timeline for Each Trial.

baseline, during the time pressure and mental demand with BAR Task. Participants received remuneration for their participation in the experiment. Figure 4 shows the participant and experiment equipment.

3.2 fNIRS Measurements

OT-R40 device was utilized to detect the hemodynamic response during the balloon analogue risk task. OT-R40 helps to keep track of cerebral activation by measuring differences in blood oxyHb and deoxyHb on the surface of the cerebrum at various points simultaneously depending on the difference in the absorption of light. In this paper, we used the fNIRS setup 10/20 system for positioning the fNIRS optodes probe which is a standardized method for positioning the optodes (light detectors) on the scalp. In this experiment, we used 52 channels, 17 emitters and 16 detectors, which are typically placed in a grid-like pattern on the scalp. The placement of the optodes is chosen based on the specific research question and the brain regions of interest. In this study, near-infrared spectroscopy (NIRS) used wavelengths between 695 and 830 nm. The hemodynamic responses of the participants were measured utilizing a custom-built 52-channel functional NIRS (fNIRS) system with sampling rates of 10 Hz. This system comprises three light sources and five detectors arranged in a 3×11 grid configuration. As shown in Fig. 5, we positioned the fNIRS channels over the bilateral frontal areas at nine distinct regions of interest (ROIs) [26], we have included the right and left DLPFC (R-DLFPC: Ch1, Ch2, Ch3, Ch4, Ch5, Ch11,

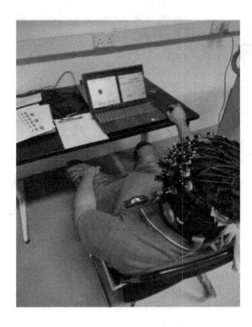

Fig. 4. fNIRS Placement and Subject During the Risky Decision-Making Task.

Ch12; L-DLPFC: Ch6, Ch7, Ch8, Ch9, Ch10, Ch19, Ch20), right and left ven-trolateral PFC (R-VLPFC: Ch22, Ch32, Ch33, Ch43; L-VLPFC: Ch31, Ch41, Ch42, Ch52), frontopolar area (FPA: Ch16, Ch26, Ch27, Ch37), R-DLPFC/FPA (Ch13, Ch14 Ch15, Ch23, Ch24, Ch25), L-DLPFC/FPA (Ch17, Ch18, Ch28, Ch29, Ch30, Ch38), and right and left orbitofrontal cortex (OFC; R-OFC: Ch34, Ch35, Ch36, Ch44, Ch45, Ch46, Ch47; L-OFC: Ch39, Ch40, Ch48, Ch49, Ch50, Ch51).

3.3   Recording Procedure

To enhance visual instruction of the activity, the display was placed roughly 85 cm in front of the participants, who were situated in comfortable chairs, as depicted in Fig. 4. A blank screen served as the baseline for the experiment for the first 30 s of the trial, to aid in the detection of changes in hemodynamics during the pre-processing stage. The participant was then permitted to move their hand and use the mouse to begin the first task, which lasted for 120 s. Subsequently, the screen transitioned to a black state, indicating a period of rest for 30 s. The following task involved the participant performing the same task under time pressure for 25 s, followed by a rest period and a black screen for 30 s. The subject had to complete the assignment under time constraints and mental stress in the final step, utilizing the N-back task for 25 s. To reduce outside noise, the data-collecting experiment was carried out in a quiet lab, Before the experiment, participants were instructed to remain relaxed for at least 5 min before the task begin.

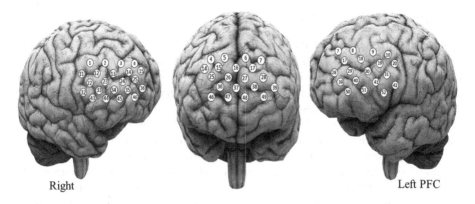

Right   Left PFC

Fig. 5. Right PFC, Middle PFC and Left PFC View of the fNIRS Channels Placement.

4   Results and Discussion

4.1   Observatory Results

In the previous section, subjects underwent three tasks: BART, BART with time pressure, and BART with time pressure and 2-back memory task. In the first task, both subjects performed well, with profits of 54$ and 44$ respectively (as shown in Table 1). However, in the second task, the subjects struggled to keep up their performance, finishing only 5 and 6 balloons respectively and earning lower profits compared to the first task. In the third task, a significant difference between the two subjects was observed. The first subject successfully pumped all balloons and earned 16$, while the second subject only managed to pump 3 balloons and earned 5$. Nevertheless, the first subject scored 50% in the memory task while the second subject scored 100%. The difference in performance was due to their switching gears to bounce back and forth between tasks and also the active attention to one task more than the other. It's noteworthy that despite a focus on increasing profits by the first subject, he earned lower profits in the third task compared to the first, where there was no time pressure or mental demands introduced.

Table 1. Participants' Performance from BART and Memory Tasks.

Participant No	Task 1	Task 2	Task 3
1	(54.20$) 10/10 balloons	(14.80$) 5/10 balloons	(16.60$) 10/10) balloons (2-Back 50% correct
2	(44.80$) 10/10 balloons	(15.20$) 6/10 balloons	(5.6$) 3/10) balloons (2-Back 100% correct)

4.2 Preliminary Results

The results in Fig. 6 and Table 1 revealed that introducing time pressure in the second task (condition II) led to increased activation in the prefrontal cortex (PFC), consistent with previous research linking task difficulty to PFC activation. The observed results suggest that the participants found the task more challenging under time pressure and focused on maximizing their gains by completing the task quickly and obtaining additional compensation. This may have resulted in a tendency to overlook the inherent risks associated with the task. The findings are consistent with previous studies, indicating that the initial results align with existing research in the field. Overall, these findings imply that PFC activity is a valid proxy for decision-making attempts in situations involving increased task difficulty, such as those encountered under time pressure. In condition III, the participants completed a task that required them to work under time constraints and complete a 2-back test, which is a task specifically designed to test working memory. Working memory is a type of short-term memory that mainly depends on the PFC, and it is responsible for integrating and manipulating newly acquired knowledge.

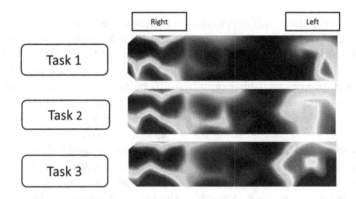

Fig. 6. Hemodynamic Response of L-PFC and OFC.

An increase in the load placed on working memory leads to an increase in cognitive activity across the PFC, especially in the dorsolateral prefrontal cortex (DLPFC) and orbitofrontal cortex (OFC). Figure 6 shows that oxyhemoglobin (oxy-Hb), represented by the dark red colored area on the left side of the brain, significantly increased in concentration in the DLPFC region during task III compared to the control condition (Task I). This substantial increase in brain activity in the PFC region indicates the extent of working memory utilization in processing the received information. The present study aimed to determine the most active region of the brain by analyzing and comparing the results obtained from a total of 52 channels. It was observed that the left PFC exhibited the highest level of activity. From the fNIRS 52 channels, channel 44 was selected as it displayed the most activation based on a rigorous t-test analysis where $p < .001$. This finding is visually depicted in Fig. 7. which demonstrates the notable differences in activation observed in channel 44 across three distinct tasks.

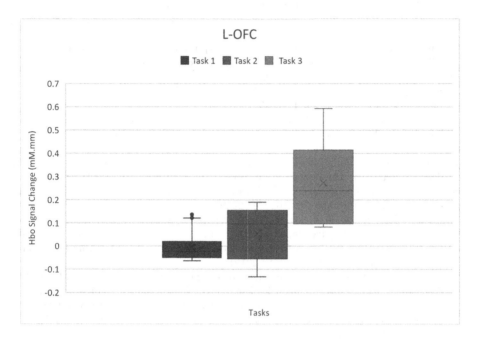

Fig. 7. t-test Analysis and the difference between the Three Tasks.

5   Conclusion

In summary, the findings of this study have illustrated the complex relationship between risk compensation, acute stress, and decision-making in risky tasks. Through a meticulously designed and executed experiment, which built upon a comprehensive review of prior research, the initial results have revealed that time pressure and mental demand affect task performance, leading to a heightened inclination towards risky decision-making. These results have important implications for understanding the underlying mechanisms of decision-making in high-stress environments.

Acknowledgment. The authors extend their appreciation to the Electrical & Electronics Engineering Department at Universiti Teknologi PETRONAS and the University of Ruse "Angel Kanchev" for their ongoing technical support and the use of their facilities.

References

1. COVID, W.: pandemic triggers 25% increase in prevalence of anxiety and depression worldwide, no. 19 (2022)
2. Moni, A.S.B.: Psychological distress, fear and coping among Malaysians during the covid-19 pandemic. PloS One **16**(9), e0257304 (2021)
3. Selye, H.: A syndrome produced by diverse nocuous agents. Nature **138**(3479), 32–32 (1936)
4. Hinds, J.A., Sanchez, E.R.: The role of the hypothalamus-pituitary-adrenal (HPA) axis in test-induced anxiety: assessments, physiological responses, and molecular details. Stresses **2**(1), 146–155 (2022)
5. Sriramprakash, S., Prasanna, V.D., Murthy, O.R.: Stress detection in working people. Procedia Comput. Sci. **115**, 359–366 (2017)
6. Panicker, S.S., Gayathri, P.: A survey of machine learning techniques in physiology based mental stress detection systems. Biocybern. Biomed. Eng. **39**(2), 444–469 (2019)
7. Moghadam, S.S., Khodadad, F.S., Khazaeinezhad, V.: An algorithmic model of decision making in the human brain. Basic Clin. Neurosci. **10**(5), 443 (2019)
8. Ayton, P., Bernile, G., Bucciol, A., Zarri, L.: The impact of life experiences on risk taking. J. Econ. Psychol. **79**, 102274 (2020)
9. Hasanzadeh, S., de la Garza, J.M.: Understanding roofer's risk compensatory behavior through passive haptics mixed-reality system. In: ASCE International Conference on Computing in Civil Engineering 2019, pp. 137–145. American Society of Civil Engineers Reston, VA (2019)
10. Casson, A.J.: Wearable EEG and beyond. Biomed. Eng. Lett. **9**(1), 53–71 (2019)
11. Al-Quraishi, M.S., Elamvazuthi, I., Tang, T.B., Al-Qurishi, M., Parasuraman, S., Borboni, A.: Detection of lower limb movements using sensorimotor rhythms. In: 2020 8th International Conference on Intelligent and Advanced Systems (ICIAS), pp. 1–5. IEEE (2021)
12. Al-Quraishi, M.S., Elamvazuthi, I., Tang, T.B., Al-Qurishi, M., Adil, S.H., Ebrahim, M.: Bimodal data fusion of simultaneous measurements of EEG and fNIRS during lower limb movements. Brain Sci. **11**(6), 713 (2021)
13. Niu, B., Li, Y., Ding, X., Shi, C., Zhou, B., Gong, J.: Neural correlates of bribe-taking decision dilemma: an fNIRS study. Brain Cogn. **166**, 105951 (2023)
14. Huang, M., et al.: Joint-channel-connectivity-based feature selection and classification on fNIRS for stress detection in decision-making. IEEE Trans. Neural Syst. Rehabil. Eng. **30**, 1858–1869 (2022)
15. Pooladvand, S., Hasanzadeh, S.: Neurophysiological evaluation of workers' decision dynamics under time pressure and increased mental demand. Autom. Constr. **141**, 104437 (2022)
16. Zhao, H., et al.: Acute stress makes women's group decisions more rational: a functional near-infrared spectroscopy (fNIRS)-based hyperscanning study. J. Neurosci. Psychol. Econ. **14**(1), 20 (2021)
17. Wanniarachchi, H., et al.: Alterations of cerebral hemodynamics and network properties induced by newsvendor problem in the human prefrontal cortex. Front. Hum. Neurosci. **14**, 598502 (2021)
18. Lee, E.J., Yun, J.H.: Moral incompetency under time constraint. J. Bus. Res. **99**, 438–445 (2019)
19. Wolff, A., Gomez-Pilar, J., Nakao, T., Northoff, G.: Interindividual neural differences in moral decision-making are mediated by alpha power and delta/theta phase coherence. Sci. Rep. **9**(1), 4432 (2019)

20. Li, Y., et al.: Hemispheric MPFC asymmetry in decision making under ambiguity and risk: an fNIRS study. Behav. Brain Res. **359**, 657–663 (2019)
21. Li, L., Cazzell, M., Zeng, L., Liu, H.: Are there gender differences in young vs. aging brains under risk decision-making? an optical brain imaging study. Brain Imaging Behav. **11**, 1085–1098 (2017)
22. Robertson, C.V., Immink, M.A., Marino, F.E.: Exogenous cortisol administration; effects on risk taking behavior, exercise performance, and physiological and neurophysiological responses. Front. Physiol. **7**, 640 (2016)
23. Cazzell, M., Li, L., Lin, Z.J., Patel, S.J., Liu, H.: Comparison of neural correlates of risk decision making between genders: an exploratory fNIRS study of the balloon analogue risk task (bart). Neuroimage **62**(3), 1896–1911 (2012)
24. Wang, P., Gu, R., Zhang, J., Sun, X., Zhang, L.: Males with low risk-taking propensity overestimate risk under acute psychological stress. Stress **24**(6), 898–910 (2021)
25. Chen, X.J., Kwak, Y.: What makes you go faster?: the effect of reward on speeded action under risk. Front. Psychol. **8**, 1057 (2017)
26. Narita, N., et al.: Prefrontal modulation during chewing performance in occlusal dysesthesia patients: a functional near-infrared spectroscopy study. Clin. Oral Invest. **23**, 1181–1196 (2019)

A Survey and Analysis on Customer Satisfaction of Bitter Melon Supplements

Kai En Mak, Sabariah Saharan, and Aida Mustapha[✉]

Faculty of Applied Sciences and Technology, Universiti Tun Hussein Onn Malaysia,
KM 1, Jalan Panchor, 84600, Pagoh Johor, Malaysia
{sabaria,aidam}@uthm.edu.my

Abstract. While the distributor concerned striving to provide quality product and service to achieve customer satisfaction with opportunities given by the rapidly developing supplement market and e-commerce, the sales of their bestselling health supplement, bitter melon supplement dropped. Research was conducted to discover the customer satisfaction level, customers' latent profile, opinion, and perspectives towards their purchase of the bitter melon supplement. 130 random samples of data collected from a total of 600 customers approached via a questionnaire. The data were analyzed with exploratory data analysis, association rule mining, and text analysis. The result indicated that the customers are mostly satisfied but knowing the ineffectiveness of the supplement will increase the expected dissatisfaction of customer by 1.68 times. It is found that the customers were mainly chased away by the low effectiveness of the product instead of customer service. Male at age of 45 to 54 are suggested to be considered in customer segmentation with emphasis on the natural ingredient when increasing the product visibility. The findings suggested that the enhancement in product quality and lower price is necessary for repositioning in terms of product and price.

Keywords: Text analysis · Association rules · Customer satisfaction

1 Introduction

The prevalence of diabetes has been growing globally, especially in Malaysia, earning a title of "Sweetest Nation in Asia" due to the 4.9% increase in the disease from 2015 to 2019 [2]. The issue causes an upsurge of health consciousness among Malaysians which contribute to the growing marketing for health supplements [1]. In the meantime, cross border e-commerce has been expanding rapidly with the development of the internet and technologies [12]. Recognizing the window of opportunity, the distributor concerned which distributes and sells various health supplements was established in 2017. They are ambitious in creating 10, 000 entrepreneurs and contribute to business growth and community stability. The distributor is committed to providing quality and flexible service to achieve customer satisfaction, as well as training services and high-quality products.

© The Author(s), under exclusive license to Springer Nature Singapore Pte Ltd. 2024
F. Hassan et al. (Eds.): AsiaSim 2023, CCIS 1912, pp. 464–478, 2024.
https://doi.org/10.1007/978-981-99-7243-2_38

One of their best-selling products is capsules of bitter melon supplement, which is 100% formulated by Momordica Charantia and are claimed to aid in both prevention and supplying internal energy, to help regulate blood glucose as well as to stimulate insulin secretion.

The success in meeting existing customers' satisfaction, bringing new customers, and enhancing customer loyalty is indicated by customer retention [21]. While the distributor is striving to meet their mission and vision, the top management of the company believed that there is an issue with the product or the agents that is causing a drop in customer satisfaction, leading to fewer repeat purchases. The company was experiencing a drop in sales of the product, by 640,000 units from 2020 to 2021. However, the root cause of the situation was unknown. Additionally, Selangor was noticed to have relatively high sales compared to other regions, having most of the customers purchasing the supplement.

The study is aimed to identify the level of customer satisfaction towards their purchase of the bitter melon supplement using Exploratory Data Analysis. Besides, the objective of the study is to determine the latent profiles of the bitter melon supplement customer using association rule mining. The study is also conducted to summarize the opinions and perspectives of customers towards the performances of the distributor using Text Analysis. The study collaborated with the company only involved those eligible customers in Selangor.

The remainder of this paper proceeds as follows. Section 2 presents the materials and methods in conducting the survey and performing the exploratory data analysis. Section 3 presents and discusses the results and finally Sect. 4 concludes the paper with some direction for future research.

2   Materials and Methods

A survey questionnaire was utilized to collect data from respondents. The questionnaire was constructed of both close-ended questions with multiple choices and Likert scale as well as open-ended questions. The details of customers in terms of personal information including phone number, gender and age and satisfaction level based on their experience in the purchase of the bitter melon supplement by the distributor was collected with close-ended questions. Meanwhile, the latter collect the corresponding opinions and perspectives of customers. Relevant secondary data from various sources such as sales records by the company and previous research were referred as well.

2.1   Sampling Methods

Random sampling was used on the purposely targeted customer in Selangor in this study with the consideration of time and cost consumption. Every individual of the customers in Selangor that purchased the product has equal probability of being included in this study. A random number of 600 customers from Selangor was generated to be respondents of the survey using Microsoft Excel. However, only 130 of them picked the calls and responded to the survey due to viral fraud with scam calls. This number of samples often produces a favorable result although it is quite small compared to the whole population [7].

2.2 Exploratory Data Analysis

Exploratory data analysis is a numerical and graphical discovery of insights and clues [18]. Descriptive statistics such as median is the best measure of central tendency focused to describe ordinal data while frequencies, fractions, and percentages for categorical data [11]. Net Promoter Score (NPS) is one of the statistics used to summarize customer satisfaction and loyalty. The computation is based on the customer rating on likelihood to recommend on a scale of 0 to 10. Scale of 0 to 6 is classified as detractors, 9 and above is promoter while the others are passive. A Net Promoter Score of more than 0 is generally considered good. The equation for Net Promoter Score (NPS) is as in Eq. 1 [5].

$$\text{NPS} = \text{percentage of promoter (9-10)} - \text{percentage of detractor (0-6)} \quad (1)$$

The visualization produced reveals trends and fulfillments [6]. The statistics and visualizations of the data from the customer provide an overview of customer satisfaction of the distributor.

2.3 Association Rules Mining

Association rule mining is a data mining approach used to determine rules or patterns that describe the relationship between items in a dataset [15]. The patterns discovered can be used to determine opportunities for targeted marketing in making sound business decisions [17]. In this study, the classic Apriori algorithm which is a "bottom-up" method in association rule mining was utilized. This classic algorithm is efficient for datasets with size which is small or moderate.

In the Apriori algorithm, a frequent itemset must have a frequent subset, and vice versa [14]. The identification of rules involves setting three parameters as minimum thresholds that the item and itemset should appear at least so that the item and itemset are considered frequent for further analysis [16]. Support is to discover frequent items while confidence to discover frequent item sets that indicate relationship strength [3]. Meanwhile, lift discovers the rules that are likely to be useful when the two sets are independent. Rules that fulfill both minimum support and minimum confidence are strong association rules. Rules that have a lift value of more than 1 indicate a positive association where A favors B's occurrence. Following [9], the equations of support, confidence and lift are shown in Eq. 2. Eq. 3, and Eq. 4, respectively.

$$\text{Support}(s) = p(A \cup B) \quad (2)$$

$$\text{Confidence}(c) = P(B|A) = \frac{\text{Support}(A \cup B)}{\text{Support}(A)} = \frac{P(A \cup B)}{P(A)} \quad (3)$$

$$\text{Lift}(l) = \frac{\text{Confidence}(c)}{\text{Support}(B)} = \frac{P(A \cup B)}{P(A)P(B)} \tag{4}$$

2.4   Text Analysis

Text analysis is a method of extracting information and insights from the unstructured data obtained from the open-ended question in the survey study [4]. To be specific, the study applied lexical analysis which focuses on the words used in a text [8]. The word frequency technique as one of its techniques was applied to identify the most frequently used words, overviewing the main themes in the text [13]. The analysis is started by text pre-processing which includes tokenization which breaks sentences into words, removal of punctuation, lower-casing, lemmatization which transform the word into base form and removal of stop words which remove unnecessary words [19]. The frequency of each word is computed. The word cloud or tag cloud is generated based on the frequency of the word, providing a visual representation of the unstructured data [10]. The more frequent terms would have a bigger font in the word cloud [20].

3   Results and Discussion

This section presents the results from exploratory data analysis and association rule mining from the survey conducted.

3.1   Exploratory Data Analysis

Gender: Figure 1 shows that most of the respondents in this survey are female, recording a portion of 63% (82 people), while the rest (37%) are male respondents with 48 people.

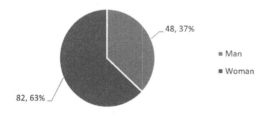

Fig. 1. Gender for respondents.

Age: Figure 2 shows that most of the respondents in this survey are aged between 55 to 64 years old, which is 44 people out of 130 respondents (34%). Only 8 respondents (6%) were aged between 25 to 34 years old. The number of respondents increases by about twofold as the age range increases from one range to another range.

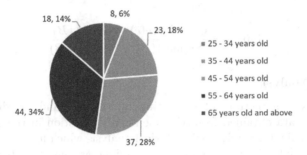

Fig. 2. Age of respondents.

Customer Acquisition Channels: Figure 3 illustrates how customers first find the bitter melon supplement. Out of 130 respondents, 71 of them (54%) found out about the product via social media feeds such as Facebook, Instagram and so on, acting mode of the customer acquisition channel. There are 22 respondents (16.9%) who found out about the product through word-of-mouth, which is from the people around them such as family, friend, or colleague. 17 respondents (13%) found out about the product from shopping platforms such as Shopee and Lazada, while 15 of them (11.5%) found out about the product from the broadcast media. Three (2%) of them knew the product from both social media feeds and people around them. One respondent (0.7%) is recorded for each search result from search engines and the combination of social media feed and broadcast media.

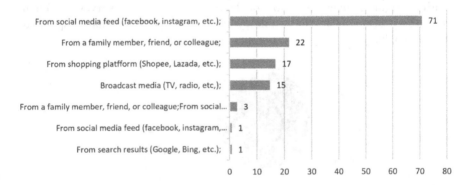

Fig. 3. Customer acquisition channels of respondents.

Customer Satisfaction: Figure 4 illustrates the response of respondents to the question on how effective is the bitter melon supplement. More than half of the respondents, including 60 respondents (46%) reflected that the product is effective, acting as mode value while another 19 respondents (14.6%) reflected that the product is very effective. The ineffectiveness of the product was strongly claimed by 16 respondents (12.3%) than the rest of 35 respondents (26.9%) who reflected that the product is ineffective. From this result, most of the respondents (60.8%), agreed with the effectiveness of the product with sample statistics of median of 3.

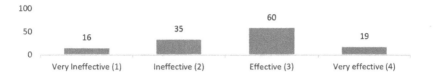

Fig. 4. Effectiveness of bitter melon supplement based on experience respondents.

Table 1 shows the frequency for teh satisfaction levels used in this study with the corresponding median score.

Table 1. Frequency of each satisfaction level (1 to 4).

Item	1	2	3	4	Median
Product Effectiveness	16	35	60	19	3
Price	24	53	53	0	2
Quality Upon Delivery	5	16	73	36	3
Customer Service	2	14	79	35	3

Figure 5 shows the response of respondents to the question what do they think about the price of the bitter melon supplement. More than half of the respondents including 53 respondents (40.8%) reflected that the product is pricey, acting as mode value while another 24 respondents (18.5%) reflected that the product is very pricey. The rest of 53 respondents felt that the price of the product is reasonable while none of them claimed it is cheap. Based on the median value at 2, it can be inferred that the product is pricey for most of the respondents.

Fig. 5. Price of the bitter melon supplement based on experience respondents.

Figure 6 illustrates that about half of the respondents (56.2%) pointed out that the quality of the product upon delivery is good while 36 respondents

(27.7%) strongly agreed that they receive the product in good condition. On the other hand, 16 respondents (12.3%) complained that the quality of delivery is not good, and 5 respondents (3.8%) received the not all well product. According to the median at value of 3, most of the respondents are satisfied with the quality of the product upon delivery.

Fig. 6. Quality upon delivery of the bitter melon based on experience respondents.

Figure 7 illustrates the response to the question how do they rate their overall experience with the customer service. Positive reviews of the customer service experience are obtained from most respondents (60.8%) while 35 respondents (26.9%) even gave the highest performance rating. However, 14 respondents (10.8%) are dissatisfied with the customer service. There are two respondents who were greatly dissatisfied with the customer service experienced.

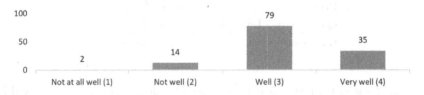

Fig. 7. Respondents' experience in customer service in the purchase of the product from the distributor concerned.

The overall satisfaction with the overall question on how satisfied they are with the bitter melon supplement is visualized as in Fig. 8. The median at value 3 shows that most of the respondents were satisfied with their purchase. In fact, customer service is not the main issue of customer satisfaction and the drop in sales. Based on the overview with the statistics and visualization, the price of the product was dissatisfied the most. The need for initiatives for improvement is followed by the aspect of effectiveness, quality upon delivery, and lastly, customer service.

Fig. 8. Overall satisfaction with the bitter melon supplement.

Meanwhile, Fig. 9 and Fig. 10 show the likelihood to repurchase and to recommend the bitter melon supplements by respondents. The need and room to improve is proved by the low likelihood to repurchase at 2 in median value, which is unlikely and having more detractors than promoters with a negative net promoter score of 25% while most of the respondents not at all likely to recommend the product.

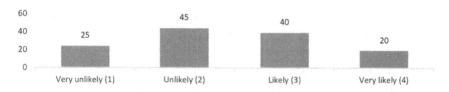

Fig. 9. Likelihood of repurchase of the bitter melon supplement by respondents.

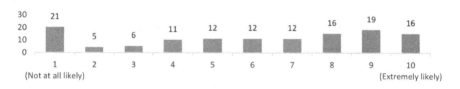

Fig. 10. Likelihood to recommend bitter melon supplement by respondents.

These findings are reflected from the Net Promoter Score (NPS) value of -25% in the following calculation. NPS is used to gauge the loyalty of a company's customer relationships and to assess whether customers are likely to recommend the company's products or services to others.

$$\text{NPS} = \left(\frac{35}{130} - \frac{67}{130} \right) \times 100\%$$
$$= (0.27 - 0.52) \times 100\%$$
$$= -25\%$$

3.2   Association Rule Mining

Table 2 shows the association rules mined from the data. The rules indicate that the dissatisfaction of respondents as a consumer of the product is significantly associated with the ineffectiveness of the product being a health supplement. There are 20% of the feedback (26 responses) that reflect both ineffectiveness and their dissatisfaction towards their purchase. The probability of getting a negative review in overall satisfaction with the knowledge that the product is ineffective is 74.3%, indicating the strong association between ineffectiveness and dissatisfaction among consumers of the bitter melon supplement.

Table 2. Association rules mined.

Antecedent	Consequent	s	c	l
Effectiveness ⇒ Ineffective	Overall Satisfaction ⇒ Dissatisfied	0.200	0.743	2.683
Gender ⇒ Male, Quality Upon Delivery ⇒ Well, Effectiveness ⇒ Effective	Overall Satisfaction ⇒ Satisfied	0.238	0.969	1.999
Quality Upon Delivery ⇒ Well, Overall Satisfaction ⇒ Satisfied	Repurchase ⇒ Likely	0.231	0.612	1.990
Quality Upon Delivery ⇒ Well, Effectiveness ⇒ Effective	Overall Satisfaction ⇒ Satisfied	0.323	0.955	1.970
Quality Upon Delivery ⇒ Well, Effectiveness ⇒ Effective	Repurchase ⇒ Likely	0.200	0.591	1.920
Effectiveness ⇒ Effective, Overall Satisfaction ⇒	Repurchase ⇒ Likely	0.223	0.569	1.848
Gender ⇒ Male, Effectiveness ⇒ Effective	Overall Satisfaction ⇒ Satisfied	0.285	0.881	1.818
Overall Satisfaction ⇒ Satisfied	Repurchase ⇒ Likely	0.269	0.555	1.806
Effectiveness ⇒ Effective	Overall Satisfaction ⇒ Satisfied	0.392	0.850	1.754
Effectiveness ⇒ Effective	Repurchase ⇒ Likely	0.223	0.533	1.733

When the ineffectiveness of the product is known, the expectation that someone the customer will be dissatisfied with the purchase increases by 168%. Other than the strong association that customer will mostly be satisfied and repurchase due to effectiveness of the health supplement and the good quality upon deliver as well as the likely of repurchase when customer is satisfied, it is worth to highlight that the gender of customers being a male increases the expectation that the customer will be satisfied in overall.

3.3 Text Analysis

Figure 11 shows the chart generated from the word frequency for the response to the question "How long have you been taking the bitter melon supplement to notice its effectiveness?". The "1 month" with the largest font shows that most of the respondents find that the product is effective after eating for 1 month. The second major period for the medicine to take effect is 2 years as experienced by customers. It is followed by around 3 to 6 months and after 9 months to 1 year while the most time taken is 4 years for the medicine to take effect.

Surprisingly, there is a customer that had not tried the product yet after purchase. Follow-up should be made regularly so that customers can benefit from their purchase. Although the time taken for the product to work is different, the analysis outcome and reviews, especially from influencers considered as word of mouth can be shared to customers to gain their confidence.

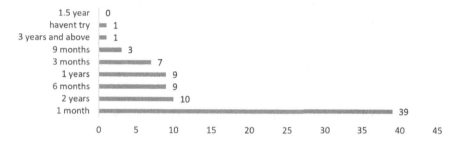

Fig. 11. Bar chart of the time taken for customers to find effectiveness of the product.

Figure 12 shows the response to the question "How long has you been taking the bitter melon supplement?" after they reflected the ineffectiveness of the product. 3 months is the most period when the customers felt that the product is not working. Based on the word frequency, except the 10 responses with 1 year and above while one being forgotten, most of the records are 2 months, followed by 4 months and 8 months. Customers especially those that feel that the product is ineffective should be followed up to facilitate them in terms such as dose needed. Understand the situation and relevant information of the customer for the continuous research and development to provide a better product.

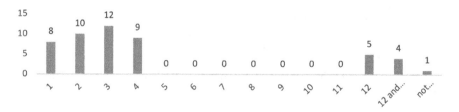

Fig. 12. Bar chart of the period of customer have been consuming the product but find no difference.

The word cloud as in Fig. 13 was generated from the response of respondents if they have any thought about having any problem or suggestion for better customer service. The data is translated and manually transformed based on the word frequencies of the raw text data to get more meaningful visualization for insight. It turns out that most of the respondents were satisfied without problem with compliments such as friendly, understanding, and effective communication. However, some of the respondents complained that their calls were not answered promptly, being not helpful. Suggestions such as training, and development of live chat were provided while some of them purchased via website and Shopee indeed without interaction with customer service, being aligned with recent research about embrace of technology among consumers.

Fig. 13. Word cloud of the feedback regarding customer service.

Similarly, at the right side of Fig. 14 gives hints about the word presented in the word cloud at the left side. With the question "What did you like most about the bitter melon supplement?", the help in reducing high blood pressure and blood sugar is the part where the respondents love the most. Other than regulating blood pressure and blood sugar that cause diabetes, the convenience of the product, ability to reduce cholesterol, maintain health, relieve pain with natural ingredients are the unique selling points the bitter melon supplement supplied in market positioning.

Fig. 14. Word cloud of the respondents' favourite part about the bitter melon supplement.

When the respondents were asked about their opinion regarding their dissatisfaction pointed out in the previous question, the low effectiveness of the product seemed to be the main problem that led to the dissatisfaction based on the funnel chart as shown in Fig. 15. Price is shown to be the main issue in common, followed by effectiveness and alternatives of the supplement. Customers pointed out that they might not repurchase the product when they have other commitments that need to be prioritized in expenses.

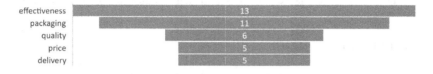

Fig. 15. Funnel chart of the respondents' feedback regarding their dissatisfaction.

Information obtained from the customer feedback is deemed to be crucial in giving a sneak peak of the market situation to take prompt action. For instance, they hope for enhancement in quality of product while being affordable so that every consumer can get a better experience. As a smart buyer, customers are aware of the alternatives available in the market and compare the products from different aspects. Therefore, the distributor should always be alert to the changes in the market and both the potential and existing competitor to stay competitive. These findings are reflected from Fig. 16, 17 and 18.

Fig. 16. Tree map of the reason that almost stops respondents from buying the product.

Fig. 17. Bar chart of the reason that stopped respondents from buying the product.

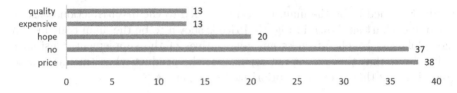

Fig. 18. Bar chart of the respondents' additional comment for their purchase.

4 Conclusions

In this research, the exploratory data applied provided an overview of the customer satisfaction level towards their purchase of the bitter melon supplement, indicating that the customers were mostly satisfied with their purchase. The mined association rules discovered the strong association between product ineffectiveness and dissatisfaction that lead to a drop in sales. A customer is 2.68 times more likely to be dissatisfied when the health supplement is ineffective, compared to the likelihood of dissatisfaction regardless of the effectiveness. The result of text analysis shows that the selling point of the product is its natural ingredient and convenience. Rather than the problem of customer service deemed by the company, the deterrent of the product is the low effectiveness and price that the customers complained about.

Male customers at age ranging 45 to 54 are recommended to be considered in customer segmentation. This segment of customers could be targeted with emphasized natural ingredients and convenience of the product while increasing the product visibility. Overall, the results gained from three different analyses are consistent where the supplement efficacy should be enhanced in product positioning to gain customer satisfaction in customer retention. An intention-to-buy scale could be included in the survey of market research to prior to product launch to understand the purchase intention in order to make further development decisions. To improve the brand equity to chosen target customers, especially male, social media should be focused for both placing and promotion positioning to the target market, so that the likelihood to purchase rises with the increased exposure that directs them to customer service and e-commerce websites and platforms. Pricing analysis is needed for price positioning. Lowering price with price position of cost-effective is suggested for the current stage of product's life cycle. Encourage customer feedback with reward in the next purchase for continuous monitoring and corresponding development to earn strong brand reputation created which results in higher customer loyalty and sustainable growth in sales. The weightage of each aspect should be investigated, and more samples are suggested for future further study for more accurate results to make informed business decisions.

Acknowledgements. This study is funded by Universiti Tun Hussein Onn Malaysia.

References

1. Dietary supplements market size and trends | growth, 2021–2028. https://www.fortunebusinessinsights.com/dietary-supplements-market-102082. Accessed 08 Jan 2023

2. Report of the second meeting of the who technical advisory group on diabetes: virtual meeting, 16–17 December 2021. https://www.who.int/publications/i/item/9789240047945. Accessed 08 Jna 2023

3. Aqra, I., et al.: Correction: a novel association rule mining approach using TID intermediate itemset. PLoS ONE **13**(5), e0196957 (2018)

4. Arya, A.: An overview of textual analysis as a research method for cultural studies. Int. J. Innovative Res. Multidisciplinary Field **6**(3), 173–177 (2020)

5. Baehre, S., O'Dwyer, M., O'Malley, L., Lee, N.: The use of net promoter score (NPS) to predict sales growth: insights from an empirical investigation. J. Acad. Mark. Sci. **50**(1), 67–84 (2022)

6. Carranza, E.J.M.: Exploratory data analysis. In: Daya Sagar, B., Cheng, Q., McKinley, J., Agterberg, F. (eds.) Encyclopedia of Mathematical Geosciences. Encyclopedia of Earth Sciences Series, pp. 1–5. Springer, Cham (2020). https://doi.org/10.1007/978-3-030-26050-7_105-1

7. Dawson, J.F.: Analysing quantitative survey data for business and management students, pp. 1–104 (2016)

8. Gonçalves Júnior, J., Sales, J.P.D., Moreno, M.M., Rolim-Neto, M.L.: The impacts of SARS-COV-2 pandemic on suicide: a lexical analysis. Front. Psychiatry **12**, 593918 (2021)

9. Ju, C., Bao, F., Xu, C., Fu, X.: A novel method of interestingness measures for association rules mining based on profit. Discrete Dyn. Nat. Soc. **2015** (2015)

10. Kabir, A.I., Ahmed, K., Karim, R.: Word cloud and sentiment analysis of amazon earphones reviews with R programming language. Informatica Economica **24**(4), 55–71 (2020)

11. Manikandan, S., et al.: Measures of central tendency: median and mode. J. Pharmacol. Pharmacother. **2**(3), 214–215 (2011)

12. Nogoev, A., Yazdanifard, R., Mohseni, S., Samadi, B., Menon, M.: The evolution and development of e-commerce market and e-cash. In: International Conference on Measurement and Control Engineering 2nd (ICMCE 2011), vol. 1, pp. 1–5 (2011)

13. Nouri, N., Zerhouni, B.: Lexical frequency effect on reading comprehension and recall. Arab World Engl. J. (AWEJ) **9** (2018)

14. Sarker, I.H.: Machine learning: algorithms, real-world applications and research directions. SN Comput. Sci. **2**(3), 160 (2021)

15. Saxena, A., Rajpoot, V.: A comparative analysis of association rule mining algorithms. In: IOP Conference Series: Materials Science and Engineering, vol. 1099, p. 012032. IOP Publishing (2021)

16. Selvarani, S., Jeyakarthic, M.: Rare itemsets selector with association rules for revenue analysis by association rare itemset rule mining approach. Recent Adv. Comput. Sci. Commun. (Formerly: Recent Patents Comput. Sci.) **14**(7), 2335–2344 (2021)

17. Sinha, A.: Implying association rule mining and market basket analysis for knowing consumer behavior and buying pattern in lockdown-a data mining approach (2021)

18. Tukey, J.W., et al.: Exploratory Data Analysis, vol. 2. Reading, MA (1977)

19. Udgave, A., Kulkarni, P.: Text mining and text analytics of research articles. PalArch's J. Archaeol. Egypt/Egyptology **17**(6), 4483–4489 (2020)

20. Yusop, N.M., Azmi, N.N., Azlan, N.: Analysis of tenants complaints using text mining. Int. J. Acad. Res. Bus. Soc. Sci. **12**(4), 1111–1119 (2022)
21. Zephan, N.: Relationship between customer satisfaction and customer loyalty. Centria University of Applied Science and Business Management (2018)

Enhancement of OWASP Monitoring System with Instant Notification

Mazlan Sazwan Syafiq⬤, Mohamed Norazlina⁽✉⁾⬤, and M. Fatin Faqihah

Infrastructure University Kuala Lumpur, Kajang, Malaysia
{sazwan,norazlina}@iukl.edu.my

Abstract. Security is a major concern for all premise properties, such as residences, industries, etc. Lack of a secure system will open a window of opportunity for theft that causes damage, loss of property, and emotional misery. This paper mainly focused on the OWASP development of an embedded live surveillance camera with instant notification. The system functions by capturing video of any motion detected in restricted areas, such as user belongings and property. The system will immediately alert the user via Telegram and email, for further action to be taken as the result proves within an average of 2.3 s of motion detection and a user alert system of 3.8 s. The monitoring system with a high alarm rate is able to integrate with the system and communicate with the user by using an embedded system of Raspberry Pi 4 Model B. The algorithms used in this research include the dynamic programming algorithm (DPA) and the application programming interface (API). The integration and embed system indicate the system efficiency of 92.65% with error tolerance below 3.1% using Support Vector Machine (SVM) as a tool to analyze system performance of True Positive (TP), True Negative (TN), False Positive (FP), and False Negative (FN).

Keywords: Internet of Things (IoT) · Security System · Industry Revolution (IR 4.0) · Artificial Intelligence · Machine Learning

1 Introduction

The demand for the current trend in a modern surveillance system that interfaces with the Internet connectivity has grown as a result of Malaysia's rise in crime cases. According to the crime index, there were 52,344 theft cases in 2020, which will make up nearly 80% of all instances [10]. Despite being widely used, closed-circuit television (CCTV) is still regarded as a passive monitoring system that requires constant and continuing supervision by individuals, requires more time, is very expensive, and frequently produces corrupted files [9].

Internet of Things (IoT) technology creates a new platform for information sharing, productivity, and modern living [1]. IoT has facilitated modernity and ease in human lifestyle. Over the past few years, people have become more dependent on IoT to fulfil tasks like work, school, research, and daily chores. Consequently, this advancement has the ability to broaden the fundamental functions of a surveillance system [5]. Any

© The Author(s), under exclusive license to Springer Nature Singapore Pte Ltd. 2024
F. Hassan et al. (Eds.): AsiaSim 2023, CCIS 1912, pp. 479–487, 2024.
https://doi.org/10.1007/978-981-99-7243-2_39

organization needs a surveillance system, used to protect people and their possessions from various risks including theft and burglary.

Researchers and academics were motivated to create a non-passive surveillance system because of the drawbacks of passive monitoring technologies described above [7]. The majority of researchers then utilized Wireless Sensor Networks (WSN) for monitoring, taking full advantage of its features and advantages [4, 5]. Due to the wireless connectivity, sensor nodes can be placed anywhere in a building, giving them the advantage of portability when being deployed [8].

The main objective is to develop a system algorithm for the ease user to monitor asset and property, this research mainly focused on the algorithm surveillance system that integrate with open-source application programming interphase (API). The response able to integrate between user and the system instantly via different platform (Telegram and email). These system implicates by auto record of motion detection and for the used as evidence for further forensic.

2 Literature Review

The Raspberry Pi 3 computer, a Universal Serial Port (USB) webcam, a Passive Infrared Sensor (PIR) sensor, ESP8266 Wi-Fi module, current sensor, and ESP8266 Wi-Fi module make up the system's hardware in this study. The Python programming language, the Arduino Integrated Development Environment (IDE), and the Raspbian operating system are all used in these articles. This paper describes a security system that sends the user a telegraph message with an series of images in the event of an intrusion. As opposed to a surveillance system, the system described in this paper uses the Message Queuing Telemetry Transport (MQTT) protocol to read the state of numerous sensors as well as monitoring.

The main user and the surveillance system are connected into a single system, the monitoring system proposed in the research proposal by researcher(s) [3] enables consumers to monitor their homes live through mobile application. Additionally, a motion sensor that detects an intruder nearby will send an email notification with a picture attachment as well as two notifications through SMS and email. This project makes use of the Raspberry Pi, the Pi camera, the PIR motion sensor, the Ultrasonic sensor, the buzzer, and the light-emitting diode (LED). The applications employed in this study include Node-Rack, broker, and ThingSpeak. The investigation by is one more study utilizing Node technology [2].

Any ingenious motion detected method will enables the Raspberry Pi to send instant image notifications is described in research [2]. On the other hand, it also discussed the method on sending emails using Transmission Control Protocol on port 55 and SMTP on port 587. In early 90s, deployment technologies of the Raspberry Pi 3-B, Passive Infrared Sensot (PIR) sensor and Pi camera.

N. Patil, S. Ambatkar, and S. Kakde 2017, proposed a semiautomated monitoring system, that triggered with motion detector. It operates by send out an email to the user after some times.

In the study by Sharma H. K. and Sharma M. (2019), explains on a video monitoring system service that sends instant notification with a short clip video. However, the approach, is unique; the paper used a Raspberry Pi B + as a video streaming server to store memory, as well as an AT-Mega 328 Arduino as a microcontroller and processor. The Raspberry Pi processing in this article applied a speed of 1.5 GHz, significantly faster than the 20 MHz speed of the Arduino used most research reported [7]. The uses of Arduino have an effect on the system's performance by slowing down operation speed.

3 Methodology

The development of the Open Web Security Project (OWASP) algorithm is used to monitor and control system between user and device (camera). Besides that, OWASP also act as a platform to send out notification and alert user of any suspicious motion identified. In order to navigate the OWASP algorithm, integration of several element such as processor, input, network, system and output should be in place and able to communicate. Since the OWASP is an open space, it do have its high security breach to prevent the user privacy.

The hardware for this system comprises of a motion sensor, a camera module, and Raspberry-Pi 4- B processor of 8 Gigabyte (GB) of Random Access Memory (RAM) [11]. The camera is integrated into the Raspberry Pi through a specific camera connector. A General-Purpose Input/Output (GPIO) pin on the Raspberry Pi connects it to the motion sensor. The GPIO4 pin, 5V, and GND on the Raspberry Pi have all been used.

Assembling component of a motion sensor and camera are attached to the Raspberry Pi. The Raspberry Pi's Python software must be launched in order to begin the camera's live feed of the residence, its contents, and its occupants. Any intruders will be caught by the property's motion sensor as they approach. Within 10 s, the camera module will start capturing the events. A Python script will send the 10 s video recording and transmit it to the user's email and Telegram. The camera module can support on live monitoring location to recognise detection based on the same script as shows in Fig. 1.

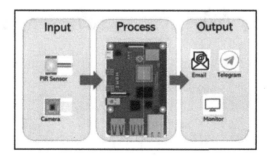

Fig. 1. System Flow.

In this research, a 32 GB storage memory are used to store approximately 2500 images or 57 h video recording by the camera, the Raspberry Pi applied Raspbian OS with support of using an SD card (Dow, 2018). The system operation affects the camera module and motion sensor are set up, as well as alert system (send out notification). For instance, if the camera identify any suspicious movement, the embsed system will automatically record with ten seconds (10) will also be stored on the micro-SD card and instantly notify the user.

In this study, application programming interface (API) is used. As an IoT standard, MQTT messaging protocol is used [12]. It serves as an exceptionally lightweight publish/subscribe message transport and is designed to connect faraway devices with little code footprint and little network bandwidth [9]. MQTT is used by many different industries, such as the automobile, manufacturing, telecommunications, oil and gas, etc. The MQTT API is most suited for this study due to a select few features. It is quick and effective, scales to many IoT devices, supports unstable networks regardless of speed, allows bi-directional communication, and last but not least, is security enabled.

This structure consists of three (3) levels of user authentication methods accessible for MQTT broker, it is important for the security and optimization of the API protocol which are user identification, username and password and client verification.

The user identification for sending Telegram notifications is hard coded in the script, whereas the username and password for sending email notifications are written and programmed by the researcher. The MQTT broker verifies the user authentication credentials along with the CONNECT packet before accepting the MQTT session. While in the CONNECT packet, used port 8883 for connection, the credentials are provided to the broker in clear text first and they are being encrypted at the transport layer with port 8883 mentioned above.

The user must first establish the connection by allowing the users' platform of Telegram ID and email credentials to be coded using Pythom C-programming code that runs on the Internet of Things (IoT) device. The API will initiate the MQTT Service, execute the script code, and finish the IoT system. The surveillance system's API structural diagram is shown in Fig. 2.

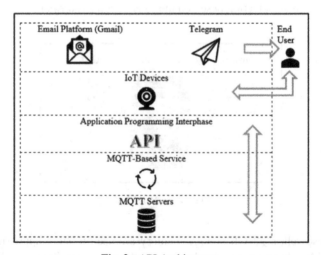

Fig. 2. API Architecture.

The OWASP is made up of Python, PHP, and JavaScript. Primarily written in Python, the script combines Telegram and Email notifications with the integration of hardware;

motion sensor and camera. Nevertheless, Python assembly language also will be an interface to notify signal to network alert user [6]. Additional software needs used in the development of the suggested system include Thonny IDE and Raspbian OS. The code was written using Visual Studio Code and Thonny. For the Python programming language, an Integrated Development Environment (IDE) called Thonny was developed for open source. On the other hand, it do have the capability to analyze step-by-step and has an integrated debugger that can be used to run in order to fix faults.

The notification alert is sent via the Email API as well as the Telegram API. An open-source, cloud-based, cross-platform instant messaging service is called Telegram. File sharing, VoIP, and a number of other capabilities are also available. In this research, a short clip of ten-second (10s) video will be transmitted over Telegram as an immediate

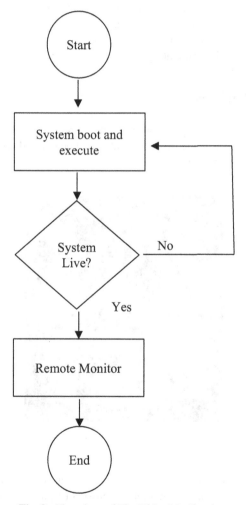

Fig. 3. Flowchart of The Video Notification.

warning if a motion sensor detects any intruder movement. Email can be used to communicate with others, exchange messages, or alert a system. Google Mail, also known as Gmail, is used for this project.

The setup of the Raspberry Pi, infrared sensor, camera V2, Telegram and email will be catagorized into five (5) critical steps. For live streaming and video capture, the Raspberry Pi 4's camera port was connected to the camera module V2. Ten seconds of video recording were set up using Python code. H.264 video recording is the industry standard, thus that is how the video will be saved by default. The Python code will convert it to MP4 due to the ease format by the user device (Android and iOS).

The system begins with the capability to feed live video of the space structure and the motion is immediately activated. Even if the sensor does not detect any motion, the video recording features will continue to function as usual. Users can access the surveillance system using any device, including smartphones, tablets, laptops, and computers, thanks to Python coding. Not all users are allowed to sign into the web application, which increases security and assurance. Only the system user is authorized to access into the dashboard. The process flow activating video recording shows in Fig. 3.

4 Result and Analysis

The camera module activates and record the ten seconds (10s) video and will be stored in memory as set up. In the same times, user also will be notified instantly by the system via Telegram and email as shows in Fig. 4.

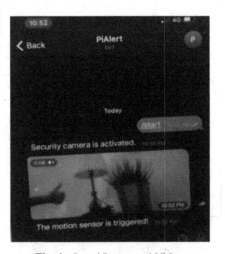

Fig. 4. Saved Image and Video.

The reliability and capability of the sensor are were examined, to ensure its performance towards real situation. The motion sensor's responses to movement and light conditions are displayed in Tables 1 and 2 below. The sensor is unable to detect subject movement at distances greater than 3 m, based on the test that has been conducted within a range of 1 to 5 m. Movement of the test subject can be detected in both light and dark environments in the presence of light. In conclusion, the motion sensor detection range and light condition restriction on subject movement is between 1 to 3 m apart. The flames can only be located within 1 to 2 m, though as shows in Table 1.

Table 1. Sensors Respond to Movement

Testing subject	Distance (Meter)				
	1.0	2.0	3.0	4.0	5.0
Human	✓	✓	✓	✗	✗
Cat	✓	✓	✓	✗	✗
Ball	✓	✓	✓	✗	✗
Fire	✓	✓	✗	✗	✗

Table 2. Sensors Respond to Light Condition

Testing subject	Light Condition	
	Light	Dark
Human	✓	✓
Cat	✓	✓
Ball	✓	✓
Fire	✓	✓

Table 3. Confusion Matrix for OWASP structure.

	Actual	
	Negative	Negative
Predictive	Positive True	Positive False
	Negative False	Negative True

Support vector machine (SVM) are used to analyze the data captured form the experimental. The SVM algorithm applied variable of Positive True (TP), Positive False (TP), Negative False (FN) and Negative True (TN). A sample shows in Tables 2 and 3 comes from different type of obstacle movement, and the result complied error value of approximately 3%, which contributes to over 97% accuracy.

The result can be derived from the test results in Table 4. Specifically, the delay between sending and receiving a telegraph or email alert is 2 s for repetitions 1–3 and 3 s for repetitions 4 and 5. Different video file sizes and the state of the internet network connection can have an impact on this disparity. The Raspberry Pi's sensor and camera continue to function even without internet access, but the video is preserved instead of being sent straight to telegram and email as an alarm. When the internet is back up and running, it will be transmitted.

Table 4. Testing Result of Notification

Testing	Time (hour: minute: second)		
	Sending	Received	Delay (sec)
1	10:18:20	10:18:22	2
2	10:18:30	10:18:32	2
3	10:19:12	10:19:14	2
4	10:19:40	10:19:42	3
5	10:20:02	10:20:04	3

5   Discussion

The Raspberry Pi starts a completely new era in terms of modern technologies. Not just for its size, but also for what it is capable of. It can be used for nearly anything because of its portability [8]. The surveillance system project serves as an example of this. With the advancement over the prior research, this study has succeeded in achieving its four goals, which were described in detail in the introduction.

The goals were to monitor people, property, and belongings, to send instant email notifications with video attachments when nearby surveillance systems detect motion, to send instant Telegram notifications when motion is detected. When the system detects motion, the research has improved to deliver an instant Telegram and email message with a 10-s video file.

Each research project has restrictions either when creating a hardware-based system or software-based system. The limitation we encountered for this research project is if more than eight (8) devices connect to the Raspberry Pi at once, the network's performance will be disrupted. This restriction does. However, have a justification. For instance, the authority granted to the surveillance system in a building, home, or other premises supposedly limited to few personnel only.

6   Conclusion

The system enhancement, the efficiency, adaptability, security, and quick response whenever a threat is detected by the monitoring system have all been maximized. The system's drawback is that it can be challenging for non-technical people to set up without a clear written manual, and it significantly relies on internet connectivity to send the warning, making it challenging for people who live in rural areas with limited internet access.

More GPIO pins in the Raspberry Pi can be utilized to this surveillance system in order to maximized its functionality and research applications. More features and applications can make daily tasks and activities easier for users. To sum up, this project can be expanded to completely employ the suggested system with cutting-edge technology like an AI recognition feature or other detection, making it a full security system.

The research project contribution is to construct a secured and integrated Internet of Things (IoT) for surveillance system. The system should be reliable and user-friendly. The two main novelties to be highlighted are:

i. The surveillance system provides passive and active surveillance so the system can alternatively be used for local monitoring or remote monitoring.
ii. The surveillance system is a secured IoT system from virtual attacker since it is protected from injection and Broken Authentication.

In conclusion, the research project is a secured surveillance system that gives users a platform to watch over their house locally within same network and remotely even without same network. The primary elements of this thesis, other than the main function are secure coding, secure development, and secure architecture. To make sure the technology operates properly, the prototype has been tested through functionality testing and security testing based on OWASP.

References

1. Ahmad, S.: Application of raspberry Pi and PIR sensor for monitoring of smart surveillance system. Int. J. Sci. Res. (IJSR) **5**(2), 736–737 (2016)
2. Weed, J., Luu, P.V., Rodriguez, S., Akhtar, S.: An AI-based web surveillance system using raspberry Pi. J. Adv. Technol. Eng. Res. **5**(6) (2019)
3. Narkhed, Y.V., Khadke, S.G.: Application of raspberry Pi and PIR sensor for monitoring of smart surveillance system. Int. J. Sci. Res. (IJSR) **5**(2), 736–737 (2016)
4. Fatima, N.: IoT based interactive home automation with SMS and Email alert system. Int. J. Res. Appl. Sci. Eng. Technol. **7**(10), 569–574 (2019)
5. Macías, E., Gualotuna, T., Suárez, A., Rivadeneira, A.: Low cost efficient delivering video surveillance service to moving guard for smart home. Sensors **8**(3), 745 (2018)
6. Yong, E.L.C., Ramlee, R., Subramaniam, S.K., Khmag, A., Rahman, A.S.: Home switching using IoT system via telegram and web user interface. Int. J. Recent Technol. Eng. **8**(26), 814–819 (2019)
7. Hasan Basri, A.H., Noorjannah Ibrahim, S., Liza Asnawi, A.: Development of web-based surveillance system for Internet of Things (IoT) application. Bull. Electr. Eng. Inf., **8**(3) (2019)
8. Pathak, G.K.: A review of IOT based SMS & Email enabled smart home automation system. Int. J. Res. Appl. Sci. Eng. Technol. **5**, 2872–2875 (2017)
9. Patil, N., Ambatkar, S., Kakde, S.: IoT based smart surveillance security system using raspberry Pi. In: 2017 International Conference on Communication and Signal Processing (ICCSP), vol. 2, no. 4, pp. 344–348 (2017)
10. Indora, S., Rani, R.: A review IoT based camera surveillance system. Int. J. Comput. Sci.Eng. **7**(6), 793–800 (2019)
11. Sucipto, S., Sasongko, M.Z.: Design prototype IoT menggunakan bot telegram berbasis text recognition. Res. J. Comput. Inf. Syst. Technol. Manage. **4**(1), 21 (2021)
12. Sharma, M., Sharma, H.K.: IoT based home security system with wireless sensors and telegram messenger. SSRN Electron. J. **1**(1), 584–588 (2019)

Correction to: Ultrasonic Sensors in Companion Robots: Navigational Challenges and Opportunities

Isaac Asante⬤, Lau Bee Theng⬤, Mark Tee Kit Tsun⬤, and Zhan Hung Chin⬤

Correction to:
Chapter 29 in: F. Hassan et al. (Eds.): *Methods and Applications for Modeling and Simulation of Complex Systems*, **CCIS 1912,**
https://doi.org/10.1007/978-981-99-7243-2_29

The original published version of this chapter, an acknowledgement and footnote was missing. This now has been added.

The updated version of this chapter can be found at
https://doi.org/10.1007/978-981-99-7243-2_29

Author Index

Printed in the United States
by Baker & Taylor Publisher Services